Springer Proceedings in Mathematics & Statistics

Volume 134

More information about this series at http://www.springer.com/series/10533

Springer Proceedings in Mathematics & Statistics

This book series features volumes composed of select contributions from workshops and conferences in all areas of current research in mathematics and statistics, including OR and optimization. In addition to an overall evaluation of the interest, scientific quality, and timeliness of each proposal at the hands of the publisher, individual contributions are all refereed to the high quality standards of leading journals in the field. Thus, this series provides the research community with well-edited, authoritative reports on developments in the most exciting areas of mathematical and statistical research today.

Mehiddin Al-Baali • Lucio Grandinetti
Anton Purnama
Editors

Numerical Analysis and Optimization

NAO-III, Muscat, Oman, January 2014

Springer

Editors
Mehiddin Al-Baali
Department of Mathematics
College of Science
Sultan Qaboos University
Muscat, Oman

Lucio Grandinetti
Faculty of Engineering
University of Calabria
Arcavacada di Rende, Italy

Anton Purnama
Department of Mathematics and Statistics
College of Science
Sultan Qaboos University
Muscat, Oman

ISSN 2194-1009 ISSN 2194-1017 (electronic)
Springer Proceedings in Mathematics & Statistics
ISBN 978-3-319-17688-8 ISBN 978-3-319-17689-5 (eBook)
DOI 10.1007/978-3-319-17689-5

Library of Congress Control Number: 2015941730

Springer Cham Heidelberg New York Dordrecht London

Printed on acid-free paper

Springer International Publishing AG Switzerland is part of Springer Science+Business Media (www.springer.com)

Preface

This special edited book series of Springer Proceedings in Mathematics and Statistics contains 13 selected papers presented at the Third International Conference on Numerical Analysis and Optimization: Theory, Methods, Applications and Technology Transfer (NAOIII-2014) held during January 5–9, 2014, at Sultan Qaboos University (SQU), Muscat, Oman. The NAO conference series is held once every 3 years at SQU: the first conference (NAO-2008) was held on April 6–8, 2008, and the second conference (NAOII-2011) was held on January 3–6, 2011. The NAO conference will hopefully become a forum where prominent mathematicians, worldwide experts and active researchers gather and meet to share their knowledge on new scientific methodologies and simulate the communication of new innovative ideas, promote scientific exchange and discuss possibilities of further cooperation, networking and promotion of mobility of senior and young researchers and research students. NAOIII-2014 was inaugurated by the Under-Secretary of the Ministry of Higher Education, Vice Chancellor of SQU and Ambassador of Italy to the Sultanate. The conference was sponsored by SQU, The Research Council of Oman, The International Center for Theoretical Physics (ICTP, Italy), German University of Technology (GUtech) in Oman, AMPL (USA), Al-Anan Press (Oman) and Al-Roya Newspaper (Oman). Twenty world leading researchers gave keynote lectures. In total, 40 international participants contributed talks. After the conference, selected contributed papers were invited to submit for publication in a special issue of the following international journals: *Optimization Methods and Software*, *International Journal of Operational Research* and *SQU Journal for Science*. More information is available at http://conference.squ.edu.om/nao. Thirteen of the keynote papers were selected for this edited proceedings volume, each of which was accepted after a stringent peer review process by independent reviewers. We wish to express our gratitude to all contributors. We are also indebted to many anonymous referees for the care taken in reviewing the papers submitted for publication.

Muscat, Oman — Mehiddin Al-Baali
Arcavacada di Rende, Italy — Lucio Grandinetti
Muscat, Oman — Anton Purnama

Contents

Invited speakers and members of the organizing committee at Sultan Qaboos University

Contributors

Mehiddin Al-Baali Department of Mathematics and Statistics, Sultan Qaboos University, Muscat, Oman

Pietro Belotti Xpress Optimization Team, FICO, Birmingham, UK

Marida Bertocchi Department of Management, Economics and Quantitative Methods, University of Bergamo, Bergamo, Italy

J.C. Butcher The University of Auckland, Auckland, New Zealand

Andrew R. Conn IBM Thomas J. Watson Research Center, Yorktown Heights, NY, USA

Yu-Hong Dai State Key Laboratory of Scientific and Engineering Computing, Chinese Academy of Sciences, Beijing, P.R. China

Leonhard Frerick FB IV – Mathematik, University of Trier, Trier, Germany

David M. Gay AMPL Optimization Inc., Albuquerque, NM, USA

Julio C. Góez GERAD and École Polytechnique de Montréal, Montreal, QC, Canada

Effat Golpar-Raboky Department of Mathematics, University of Qom, Qom, Iran

Nick I.M. Gould Computational Science and Engineering Department, Rutherford Appleton Laboratory, Chilton, Oxfordshire, UK

M. Hintermüller Department of Mathematics, Humboldt-Universität zu Berlin, Berlin, Germany

Lior Horesh IBM Thomas J. Watson Research Center, Yorktown Heights, NY, USA

Eduardo A. Jimenez Shell, Katy, TX, USA

C. Löbhard Department of Mathematics, Humboldt-Universität zu Berlin, Berlin, Germany

Ding Ma Department of Management Science and Engineering, Stanford University, Stanford, CA, USA

Nezam Mahdavi-Amiri Faculty of Mathematical Sciences, Sharif University of Technology, Tehran, Iran

Jakub Mareček IBM Research – Ireland, Dublin, Ireland

Dominique Orban Département de Mathématiques et Génie Industriel, GERAD and École Polytechnique de Montréal, Montréal, QC, Canada

Paolo Pisciella Department of Management, Economics and Quantitative Methods, University of Bergamo, Bergamo, Italy

Imre Pólik SAS Institute, Cary, NC, USA

Ted K. Ralphs Department of Industrial and Systems Engineering, Lehigh University, Bethlehem, PA, USA

Peter Richtárik School of Mathematics, University of Edinburgh, Edinburgh, UK

Kees Roos Delft University of Technology, Delft, Netherlands

Ekkehard W. Sachs FB IV – Mathematik, University of Trier, Trier, Germany

Michael A. Saunders Department of Management Science and Engineering, Stanford University, Stanford, CA, USA

Martin Takáč Department of Industrial & Systems Engineering, Lehigh University, Bethlehem, PA, USA

M.H. Tber Faculté des Sciences et Technique, Université Sultan Moulay Slimane, Béni-Mellal, Morocco

Tamás Terlaky Department of Industrial and Systems Engineering, Lehigh University, Bethlehem, PA, USA

Philippe L. Toint Department of Mathematics, University of Namur, Namur, Belgium

Gijs M. van Essen Shell, Katy, TX, USA

Maria Teresa Vespucci Department of Management, Economics and Quantitative Methods, University of Bergamo, Bergamo, Italy

Xiaoqi Yang Department of Applied Mathematics, The Hong Kong Polytechnic University, Kowloon, Hong Kong

Lukas A. Zimmer FB IV – Mathematik, University of Trier, Trier, Germany

A Conic Representation of the Convex Hull of Disjunctive Sets and Conic Cuts for Integer Second Order Cone Optimization

Pietro Belotti, Julio C. Góez, Imre Pólik, Ted K. Ralphs, and Tamás Terlaky

Abstract We study the convex hull of the intersection of a convex set E and a disjunctive set. This intersection is at the core of solution techniques for *Mixed Integer Convex Optimization*. We prove that if there exists a cone K (resp., a cylinder C) that has the same intersection with the boundary of the disjunction as E, then the convex hull is the intersection of E with K (resp., C).

The existence of such a cone (resp., a cylinder) is difficult to prove for general conic optimization. We prove existence and unicity of a second order cone (resp., a cylinder), when E is the intersection of an affine space and a second order cone (resp., a cylinder). We also provide a method for finding that cone, and hence the convex hull, for the continuous relaxation of the feasible set of a Mixed Integer Second Order Cone Optimization (MISOCO) problem, assumed to be the intersection of an ellipsoid with a general linear disjunction. This cone provides a new conic cut for MISOCO that can be used in branch-and-cut algorithms for MISOCO problems.

Keywords Conic cuts • Mixed integer optimization • Second order cone optimization

Subject Classification: 90C10, 90C11, 90C20

P. Belotti
Xpress Optimization Team, FICO, Birmingham B37 7GN, UK
e-mail: pietrobelotti@fico.com

J.C. Góez
GERAD and École Polytechnique de Montréal, Montreal, QC H3C 3A7, Canada
e-mail: jgoez1@gmail.com

I. Pólik
SAS Institute, 100 SAS Campus Drive, Cary, NC 27513-2414, USA
e-mail: imre@polik.net

T.K. Ralphs • T. Terlaky (✉)
Department of Industrial and Systems Engineering, Lehigh University, 200 West Packer Ave, Bethlehem, PA 18015, USA
e-mail: ted@lehigh.edu; terlaky@lehigh.edu; tat208@lehigh.edu

M. Al-Baali et al. (eds.), *Numerical Analysis and Optimization*, Springer Proceedings in Mathematics & Statistics 134, DOI 10.1007/978-3-319-17689-5_1

1 Introduction

We consider the very general class of Mixed Integer Convex Optimization problems, which can be formulated as $\min\{c^\top x : x \in \mathscr{E}, x \in \mathbb{Z}^p \times \mathbb{R}^{n-p}\}$, where $\mathscr{E}$ is a closed convex set. Solving such a problem often requires finding the convex hull of the intersection of $\mathscr{E}$ with a disjunction $\mathscr{A} \cup \mathscr{B}$, where $\mathscr{A}$ and $\mathscr{B}$ are two half-spaces. In the first part of this paper, using proper disjointness, nonempty, and boundedness assumptions on the intersection $\mathscr{E} \cap (\mathscr{A} \cup \mathscr{B})$, we prove that $\mathrm{conv}(\mathscr{E} \cap (\mathscr{A} \cup \mathscr{B}))$ is the intersection of $\mathscr{E}$ with an appropriate cone $\mathscr{K}$.

In the second part of this paper, we apply our result to a specific subclass of optimization problems where the set $\mathscr{E}$ is the intersection of an affine space and a second order cone. In order to establish a mindset that encompasses both mixed integer convex problems and mixed integer conic problems, we explicitly describe $\mathscr{E}$ as the intersection of a cone and an affine subspace. Note that any mixed integer convex problem can be described as a mixed integer conic optimization (MICO) problem and vice versa, since the former is a superset of the latter and any convex problem can be turned into a conic one by adding an auxiliary variable. Therefore, we consider problems of the form:

$$\begin{aligned} &\text{minimize: } c^\top x \\ &\text{subject to: } Ax = r \qquad\qquad \text{(MICO)} \\ &\qquad\qquad x \in \mathscr{K} \\ &\qquad\qquad x \in \mathbb{Z}^l \times \mathbb{R}^{n-l}, \end{aligned} \tag{1}$$

where $A \in \mathbb{R}^{m\times n}$, $c \in \mathbb{R}^n$, $r \in \mathbb{R}^m$, $\mathscr{K}$ is a convex cone, and the rows of A are linearly independent.

MICO problems comprise a wide range of discrete optimization problems. A very important class of MICO is the class of mixed integer second order cone optimization (MISOCO) problems, which find applications in engineering, finance, and inventory problems [3, 10, 16, 20]. Theoretically, the integrality constraint can be tackled by means of a generic branch-and-bound algorithm. However, experience with mixed integer linear optimization (MILO) has shown that the development of methods for generating valid inequalities for the problem can improve the efficiency of the algorithm significantly [12]. The aim of this paper is the development of conic cuts for MICO problems.[1]

MICO problems are a class of non-convex optimization problems in which the non-convexity comes from the integrality of a subset of variables. Such non-convexity can be dealt with by means of disjunctive methods, which partition the set of feasible solutions into two or more feasible subsets. Disjunctive methods

[1] A cone is called a *conic cut* if it cuts off some non-integer solutions but none of the feasible integer solutions.

in mixed integer linear optimization have been studied extensively during the past decades [4, 5, 13, 17]. The contribution of this paper is twofold. First, we introduce conditions for the existence of a conic inequality arising as a disjunctive inequality for the general case of MICO that yields the convex hull of the intersection between a convex set and a disjunctive set (defined below). Second, we describe a procedure to find such a cut in the MISOCO case. The latter result allows us to generate second order cones for tightening the continuous relaxation of the MISOCO problem.

This paper is organized as follows. In Section 2 we present a brief review of the previous work done in MICO. Then, in Section 3 we derive conditions for the existence and unicity of the convex hull of the intersection between a disjunctive set and a closed convex set. In Section 4 we consider the special case of MISOCO: we introduce the disjunctive conic cut and a procedure to find it. We then compare our disjunctive cut with the conic cut introduced in [1] in Section 5. We provide some concluding remarks in Section 6.

Notation

Sets are denoted by script capital letters, matrices by capital letters, vectors by lowercase letters, and scalars by Greek letters. For a matrix M, M_{ij} is the (i,j) element, while M_j is the jth column. For a vector v, its ith component is denoted as v_i. For any two sets $\mathscr{S}_1$ and $\mathscr{S}_2$, the expression $\mathscr{S}_1 \subset \mathscr{S}_2$ is used to denote that $\mathscr{S}_1$ is a proper subset of $\mathscr{S}_2$. The notation $\text{ri}(\bullet)$ is used to refer to the relative interior of a set.

2 Literature Review

There have been several attempts to extend some of the techniques developed for MILO to the case of MICO. For the MISOCO case, one approach uses outer linear approximations of second order cones. Vielma et al. [25] used the polynomial-size polyhedral relaxation introduced by Ben-Tal and Nemirovski [9] in their "lifted linear programming" branch-and-bound algorithm for MISOCO problems. Krokhmal and Soberanis [19] generalized this approach for integer p-order conic optimization. Drewes [15] presented subgradient-based linear outer approximations for the second order cone constraints. This allows one to approximate the MISOCO problem by a mixed integer linear problem in a hybrid outer approximation branch-and-bound algorithm.

Stubbs and Mehrotra [24] generalized the lift-and-project algorithm of Balas et al. [6] for 0-1 MILO to 0-1 mixed integer convex problems. Later, Çezik and Iyengar [11] investigated the generation of valid convex cuts for 0-1 MICO problems and discussed how to extend the Chvátal-Gomory procedure for generating linear cuts for MICO problems and the extension of *lift-and-project* techniques for MICO problems. In particular, they showed how to generate linear and convex quadratic valid inequalities using the relaxation obtained by a project procedure. Later, Drewes [15] reviews the ideas proposed in [11] and [24] and applies them to MISOCO.

Atamtürk and Narayanan [1, 2] proposed two procedures for MISOCO problems that generate valid second order conic cuts. They first studied a generic lifting procedure for MICO [2], and then [1] extended the *mixed integer rounding* [22] procedure to the MISOCO case. The main idea in [1] is to reformulate a second order conic constraint using a set of two-dimensional second order cones. In this new reformulation the set of inequalities are called *polyhedral second-order conic constraint*. The authors used polyhedral analysis for studying these inequalities separately. This allowed the derivation of a mixed integer rounding procedure, which yields a *nonlinear conic mixed integer rounding*. A generalization of the use of polyhedral second-order conic constraints is presented by Masihabadi et al. [21].

Dadush et al. [14] studied the *split closure* of a strictly convex body and present a *conic quadratic inequality*. The conic quadratic inequality is introduced to present an example of a non-polyhedral split closure. In particular, the authors showed that it is necessary to consider conic quadratic inequalities in order to be able to describe the split closure of an ellipsoid. This independently obtained *conic quadratic inequality* coincides with the conic cut for MISOCO problems presented in Section 4.1.

3 The Convex Hull of a Disjunctive Convex Set

We focus on the convex hull of the intersection of a full-dimensional closed convex set $\mathscr{E} \subseteq \mathbb{R}^n$, $n > 1$ with a *disjunctive set*. Consider a *disjunctive set* of the form

$$\mathscr{A} \cup \mathscr{B}, \tag{2}$$

where $\mathscr{A} = \{x \in \mathbb{R}^n \mid a^\top x \geq \alpha\}$ and $\mathscr{B} = \{x \in \mathbb{R}^n \mid b^\top x \leq \beta\}$ are two half-spaces with $a, b \in \mathbb{R}^n$, and (a, α), (b, β) are not proportional, i.e., $\nexists \eta \in \mathbb{R}$ such that $a = \eta b$, $\alpha = \eta \beta$. This section presents a characterization of the convex hull of the set $\mathscr{E} \cap (\mathscr{A} \cup \mathscr{B})$.

Let $\mathscr{A}^= = \{x \in \mathbb{R}^n \mid a^\top x = \alpha\}$ and $\mathscr{B}^= = \{x \in \mathbb{R}^n \mid b^\top x = \beta\}$ denote the boundary hyperplanes of the half-spaces $\mathscr{A}$ and $\mathscr{B}$, respectively. Throughout this paper, we assume the following about the sets $\mathscr{E}$, $\mathscr{A}$, and $\mathscr{B}$:

Assumption 1. *$\mathscr{A} \cap \mathscr{B} \cap \mathscr{E}$ is empty.*

Assumption 2. *$\mathscr{E} \cap \mathscr{A}^=$ and $\mathscr{E} \cap \mathscr{B}^=$ are nonempty and bounded.*

3.1 Disjunctive Conic Cut

Let us recall the definition of a convex cone, as given by Barvinok [7, page 65].

Definition 1 (Convex Cone). A set $\mathscr{K} \subseteq \mathbb{R}^n$ is a convex cone if $0 \in \mathscr{K}$ and if for any two points $x, y \in \mathscr{K}$ and for any $\theta, \vartheta \geq 0$, we have $z = \theta x + \vartheta y \in \mathscr{K}$.

Remark 1. Observe that we can define a set $\hat{\mathscr{K}}$ as a translated cone if there exists a vector $x^* \in \hat{\mathscr{K}}$, called the *vertex* of $\hat{\mathscr{K}}$, such that for any $\theta, \vartheta \geq 0$ and $x, y \in \hat{\mathscr{K}}$, $x^* + (\theta(x - x^*) + \vartheta(y - x^*)) \in \hat{\mathscr{K}}$. One can use the translation $\mathscr{K} = \{y \in \mathbb{R}^n \mid y = x - x^*, x \in \hat{\mathscr{K}}\}$ to get a cone $\mathscr{K}$ in the sense of Definition 1. Although translated cones arise naturally in this setting, we assume w.l.o.g. that all cones have a vertex at the origin unless otherwise specified.

Definition 2. A closed convex cone $\mathscr{K} \subset \mathbb{R}^n$ with $\dim(\mathscr{K}) > 1$ is called a *disjunctive conic cut* (DCC) for the set $\mathscr{E}$ and the disjunctive set $\mathscr{A} \cup \mathscr{B}$ if

$$\operatorname{conv}(\mathscr{E} \cap (\mathscr{A} \cup \mathscr{B})) = \mathscr{E} \cap \mathscr{K}.$$

The following proposition gives a sufficient condition for a convex cone $\mathscr{K}$ to be a disjunctive conic cut for the set $\mathscr{E} \cap (\mathscr{A} \cup \mathscr{B})$.

Proposition 1. *A convex cone $\mathscr{K} \subset \mathbb{R}^n$ with $\dim(\mathscr{K}) > 1$ is a DCC for $\mathscr{E}$ and the disjunctive set $\mathscr{A} \cup \mathscr{B}$ if*

$$\mathscr{K} \cap \mathscr{A}^= = \mathscr{E} \cap \mathscr{A}^= \quad \textit{and} \quad \mathscr{K} \cap \mathscr{B}^= = \mathscr{E} \cap \mathscr{B}^=. \tag{3}$$

Figure 1 illustrates Proposition 1, where the set $\mathscr{E} \subset \mathbb{R}^3$ is the epigraph of a paraboloid. Before proving Proposition 1, we first provide a set of lemmas that will make the proof more compact. To begin, let us recall the definition of a *base of a cone* presented by Barvinok [7, page 66].

Definition 3 (Base of a Cone). Let $\mathscr{K} \subset \mathbb{R}^n$ be a convex cone. A set $\mathscr{L} \subset \mathscr{K}$ is called a *base* of $\mathscr{K}$ if $0 \notin \mathscr{L}$ and for every point $u \in \mathscr{K}$, $u \neq 0$, there is a unique $v \in \mathscr{L}$ and $\lambda > 0$ such that $u = \lambda v$.

We can use Definition 3 to state Lemma 1, which shows a key relationship between the cone $\mathscr{K}$ and the hyperplanes $\mathscr{A}^=$ and $\mathscr{B}^=$.

Lemma 1. *Consider a half-space $\mathscr{G} = \{x \in \mathbb{R}^n \mid g^\top x \leq \rho\}$. Assume that $\mathscr{E} \cap \mathscr{G}^=$ is nonempty, bounded, and does not contain the origin 0. If there exists a convex cone $\mathscr{K} \subseteq \mathbb{R}^n$, with $\dim(\mathscr{K}) > 1$ and $\mathscr{K} \cap \mathscr{G}^= = \mathscr{E} \cap \mathscr{G}^=$, then $\mathscr{E} \cap \mathscr{G}^=$ is a base of $\mathscr{K}$.*

Proof. From the assumptions in the lemma, we have that $0 \notin \mathscr{K} \cap \mathscr{G}^= = \mathscr{E} \cap \mathscr{G}^=$. We may assume w.l.o.g. that $0 \in \mathscr{G}$. First, since $\mathscr{K} \cap \mathscr{G}^= = \mathscr{E} \cap \mathscr{G}^=$ is bounded we know that there exists no ray of $\mathscr{K}$ parallel to $\mathscr{G}^=$. Now, let us suppose that $\mathscr{E} \cap \mathscr{G}^=$ is not a base for $\mathscr{K}$. From Definition 3 we know that there must exist a point x such that $x \in \mathscr{K}$ but there exists no point $\hat{x} \in \mathscr{E} \cap \mathscr{G}^=$ for uniquely representing x as $\lambda \hat{x}$ for

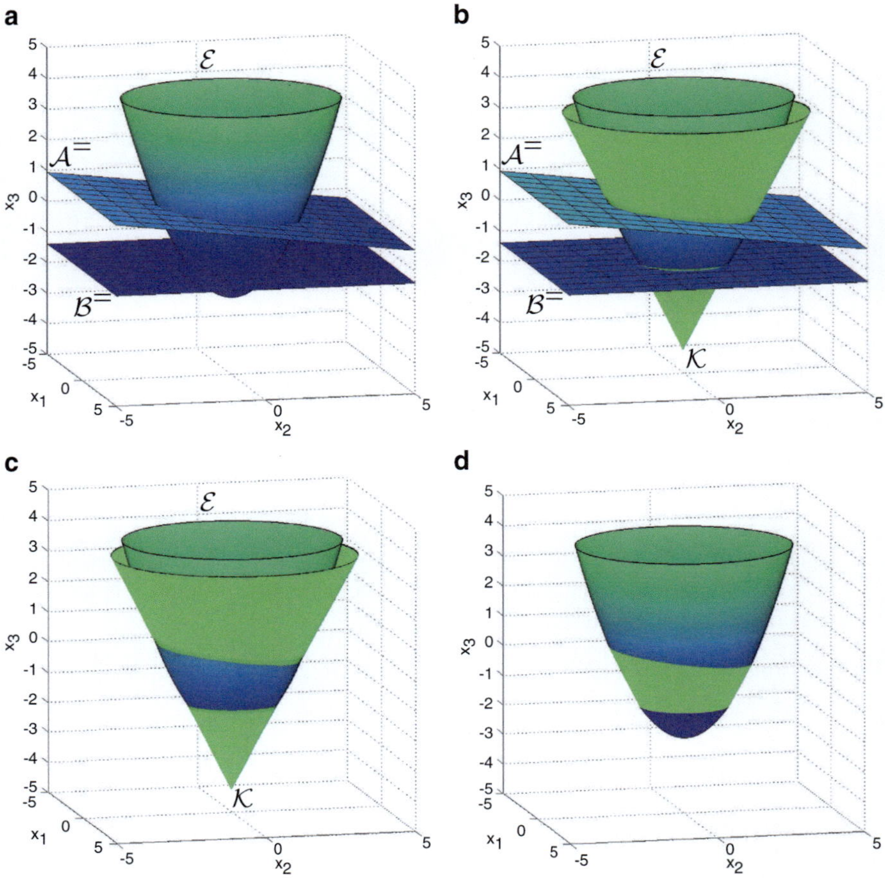

Fig. 1 Illustration of a disjunctive conic cut as specified in Proposition 1. (**a**) $\mathscr{A}^{=}$, $\mathscr{B}^{=}$, and $\mathscr{E}$. (**b**) The cone yielding $\mathrm{conv}(\mathscr{E} \cap (\mathscr{A} \cup \mathscr{B}))$. (**c**) $\mathscr{E} \cap \mathscr{K}$. (**d**) $\mathrm{conv}(\mathscr{E} \cap (\mathscr{A} \cup \mathscr{B}))$

some $\lambda > 0$. Then, there is a ray in $\mathscr{K}$ parallel to the hyperplane $\mathscr{G}^{=}$. This implies that the set $\mathscr{K} \cap \mathscr{G}^{=}$ is unbounded, which contradicts the boundedness assumption of $\mathscr{E} \cap \mathscr{G}^{=}$. Therefore, $\mathscr{E} \cap \mathscr{G}^{=}$ is a base for $\mathscr{K}$. $\square$

The result of Lemma 1 allows us to show that $\mathscr{K}$ is a pointed cone, which is an important result for our further development.

Lemma 2. *Any convex cone $\mathscr{K}$ satisfying Lemma 1 must be pointed.*

Proof. Assume that $\mathscr{K}$ is not pointed. This means that $\mathscr{K}$ contains a line. Hence, there exist two vectors $\hat{r}, \bar{r} \in \mathscr{K} \setminus \{0\}$ such that $\hat{r} = -\bar{r}$. Additionally, we have that $\mu\hat{r} + \nu\bar{r} \in \mathscr{K}$, for any $\mu, \nu > 0$. Now, since $\mathscr{E} \cap \mathscr{G}^{=}$ is a base of $\mathscr{K}$, there exists a point $\hat{x} \in \mathscr{E} \cap \mathscr{G}^{=}$ in the ray defined by $\hat{r}$ such that $\hat{x} = \mu\hat{r}$, for some $\mu > 0$. Similarly, there exists a point $\bar{x} \in \mathscr{E} \cap \mathscr{G}^{=}$ in the ray defined by $\bar{r}$ and $\nu > 0 \in \mathbb{R}$ such that $\bar{x} = \nu\bar{r}$.

Given that $\mathscr{G}^=$ is an affine set, we have

$$\gamma\hat{x} + (1-\gamma)\bar{x} \in \mathscr{G}^=, \quad \forall \gamma \in \mathbb{R}.$$

Expressing $\hat{x}$ and $\bar{x}$ in term of $\hat{r}$ and $\bar{r}$ gives

$$\begin{aligned}\gamma\hat{x} + (1-\gamma)\bar{x} &= \gamma(\mu\hat{r}) + (1-\gamma)(\nu\bar{r}) \\ &= -\gamma(\mu\bar{r}) + (1-\gamma)(\nu\bar{r}) \\ &= \nu\bar{r} - \gamma(\mu+\nu)\bar{r}.\end{aligned}$$

Hence, if $\gamma = 0$, then $\nu\bar{r} \in \mathscr{K}$. On the other hand, if $\gamma < 0$, we get that $\nu\bar{r} - \gamma(\mu+\nu)\bar{r} \in \mathscr{K}$, since it is a point on the ray defined by $\bar{r}$. Finally, if $\gamma > 0$, then $\nu\bar{r} - \gamma(\mu+\nu)\bar{r} = \nu\bar{r} + \gamma(\mu+\nu)\hat{r} \in \mathscr{K}$, since it is a positive combination of two points in the cone $\mathscr{K}$. Hence, $\mathscr{K} \cap \mathscr{G}^=$ contains a whole line, which contradicts the assumption that $\mathscr{K} \cap \mathscr{G}^=$ is bounded. □

We can now prove that the vertex of the cone $\mathscr{K}$ belongs exclusively to either $\mathscr{A}$ or $\mathscr{B}$. Observe that this does not mean that the set $\mathscr{A} \cap \mathscr{B}$ is empty, but that the vertex of $\mathscr{K}$ is not contained in it even when $\mathscr{A} \cap \mathscr{B}$ is nonempty.

Lemma 3. *Let $\mathscr{K} \subseteq \mathbb{R}^n$ be a convex cone, with* $\dim(\mathscr{K}) > 1$, *such that $\mathscr{E} \cap \mathscr{A}^= = \mathscr{K} \cap \mathscr{A}^=$ and $\mathscr{E} \cap \mathscr{B}^= = \mathscr{K} \cap \mathscr{B}^=$. Then the origin $x^* = 0$ is either in $\mathscr{A}$ or in $\mathscr{B}$, but not in $\mathscr{A} \cap \mathscr{B}$.*

Proof. First, consider the case when $x^* \in \mathscr{A}^=$. Then, we have that $x^* \in \mathscr{E} \cap \mathscr{A}^=$, since $\mathscr{E} \cap \mathscr{A}^= = \mathscr{K} \cap \mathscr{A}^=$. Hence, from Assumption 1, we have that $x^* \notin \mathscr{B}$. Similarly, we have that if $x^* \in \mathscr{B}^=$, then $x^* \notin \mathscr{A}$.

Second, assume that neither $\mathscr{A}^=$ nor $\mathscr{B}^=$ contain x^*. By Lemma 1 and Assumption 2 we have that $\mathscr{E} \cap \mathscr{A}^=$ and $\mathscr{E} \cap \mathscr{B}^=$ are bases of the cone $\mathscr{K}$. Additionally, by Lemma 2 we know that the cone $\mathscr{K}$ is pointed. Let x be a unit length vector defining a ray of $\mathscr{K}$. Then, there are two points $\hat{x} \in \mathscr{E} \cap \mathscr{A}^=$ and $\bar{x} \in \mathscr{E} \cap \mathscr{B}^=$ such that $\hat{x} = \mu x$ and $\bar{x} = \nu x$ for some $\mu, \nu > 0$.

We prove first that $x^* \in \mathscr{A} \cup \mathscr{B}$. Let us assume to the contrary that $x^* \in \bar{\mathscr{A}} \cap \bar{\mathscr{B}}$, where the bar denotes the complement set. Let $y = \gamma x$ for $\gamma \geq 0$ be a point in the ray defined by x. Then, for any $\gamma < \min\{\nu, \mu\}$ we have that $y \in \bar{\mathscr{A}} \cap \bar{\mathscr{B}}$, and w.l.o.g. we may assume that $\nu < \mu$. Note that we cannot have $\nu = \mu$ as, by Assumption 1, $\mathscr{A} \cap \mathscr{B} \cap \mathscr{E} = \emptyset$. Additionally, for any $\gamma \geq \nu$ we have that $y \in \mathscr{B}$, so the point $\hat{x}$ is contained in the half-space $\mathscr{B}$, and $\mathscr{A} \cap \mathscr{B} \cap \mathscr{E} \neq \emptyset$, which contradicts Assumption 1.

Now, we prove that $x^* \notin \mathscr{A} \cap \mathscr{B}$. Let us assume to the contrary that $x^* \in \mathscr{A} \cap \mathscr{B}$, and let $y = \gamma\bar{x} + (1-\gamma)\hat{x}$ for some $0 \leq \gamma \leq 1$. Then, we have that $y \in \mathscr{A}$ or $y \in \mathscr{B}$. When $\nu < \mu$, set $\gamma = 1$ such that $y = \bar{x}$ and we have $y \in \mathscr{A} \cap \mathscr{B}^= \cap \mathscr{E}$. Similarly, when $\mu < \nu$, set $\gamma = 0$ such that $y = \hat{x}$ and we have $y \in \mathscr{A}^= \cap \mathscr{B} \cap \mathscr{E}$. Hence, $x^* \in \mathscr{A} \cap \mathscr{B}$ implies $\mathscr{A} \cap \mathscr{B} \cap \mathscr{E} \neq \emptyset$, which contradicts Assumption 1. □

We are able now to show that $\mathscr{E} \cap (\mathscr{A} \cup \mathscr{B}) \subset \mathscr{K}$. This will facilitate the proof of the relation $\operatorname{conv}(\mathscr{E} \cap (\mathscr{A} \cup \mathscr{B})) \subseteq \mathscr{E} \cap \mathscr{K}$.

Lemma 4. *Let* $\mathscr{K} \subseteq \mathbb{R}^n$ *be a convex cone, with* $\dim(\mathscr{K}) > 1$*, for which* (3) *holds. Then:*

$$(\mathscr{E} \cap \mathscr{A}) \subset \mathscr{K} \quad \textit{and} \quad (\mathscr{E} \cap \mathscr{B}) \subset \mathscr{K}.$$

Proof. We start showing that if $\mathscr{E} \cap \mathscr{A}^=$ is a single point, then $\mathscr{E} \cap \mathscr{A} \subseteq \mathscr{K}$. First, if $\mathscr{E} \cap \mathscr{A}^=$ is a single point, then $0 \in \mathscr{E} \cap \mathscr{A}^=$, otherwise $\dim(\mathscr{K}) = 1$. The last statement follows from the assumption $\mathscr{E} \cap \mathscr{A}^= = \mathscr{K} \cap \mathscr{A}^=$. Henceforth, we obtain in this case that $\mathscr{E} \cap \mathscr{A}^= = \{0\}$. Now, it is clear that $\mathscr{E} \cap \mathscr{A}^= \subseteq \mathscr{E} \cap \mathscr{A}$, hence we need to show that $\mathscr{E} \cap \mathscr{A} \subseteq \mathscr{E} \cap \mathscr{A}^=$. Assume to the contrary that there exists a point $x \in \mathscr{E} \cap \mathscr{A}$ such that $x \notin \mathscr{E} \cap \mathscr{A}^=$. Additionally, consider a point $y \in \mathscr{E} \cap \mathscr{B}^=$ such that the vertex 0 is not contained in the line induced by x and y, which implies that $y \notin \mathscr{E} \cap \mathscr{A}$. Note that the point y exists because $\dim(\mathscr{K}) = 1$. Then x, y are in $\mathscr{E}$, and $a^\top x > 0 > a^\top y$. Additionally, from convexity of $\mathscr{E}$ we have that for any $0 \leq \gamma \leq 1$ the point $\gamma x + (1-\gamma)y \in \mathscr{E}$. Thus, there exists a convex combination z of x and y with $a^\top z = 0$, i.e. z in $(\mathscr{E} \cap \mathscr{A}^=) \setminus \{0\}$, which contradicts that $\mathscr{E} \cap \mathscr{A}^=$ is a single point. Henceforth, we obtain that $\mathscr{E} \cap \mathscr{A} = \mathscr{E} \cap \mathscr{A}^= \subseteq \mathscr{K}$. Similarly, if $\mathscr{E} \cap \mathscr{B}^=$ is a single point, then one can show that $\mathscr{E} \cap \mathscr{B} = \mathscr{E} \cap \mathscr{B}^= \subseteq \mathscr{K}$. Note that $\mathscr{E} \cap \mathscr{A}^=$ and $\mathscr{E} \cap \mathscr{B}^=$ cannot be single points simultaneously, which follows from Assumption 1 and $\dim(\mathscr{K}) > 1$.

Now, we prove that if $\mathscr{E} \cap \mathscr{A}^=$ is not a single point then $(\mathscr{E} \cap \mathscr{A}) \subseteq \mathscr{K}$. Let us assume to the contrary that there exists a vector x such that $x \in (\mathscr{E} \cap \mathscr{A})$ but $x \notin \mathscr{K}$. First, by the separation theorem,[2] there exists a hyperplane $\mathscr{H}$ separating x and $\mathscr{K}$ that contains a ray of $\mathscr{K}$, denoted by $\mathscr{K}_r$, and does not contain x. Here the assumption of $\dim(\mathscr{K}) > 1$ is needed, since the hyperplane $\mathscr{H}$ does not exist when $\dim(\mathscr{K}) = 1$.

From Lemma 3 we know that $0 \in \mathscr{A}$ or $0 \in \mathscr{B}$. On the one hand, if $0 \notin \mathscr{E} \cap \mathscr{B}^=$, then it follows from (3), Assumption 2, Lemma 1, that the set $\mathscr{E} \cap \mathscr{B}^=$ is a base for the cone $\mathscr{K}$. Hence, there exists a vector $w \in \mathscr{E} \cap \mathscr{B}^=$ that defines the ray $\mathscr{K}_r$. On the other hand, if $0 \in \mathscr{E} \cap \mathscr{B}^=$, then we know that $\mathscr{E} \cap \mathscr{B}^= = \{0\}$. In this case, one can take $w = 0$, since $0 \in \mathscr{K}_r$.

Given that the set $\mathscr{E}$ is convex, $\lambda x + (1-\lambda)w \in \mathscr{E}$ for all $0 \leq \lambda \leq 1$. On the other hand, since w is a point on a face of $\mathscr{K}$, we have that $\lambda x + (1-\lambda)w \notin \mathscr{K}$ for $0 < \lambda \leq 1$. Furthermore, since $x \in (\mathscr{E} \cap \mathscr{A})$ and $\mathscr{A} \cap \mathscr{B} \cap \mathscr{E} = \emptyset$, we have $a^\top x \geq \alpha$ and $a^\top w < \alpha$. Hence, from the equation $a^\top(\lambda x + (1-\lambda)w) = \lambda a^\top x + (1-\lambda)a^\top w$, there exists a $\lambda \in (0,1]$ such that $a^\top(\lambda x + (1-\lambda)w) = \alpha$. Therefore, there is a

[2] Lemma 8.2 in Barvinok [7, page 65] and Theorems 11.3 and 11.7 in Rockafeller [23, pages 97 and 100].

vector $\hat{x} = \lambda x + (1-\lambda)w$ for some $\lambda \in (0,1]$, such that $\hat{x} \in \mathscr{E} \cap \mathscr{A}^=$, but $\hat{x} \notin \mathscr{K}$, which contradicts condition (3). Hence, $(\mathscr{E} \cap \mathscr{A}) \subseteq \mathscr{K}$. Analogously, one can prove that $(\mathscr{E} \cap \mathscr{B}) \subseteq \mathscr{K}$ when $(\mathscr{E} \cap \mathscr{B})$ is not a single point.

Recall that the sets $\mathscr{E} \cap \mathscr{A}^=$ and $\mathscr{E} \cap \mathscr{B}^=$ are disjoint and nonempty. Then, by condition (3) we have that $\mathscr{E} \cap \mathscr{A} \neq \mathscr{K}$ and $\mathscr{E} \cap \mathscr{B} \neq \mathscr{K}$, and the result of the lemma follows. □

Now we present the proof of Proposition 1.

Proof (Proof of Proposition 1). We first prove that $\text{conv}(\mathscr{E} \cap (\mathscr{A} \cup \mathscr{B})) = \mathscr{E} \cap \mathscr{K}$. Consider a point $x \in (\mathscr{E} \cap \mathscr{A}) \cup (\mathscr{E} \cap \mathscr{B})$. Then, from Lemma 4 we have that $x \in \mathscr{E} \cap \mathscr{K}$. Now, consider any two points $x, y \in (\mathscr{E} \cap \mathscr{A}) \cup (\mathscr{E} \cap \mathscr{B})$. Then, since both $\mathscr{K}$ and $\mathscr{E}$ are convex, for any $0 \leq \lambda \leq 1$ we have $\lambda x + (1-\lambda)y \in \mathscr{E} \cap \mathscr{K}$. Hence, $\text{conv}(\mathscr{E} \cap (\mathscr{A} \cup \mathscr{B})) \subseteq \mathscr{E} \cap \mathscr{K}$.

Consider a point $x \in \mathscr{E} \cap \mathscr{K}$. First, if $x \in \mathscr{E} \cap \mathscr{A}$ or $x \in \mathscr{E} \cap \mathscr{B}$, we have that $x \in \text{conv}(\mathscr{E} \cap (\mathscr{A} \cup \mathscr{B}))$. Assume then that $x \notin (\mathscr{E} \cap \mathscr{A}) \cup (\mathscr{E} \cap \mathscr{B})$, which implies $x \in (\bar{\mathscr{A}} \cap \bar{\mathscr{B}} \cap \mathscr{K})$. Furthermore, there are two vectors $\hat{x} \in \mathscr{E} \cap \mathscr{A}^=$ and $\bar{x} \in \mathscr{E} \cap \mathscr{B}^=$ such that, for some $\mu, \nu \geq 0$, $\hat{x} = \mu x$ and $\bar{x} = \nu x$. This last statement follows from Lemma 1 directly if $\mathscr{E} \cap \mathscr{A}^=$ and $\mathscr{E} \cap \mathscr{B}^=$ are not single points. Now, if one of the intersections is a single point, in the proof of Lemma 4 we showed that such intersection must be the vertex of $\mathscr{K}$. In this case, the statement follows from applying Lemma 1 to the remaining intersection. From Lemma 3, the vertex of the cone is either in $\mathscr{A}$ or $\mathscr{B}$ but not in both. Assume w.l.o.g. that the vertex of the cone is in $\mathscr{B}$. Then, $\nu < 1 < \mu$ and there exists a $\gamma \in (0,1)$ such that $\gamma\nu + (1-\gamma)\mu = 1$. Hence, we can write

$$\begin{aligned}
\gamma\bar{x} + (1-\gamma)\hat{x} &= \gamma\nu x + (1-\gamma)\mu x \\
&= (\gamma\nu + (1-\gamma)\mu)x \\
&= x.
\end{aligned}$$

Therefore, x can be expressed as a convex combination of two points in $(\mathscr{E} \cap \mathscr{A}^=) \cup (\mathscr{E} \cap \mathscr{B}^=)$. Hence, any point $x \in (\mathscr{E} \cap \mathscr{K})$ can be written as a convex combination of two points in $(\mathscr{E} \cap \mathscr{A}) \cup (\mathscr{E} \cap \mathscr{B})$. Thus, $(\mathscr{E} \cap \mathscr{K}) \subseteq \text{conv}(\mathscr{E} \cap (\mathscr{A} \cup \mathscr{B}))$. Finally, since the subset relation is valid in both directions, this proves that $(\mathscr{E} \cap \mathscr{K}) = \text{conv}(\mathscr{E} \cap (\mathscr{A} \cup \mathscr{B}))$. Finally, since $(\mathscr{E} \cap \mathscr{A}^=)$ and $(\mathscr{E} \cap \mathscr{B}^=)$ are compact sets, then it follows from Lemma 1 and Lemma 8.6 in Barvinok [7, page 67] that $\mathscr{K}$ is closed. □

We close the analysis by showing that if a cone $\mathscr{K}$ exists satisfying the conditions of Proposition 1 exist, then it is unique.

Lemma 5. *If a closed convex cone* $\mathscr{K}$ *exists, with* $\dim(\mathscr{K}) > 1$, *satisfying property* (3), *then* $\mathscr{K}$ *is unique.*

Proof. Assume to the contrary that there are two different cones $\mathscr{K}_1$ and $\mathscr{K}_2$ that satisfy property (3). Let $v^1 \in \mathscr{K}_1$ be the vertex of $\mathscr{K}_1$ and $v^2 \in \mathscr{K}_2$ be the vertex of $\mathscr{K}_2$. Now, we may assume w.l.o.g. that $v^1 = 0$.

First, we prove that if either $\mathscr{E} \cap \mathscr{A}^=$ or $\mathscr{E} \cap \mathscr{B}^=$ is a single point, then $\mathscr{K}_1 = \mathscr{K}_2$. Since $\dim(\mathscr{K}_1) > 1$ and $\dim(\mathscr{K}_2) > 1$, we have that $\mathscr{E} \cap \mathscr{A}^=$ and $\mathscr{E} \cap \mathscr{B}^=$ cannot be both single point sets. Let $u \in \mathscr{E}$, and assume that $\mathscr{E} \cap \mathscr{A}^= = \{u\}$, then $\mathscr{K}_1 \cap \mathscr{A}^= = \{u\}$ and $\mathscr{K}_2 \cap \mathscr{A}^= = \{u\}$. Now, if $u \neq v^1$, then we have that $\mathscr{K}_1 = \{\theta u \mid \theta \geq 0\}$, which implies that the set $\mathscr{E} \cap \mathscr{B}^=$ is a single point. Thus, we have that $u = v^1$. On the other hand, if $u \neq v^2$, then we have that $\mathscr{K}_2 = \{y \in \mathbb{R}^n \mid y = v^2 + \theta(u - v^2), \theta \geq 0\}$, which also implies that the set $\mathscr{E} \cap \mathscr{B}^=$ is a single point. Hence, we have that $u = v^2$. Therefore, we have that $v^1 = v^2$, and since $\mathscr{E} \cap \mathscr{B}^= = \mathscr{K}_1 \cap \mathscr{B}^= = \mathscr{K}_2 \cap \mathscr{B}^=$ and $\mathscr{K}_2 \cap \mathscr{B}^=$ is a base for $\mathscr{K}_1$ and $\mathscr{K}_2$, we obtain that $\mathscr{K}_1 = \mathscr{K}_2$. A symmetric argument would show that $\mathscr{K}_1 = \mathscr{K}_2$ if $\mathscr{E} \cap \mathscr{B}^= = \{z\}$.

Second, we show that if $\{v^1, v^2\} \cap (\mathscr{A}^= \cup \mathscr{B}^=) = \emptyset$, then $v^1 \in \mathscr{K}_2$ and $v^2 \in \mathscr{K}_1$. Assume to the contrary that $v^1 \notin \mathscr{K}_2$. Here use a similar argument to the one in the proof of Lemma 4. Note that $\dim(\mathscr{K}_2) > 1$. By the separation theorem, there exists a hyperplane $\mathscr{H}$ separating v^1 and $\mathscr{K}_2$ properly such that $v^1 \notin \mathscr{H}$. From Lemma 1, we know that the sets $\mathscr{E} \cap \mathscr{A}^=$ and $\mathscr{E} \cap \mathscr{B}^=$ are bases for $\mathscr{K}_2$. Hence, there exists a vector $w \in \mathscr{E} \cap \mathscr{B}^=$ such that the extreme ray $\mathscr{R}_w = \{v^2 + \gamma(w - v^2) \mid \gamma \geq 0\}$ of $\mathscr{K}_2$ is in $\mathscr{H}$. Additionally, by Lemma 3 we have that v^1 is either in $\mathscr{A}$ or $\mathscr{B}$ but not in $\mathscr{A} \cap \mathscr{B}$. Let us assume that $v^1 \in \mathscr{A}$. Given that $\mathscr{K}_1$ is convex, $\lambda v^1 + (1 - \lambda)w \in \mathscr{K}_1$ for all $0 \leq \lambda \leq 1$. On the other hand, since w is a vector on an exposed face of $\mathscr{K}_2$, for $0 < \lambda \leq 1$ we have $\lambda v^1 + (1 - \lambda)w \notin \mathscr{K}_2$. Furthermore, since $v^1 \in \mathscr{A}$, by Assumption 1 we have $a^\top v^1 \geq \alpha$ and $a^\top w < \alpha$. Hence, from the equation $a^\top(\lambda v^1 + (1 - \lambda)w) = \lambda a^\top v^1 + (1 - \lambda)a^\top w$, we may obtain $0 < \lambda \leq 1$ such that $a^\top(\lambda v^1 + (1 - \lambda)w) = \alpha$. Therefore, there exists a vector $u = \lambda v^1 + (1 - \lambda)w$ for some $0 < \lambda \leq 1$, such that $u \in \mathscr{K}_1 \cap \mathscr{A}^=$, but $u \notin \mathscr{K}_2$, which contradicts $\mathscr{K}_1 \cap \mathscr{A}^= = \mathscr{K}_2 \cap \mathscr{A}^=$. When we assume $v^1 \in \mathscr{B}$, we get a contradiction in a similar manner. Hence, we obtain that $v^1 \in \mathscr{K}_2$. Using a similar argument one can prove that $v^2 \in \mathscr{K}_1$.

Third, we show that if $v^1 \neq v^2$, then they cannot be both in $\mathscr{A}$ or in $\mathscr{B}$. Assume to the contrary that $v^1 \in \mathscr{A}$ and $v^2 \in \mathscr{A}$. Note that if $\alpha > 0$, then we have $v^1 \notin \mathscr{A}$, thus we assume that $\alpha \leq 0$. On the one hand, since $v^1 \in \mathscr{K}_2$ we have that $\mathscr{R}_{v^1} = \{(1 - \theta)v^2 \mid \theta \geq 0\} \subseteq \mathscr{K}_2$. Hence, if $a^\top v^2 \leq 0$, then $\mathscr{R}_{v^1} \subseteq \mathscr{A}$ which implies that $\mathscr{R}_{v^1}$ is parallel to $\mathscr{A}^=$, and we obtain that $\mathscr{A}^= \cap \mathscr{K}_2$ is unbounded. On the other hand, since $v^2 \in \mathscr{K}_1$ we have that $\mathscr{R}_{v^2} = \{\theta v^2 \mid \theta \geq 0\} \subseteq \mathscr{K}_1$. Hence, if $a^\top v^2 \geq 0$, then $\mathscr{R}_{v^2} \subseteq \mathscr{A}$, which implies that $\mathscr{R}_{v^2}$ is parallel to $\mathscr{A}^=$, and we obtain that $\mathscr{A}^= \cap \mathscr{K}_1$ is unbounded. Hence, if $v^1 \in \mathscr{A}$ and $v^2 \in \mathscr{A}$, then we obtain a contradiction to Assumption 2. Similarly, we can prove that v^1 and v^2 cannot be simultaneously in $\mathscr{B}$.

Finally, we show that if v^1 and v^2 are in different half-spaces and $\{v^1, v^2\} \cap (\mathscr{A}^= \cup \mathscr{B}^=) = \emptyset$, then this contradicts the assumption that $\mathscr{K}_1 \cap \mathscr{A}^= = \mathscr{K}_2 \cap \mathscr{A}^=$ and $\mathscr{K}_1 \cap \mathscr{B}^= = \mathscr{K}_2 \cap \mathscr{B}^=$. Assume that $v^1 \in \mathscr{A}$ and $v^2 \in \mathscr{B}$. Recall that in this case $v^1 \in \mathscr{K}_2$ and $v^2 \in \mathscr{K}_1$, thus the set $\mathscr{R}_{v^1} = \{(1 - \theta)v^2 \mid \theta \geq 0\} \subseteq \mathscr{K}_2$ and $\mathscr{R}_{v^2} = \{\theta v^2 \mid \theta \geq 0\} \subseteq \mathscr{K}_1$. Now, since $\dim(\mathscr{K}_1) > 1$ and $\mathscr{B}^= \cap \mathscr{K}_1$ is a base of $\mathscr{K}_1$, there is at least one extreme ray $\mathscr{R}_w = \{\gamma w \mid \gamma \geq 0\}$ of $\mathscr{K}_1$ such that $v^2 \notin \mathscr{R}_w$ and

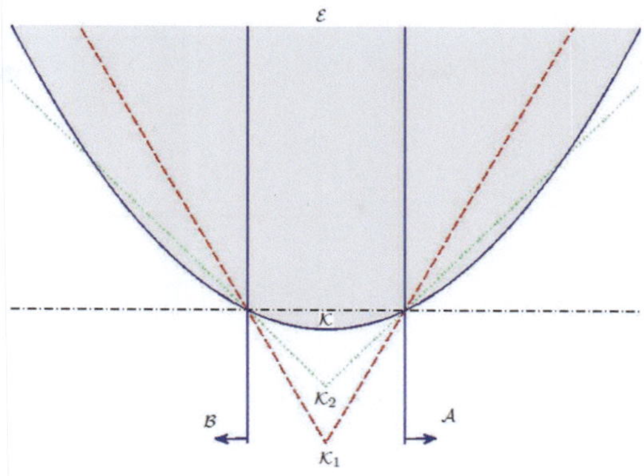

Fig. 2 Example of unbounded intersections

$w \in \mathscr{K}_1 \cap \mathscr{B}^= = \mathscr{K}_2 \cap \mathscr{B}^=$ is a vector in the boundary of $\mathscr{K}_1$. Now, if $w \in \text{ri}(\mathscr{K}_2)$, then, since $\mathscr{K}_2 \cap \mathscr{B}^=$ is bounded and is a base of $\mathscr{K}_2$, we have that $w \in \text{ri}(\mathscr{K}_2 \cap \mathscr{B}^=)$. Thus, in this case there exists a vector $u \in \mathscr{K}_2 \cap \mathscr{B}^=$ such that $u \notin \mathscr{K}_1 \cap \mathscr{B}^=$, which contradicts $\mathscr{K}_1 \cap \mathscr{B}^= = \mathscr{K}_2 \cap \mathscr{B}^=$.

Assume now that w is a vector on the boundary of $\mathscr{K}_2$. Since $w \in \mathscr{K}_2$, we have that $\{v^2 + \gamma(w - v^2) \mid \gamma \geq 0\} \subseteq \mathscr{K}_2$. Moreover, since $v^2 \in \mathscr{B}$ and $w \in \mathscr{B}^=$, there exists a $\hat{\gamma} > 1$ such that $a^\top(v^2 + \hat{\gamma}(w - v^2)) = \alpha$. However, since w is on the extreme ray $\mathscr{R}_w$ of $\mathscr{K}_1$ and $v^2 \notin \mathscr{R}_w$, then the vector $(v^2 + \hat{\gamma}(w - v^2)) \notin \mathscr{K}_1$. This contradicts the assumption $\mathscr{K}_1 \cap \mathscr{A}^= = \mathscr{K}_2 \cap \mathscr{A}^=$. A symmetric argument is valid if we assume that $v^1 \in \mathscr{B}$ and $v^2 \in \mathscr{A}$. Hence, since v^1 and v^2 cannot be in different half-spaces, then $v^1 = v^2$. In conclusion, we have that $\mathscr{K}_1 = \mathscr{K}_2$, since $\mathscr{E} \cap \mathscr{A}^=$ and $\mathscr{E} \cap \mathscr{B}^=$ are bases for $\mathscr{K}_1$ and $\mathscr{K}_2$, which proves that the disjunctive conic cut is unique. □

Figure 2 illustrates how Lemma 5 fails when the intersections $\mathscr{E} \cap \mathscr{A}^=$ or $\mathscr{E} \cap \mathscr{B}^=$ are unbounded. In this case, one can see that $\mathscr{K} \cap \mathscr{E}$ is the convex hull of $\mathscr{E} \cap (\mathscr{A} \cup \mathscr{B})$. The other two cones $\mathscr{K}_1$ and $\mathscr{K}_2$ have the same intersections with $\mathscr{A}^=$ and $\mathscr{B}^=$ as the convex set $\mathscr{E}$. However, the intersections $\mathscr{K}_1 \cap \mathscr{E}$ and $\mathscr{K}_2 \cap \mathscr{E}$ fail to give $\text{conv}(\mathscr{E} \cap (\mathscr{A} \cup \mathscr{B}))$. Indeed, $\mathscr{K}_1$ and $\mathscr{K}_2$ are not even valid for the convex hull.

Another important case to consider here is when the set $\mathscr{E} \cap (\mathscr{A} \cup \mathscr{B})$ is of dimension $n = 1$. Figure 3a illustrates this case. In particular, we can see that the uniqueness in Lemma 5 fails in this case too. Observe the cone $\mathscr{K}_1$ in Figure 3b and the cone $\mathscr{K}_2$ in Figure 3c, which are given by two half-lines. These two cones have the same intersections with $\mathscr{A}^=$ and $\mathscr{B}^=$ as the set $\mathscr{E}$. However, the intersections $\mathscr{E} \cap \mathscr{K}_1$ and $\mathscr{E} \cap \mathscr{K}_2$ differ from $\text{conv}(\mathscr{E} \cap (\mathscr{A} \cup \mathscr{B}))$. In this case, the cone $\mathscr{K}$ in Figure 3c, given by a line, is such that $\mathscr{E} \cap \mathscr{K} = \text{conv}(\mathscr{E} \cap (\mathscr{A} \cup \mathscr{B}))$.

3.2 Disjunctive Cylindrical Cut

Let us now present the definition of a convex cylinder.

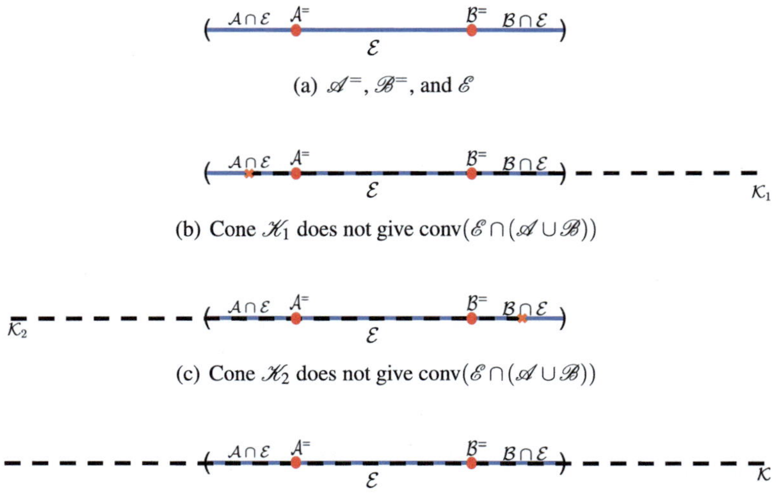

(a) $\mathscr{A}^{=}$, $\mathscr{B}^{=}$, and $\mathscr{E}$

(b) Cone $\mathscr{K}_1$ does not give $\text{conv}(\mathscr{E} \cap (\mathscr{A} \cup \mathscr{B}))$

(c) Cone $\mathscr{K}_2$ does not give $\text{conv}(\mathscr{E} \cap (\mathscr{A} \cup \mathscr{B}))$

(d) $\mathscr{K} \cap \mathscr{E} = \text{conv}(\mathscr{E} \cap (\mathscr{A} \cup \mathscr{B}))$

Fig. 3 Example when the set $\mathscr{E} \cap (\mathscr{A} \cup \mathscr{B})$ has dimension $n = 1$. (**a**) $\mathscr{A}^{=}$, $\mathscr{B}^{=}$, and $\mathscr{E}$. (**b**) Cone $\mathscr{K}_1$ does not give $\text{conv}(\mathscr{E} \cap (\mathscr{A} \cup \mathscr{B}))$. (**c**) Cone $\mathscr{K}_2$ does not give $\text{conv}(\mathscr{E} \cap (\mathscr{A} \cup \mathscr{B}))$. (**d**) $\mathscr{K} \cap \mathscr{E} = \text{conv}(\mathscr{E} \cap (\mathscr{A} \cup \mathscr{B}))$

Definition 4 (Convex Cylinder). Let $\mathscr{D} \subseteq \mathbb{R}^n$ be a convex set and $d_0 \in \mathbb{R}^n$ a vector. Then, the set $\mathscr{C} = \{x \in \mathbb{R}^n | x = d + \sigma d_0, d \in \mathscr{D}, \sigma \in \mathbb{R}\}$ is a convex cylinder in $\mathbb{R}^n$.

Definition 5. A closed convex cylinder $\mathscr{C}$ is a *disjunctive cylindrical cut* for the set $\mathscr{E}$ and the disjunctive set $\mathscr{A} \cup \mathscr{B}$ if

$$\text{conv}\,(\mathscr{E} \cap (\mathscr{A} \cup \mathscr{B})) = \mathscr{C} \cap \mathscr{E}.$$

The following proposition gives a sufficient condition for a convex cylinder $\mathscr{C}$ to be a disjunctive cylindrical cut. The result and proofs for the cylinder case are similar to the cone case, still we provide them for completeness.

Proposition 2. *A convex cylinder $\mathscr{C}$ is a disjunctive cylindrical cut for $\mathscr{E}$ and the disjunctive set $\mathscr{A} \cup \mathscr{B}$ if*

$$\mathscr{C} \cap \mathscr{A}^{=} = \mathscr{E} \cap \mathscr{A}^{=} \quad \textit{and} \quad \mathscr{C} \cap \mathscr{B}^{=} = \mathscr{E} \cap \mathscr{B}^{=}. \tag{4}$$

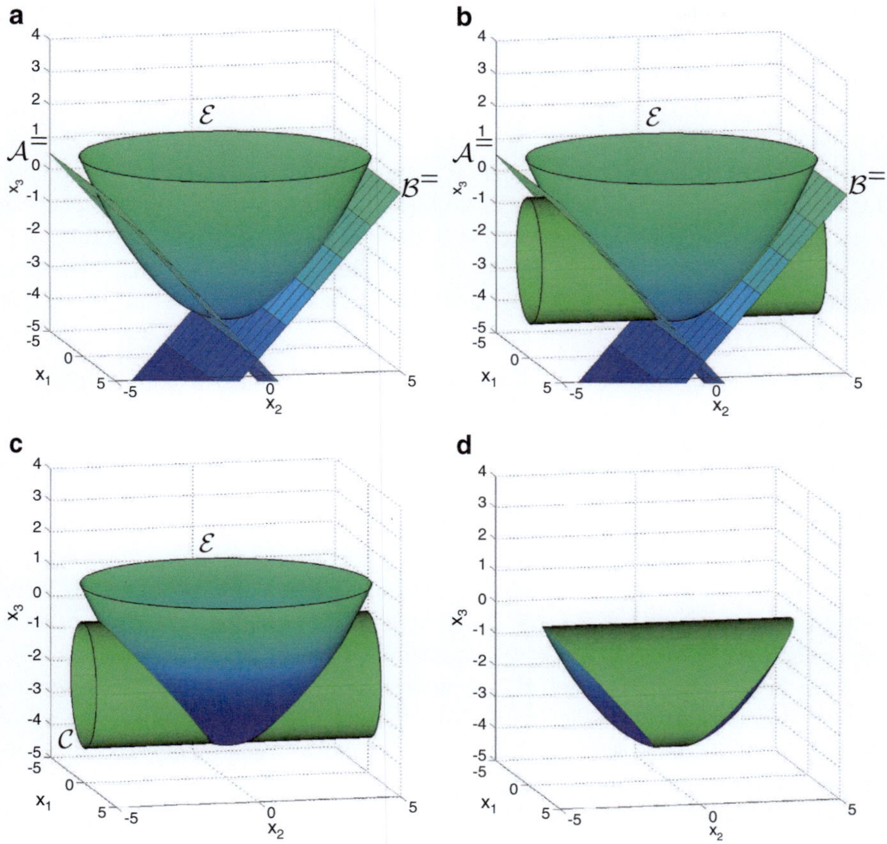

Fig. 4 Illustration of a disjunctive cylindrical cut as specified in Proposition 2. (**a**) $\mathscr{A}^=$, $\mathscr{B}^=$, and $\mathscr{E}$. (**b**) The cylinder $\mathscr{C}$ yielding $\text{conv}(\mathscr{E} \cap (\mathscr{A} \cup \mathscr{B}))$. (**c**) $\mathscr{E} \cap \mathscr{C}$. (**d**) $\text{conv}(\mathscr{E} \cap (\mathscr{A} \cup \mathscr{B}))$

Figure 4 illustrates Proposition 2, where the set $\mathscr{E}$ is the epigraph of a paraboloid. Before proving Proposition 2 we first provide a set of lemmas that will ease to understand the proof. First, let us define the *base* of a cylinder in a similar way as the base of a cone is defined in [7].

Definition 6 (Base of a Cylinder). Let $\mathscr{C} \subset \mathbb{R}^n$ be a convex cylinder. A set $\mathscr{D} \subset \mathscr{C}$ is called a *base* of $\mathscr{C}$ if, for every point $x \in \mathscr{C}$, there is a unique $d \in \mathscr{D}$ and $\sigma \in \mathbb{R}$ such that $x = d + \sigma d_0$.

Lemma 6. *Consider a half-space* $\mathscr{G} = \{x \in \mathbb{R}^n \mid g^\top x \leq \rho\}$. *Assume that* $\mathscr{E} \cap \mathscr{G}^=$ *is nonempty and bounded. If* $\mathscr{C} \cap \mathscr{G}^= = \mathscr{E} \cap \mathscr{G}^=$, *then* $\mathscr{E} \cap \mathscr{G}^=$ *is a base for* $\mathscr{C}$.

Proof. Let $\mathscr{C}$ be a cylinder such that $\mathscr{C} \cap \mathscr{G}^= = \mathscr{E} \cap \mathscr{G}^=$. Observe that if $g^\top d_0 = 0$ then for any $\hat{x} \in \mathscr{C} \cap \mathscr{G}^=$ we have that $\{y \in \mathbb{R}^n \mid y = \hat{x} + \sigma d_0, \sigma \in \mathbb{R}\} \subseteq \mathscr{C} \cap \mathscr{G}^=$, which is an unbounded set. Hence, $g^\top d_0 \neq 0$ because $\mathscr{C} \cap \mathscr{G}^= = \mathscr{E} \cap \mathscr{G}^=$ is bounded.

Now, let us assume that $\mathscr{E} \cap \mathscr{G}^=$ is not a base for $\mathscr{C}$. Then, from Definition 6 we know that there exists a point $x \in \mathscr{C}$ such that there exists no point $\bar{x} \in \mathscr{E} \cap \mathscr{G}^=$ that represents x as $\bar{x} + \sigma d_0$ for some $\sigma \in \mathbb{R}$. Thus, $\{y \in \mathbb{R}^n \mid y = x + \sigma d_0, \sigma \in \mathbb{R}\} \cap \mathscr{E} \cap \mathscr{G}^= = \emptyset$. However, with $\hat{\sigma} = (\rho - g^\top x)/g^\top d_0$ we obtain that $x + \hat{\sigma} d_0 \in \mathscr{C} \cap \mathscr{G}^= = \mathscr{E} \cap \mathscr{G}^=$ whenever $g^\top d_0 \neq 0$. Therefore, the relation $\{y \in \mathbb{R}^n \mid y = x + \sigma d_0, \sigma \in \mathbb{R}\} \cap \mathscr{E} \cap \mathscr{G}^= = \emptyset$ is true only if $g^\top d_0 = 0$. Hence, if $\mathscr{E} \cap \mathscr{G}^=$ is not a base for $\mathscr{C}$, then we have that $\mathscr{E} \cap \mathscr{G}^=$ is unbounded, which contradicts the boundedness assumption of $\mathscr{E} \cap \mathscr{G}^=$. Now, if $g^\top d_0 \neq 0$, then given an $x \in \mathscr{C}$ we have that $\{y \in \mathbb{R}^n \mid y = x + \sigma d_0, \sigma \in \mathbb{R}\} \cap \mathscr{E} \cap \mathscr{G}^= = x + \hat{\sigma} d_0$. Hence, for $\bar{x} = x - \hat{\sigma} d_0$, we obtain that x is uniquely defined by $\bar{x}\mathscr{E} \cap \mathscr{G}^=$ and $-\hat{\sigma}$ as $x = \bar{x} - \hat{\sigma} d_0$. Therefore, $\mathscr{E} \cap \mathscr{G}^=$ is a base for $\mathscr{C}$. □

The next lemma states the relationship between the cylinder $\mathscr{C}$ and the intersections of $\mathscr{E}$ with the half-spaces $\mathscr{A}$ and $\mathscr{B}$.

Lemma 7. *Let $\mathscr{C} \subset \mathbb{R}^n$ be a convex cylinder $\mathscr{C}$, for which* (4) *holds. Then*

$$(\mathscr{E} \cap \mathscr{A}) \subset \mathscr{C} \quad \textit{and} \quad (\mathscr{E} \cap \mathscr{B}) \subset \mathscr{C}.$$

Proof. We prove first that $(\mathscr{E} \cap \mathscr{A}) \subseteq \mathscr{C}$. Let us assume to the contrary that there exists $x \in (\mathscr{E} \cap \mathscr{A})$ such that $x \notin \mathscr{C}$. First, by the separation theorem, there exists a hyperplane $\mathscr{H} = \{y \in \mathbb{R}^n \mid h^\top y = \eta\}$ separating x from $\mathscr{C}$. From the definition of $\mathscr{C}$ we have that $h^\top d_0 = 0$. Now, let $\mathscr{H}$ be a supporting hyperplane of $\mathscr{C}$, which implies that $\mathscr{H} \cap \mathscr{C}$ is an exposed face of $\mathscr{C}$. Note that for any $\hat{y} \in \mathscr{H} \cap \mathscr{C}$ the inclusion $\{y \in \mathbb{R}^n \mid y = \hat{y} + \sigma d_0, \sigma \in \mathbb{R}\} \subseteq \mathscr{H} \cap \mathscr{C}$ must hold. Additionally, according to Definition 6, by Assumption 2, and Lemma 6, the sets $\mathscr{E} \cap \mathscr{A}^=$ and $\mathscr{E} \cap \mathscr{B}^=$ are bases for $\mathscr{C}$. Hence, there exists a point $w \in \mathscr{E} \cap \mathscr{B}^=$ such that $w \in \mathscr{H}$, and w is in an exposed face of $\mathscr{C}$.

Convexity of $\mathscr{E}$ implies $\lambda x + (1-\lambda) w \in \mathscr{E}$ for any $\lambda \in [0,1]$. On the other hand, the point w is in an exposed face of $\mathscr{C}$, so $\lambda x + (1-\lambda) w \notin \mathscr{C}$ for $0 < \lambda \leq 1$. Since $x \in (\mathscr{E} \cap \mathscr{A})$ and $\mathscr{A} \cap \mathscr{B} \cap \mathscr{E} = \emptyset$, we have that $a^\top x \geq \alpha$ and $a^\top w < \alpha$. Hence, from the equation $a^\top(\lambda x + (1-\lambda) w) = \lambda a^\top x + (1-\lambda) a^\top w$, there must exist a value $0 < \lambda \leq 1$ such that $a^\top(\lambda x + (1-\lambda) w) = \alpha$. Therefore, for some $0 < \lambda \leq 1$ there is a point $\hat{x} = \lambda x + (1-\lambda) w$, such that $\hat{x} \in \mathscr{E} \cap \mathscr{A}^=$, but $\hat{x} \notin \mathscr{C}$, which contradicts condition (4). Hence, $(\mathscr{E} \cap \mathscr{A}) \subseteq \mathscr{C}$. One can prove $(\mathscr{E} \cap \mathscr{B}) \subseteq \mathscr{C}$ analogously.

Recall that the sets $\mathscr{E} \cap \mathscr{A}^=$ and $\mathscr{E} \cap \mathscr{B}^=$ are disjoint and nonempty. Then, condition (4) implies that $\mathscr{E} \cap \mathscr{A} \neq \mathscr{C}$ and $\mathscr{E} \cap \mathscr{B} \neq \mathscr{C}$, and the result of the lemma follows. □

Now we can present the proof of Proposition 2.

Proof (Proof of Proposition 2). First, consider a vector $x \in (\mathscr{E} \cap \mathscr{A}) \cup (\mathscr{E} \cap \mathscr{B})$. Then, Lemma 7 implies that $x \in \mathscr{E} \cap \mathscr{C}$. Consider any two points $x, y \in (\mathscr{E} \cap \mathscr{A}) \cup (\mathscr{E} \cap \mathscr{B})$. Then, since both $\mathscr{C}$ and $\mathscr{E}$ are convex, for all $0 \leq \lambda \leq 1$ the convex combination $\lambda x + (1-\lambda) y \in \mathscr{E} \cap \mathscr{C}$. Hence, $\text{conv}(\mathscr{E} \cap (\mathscr{A} \cup \mathscr{B})) \subseteq (\mathscr{E} \cap \mathscr{C})$.

Consider now a point $x \in (\mathscr{E} \cap \mathscr{C})$. First, if $x \in (\mathscr{E} \cap \mathscr{A})$ or $x \in (\mathscr{E} \cap \mathscr{B})$, we have that $x \in \text{conv}(\mathscr{E} \cap (\mathscr{A} \cup \mathscr{B}))$. Suppose then that $x \notin (\mathscr{E} \cap \mathscr{A}) \cup (\mathscr{E} \cap \mathscr{B})$.

Then, $x \in (\bar{\mathscr{A}} \cap \bar{\mathscr{B}} \cap \mathscr{C})$. Furthermore, by Lemma 6 there are two vectors $\hat{x} \in \mathscr{E} \cap \mathscr{A}^=$ and $\bar{x} \in \mathscr{E} \cap \mathscr{B}^=$ such that $x = \hat{x} + \mu d_0$ and $x = \bar{x} + \nu d_0$, for some $\mu, \nu \in \mathbb{R}$. Note that μ and ν must have opposite signs, since $x \notin (\mathscr{E} \cap \mathscr{A}) \cup (\mathscr{E} \cap \mathscr{B})$, otherwise one of the equalities $x = \hat{x} + \mu d_0$ or $x = \bar{x} + \nu d_0$ would not be true. Let us assume w.l.o.g. that $\nu > 0$ and $\mu < 0$. Then, we have that $x = \lambda\hat{x} + (1-\lambda)\bar{x}$, where $\lambda = \nu/(\nu - \mu)$ and $0 < \lambda < 1$. In other words, x is a convex combination of $\hat{x}$ and $\bar{x}$. Since x is an arbitrary point we have that any point $x \in (\mathscr{E} \cap \mathscr{C})$ can be written as a convex combination of two points in $(\mathscr{E} \cap \mathscr{A}) \cup (\mathscr{E} \cap \mathscr{B})$. As a conclusion, we have that $(\mathscr{E} \cap \mathscr{C}) \subseteq \text{conv}(\mathscr{E} \cap (\mathscr{A} \cup \mathscr{B}))$. Finally, since $(\mathscr{E} \cap \mathscr{A}^=)$ and $(\mathscr{E} \cap \mathscr{B}^=)$ are compact sets, then it follows from Lemmas 6 and 11 that $\mathscr{C}$ is closed. □

Lemma 8. *If a convex cylinder $\mathscr{C}$ with property* (4) *exists, then $\mathscr{C}$ is unique.*

Proof. Assume that there exist two different cylinders $\mathscr{C}_1 = \{x \in \mathbb{R}^n \mid x = d^1 + \gamma d_0^1, d^1 \in \mathscr{D}_1, \gamma \in \mathbb{R}\}$ and $\mathscr{C}_2 = \{x \in \mathbb{R}^n \mid x = d^2 + \sigma d_0^2, d^2 \in \mathscr{D}_2, \sigma \in \mathbb{R}\}$ that satisfy Definition 5. Then, we have that $\mathscr{C}_1 \cap \mathscr{A}^= = \mathscr{C}_2 \cap \mathscr{A}^=$ and $\mathscr{C}_1 \cap \mathscr{B}^= = \mathscr{C}_2 \cap \mathscr{B}^=$.

Given that $\mathscr{C}_1 \neq \mathscr{C}_2$ there must exist a point $\hat{x}$ that belongs only to one cylinder, and w.l.o.g. we assume that $\hat{x} \in \mathscr{C}_1$ and $\hat{x} \notin \mathscr{C}_2$. Observe that if $\hat{x} \in \mathscr{A} \cap \mathscr{B}$, then there exists a point $\bar{x} \in \mathscr{C}_1$ such that either $\bar{x} \in \mathscr{A}^= \cap \mathscr{B}$ or $\bar{x} \in \mathscr{A} \cap \mathscr{B}^=$, which implies that $\bar{x} \in \mathscr{E} \cap \mathscr{A} \cap \mathscr{B}$, contradicting Assumption 1.

Let us begin assuming that $\hat{x} \in \bar{\mathscr{A}} \cap \bar{\mathscr{B}}$. Then, given that $\mathscr{E} \cap \mathscr{A}^=$ is a base for both cylinders there exists a $\hat{\gamma} \in \mathbb{R}$ such that $\hat{x} = \hat{d}^1 + \hat{\gamma} d_0^1$ for some $\hat{d}^1 \in \mathscr{E} \cap \mathscr{A}^= = \mathscr{C}_1 \cap \mathscr{A}^= = \mathscr{C}_2 \cap \mathscr{A}^=$. On the other hand, since $\mathscr{E} \cap \mathscr{B}^=$ is a base for $\mathscr{C}_1$, there exists $\bar{\gamma} \in \mathbb{R}$ such that $\hat{x} = \bar{d}^1 + \bar{\gamma} d_0^1$ for some $\bar{d}^1 \in \mathscr{E} \cap \mathscr{B}^= = \mathscr{C}_1 \cap \mathscr{B}^=$. Hence, $\hat{x} = \lambda \bar{d}^1 + (1-\lambda)\hat{d}^1$ where $\lambda = \hat{\gamma}/(\hat{\gamma} - \bar{\gamma}) \leq 1$, since $\hat{\gamma}$ and $\bar{\gamma}$ must have opposite signs. Additionally, given that the two cylinders are convex we get that $\bar{d}^1 \notin \mathscr{C}_2$. Then, $\mathscr{C}_1 \cap \mathscr{B}^= \neq \mathscr{C}_2 \cap \mathscr{B}^=$, which is a contradiction.

Let us assume now that $\hat{x} \in \mathscr{A}$ and $\hat{x} \notin \mathscr{B}$. By the separation theorem, there exists a hyperplane $\mathscr{H} = \{x \in \mathbb{R}^n \mid h^\top x = \eta\}$ separating $\hat{x}$ from $\mathscr{C}_2$. By the definition of a cylinder, we have $h^\top d_0^2 = 0$. Now, let $\mathscr{H}$ be a supporting hyperplane of $\mathscr{C}_2$, which implies that $\mathscr{H} \cap \mathscr{C}_2$ is an exposed face of $\mathscr{C}_2$. Note that for any $\hat{y} \in \mathscr{H} \cap \mathscr{C}_2$ we have that $\{y \in \mathbb{R}^n \mid y = \hat{y} + \sigma d_0^2, \sigma \in \mathbb{R}\} \subseteq \mathscr{H} \cap \mathscr{C}_2$. Additionally, we know that the sets $\mathscr{E} \cap \mathscr{A}^=$ and $\mathscr{E} \cap \mathscr{B}^=$ are bases for $\mathscr{C}_2$. Hence, there exists a point $w \in \mathscr{C}_2 \cap \mathscr{E} = \mathscr{E} \cap \mathscr{B}^=$ such that $w \in \mathscr{H}$, and w is in an exposed face of $\mathscr{C}_2$.

Convexity of $\mathscr{C}_1$ implies that for any $\lambda \in [0,1]$, $\lambda\hat{x} + (1-\lambda)w \in \mathscr{C}_1$. On the other hand, since $w \in \mathscr{H}$ is a point in an exposed face of $\mathscr{C}_2$, $\lambda\hat{x} + (1-\lambda)w \notin \mathscr{C}_2$ for $0 < \lambda \leq 1$. Since $\hat{x} \in \mathscr{A} \cap \mathscr{C}_1$ and $\mathscr{A} \cap \mathscr{B} \cap \mathscr{C}_1 = \emptyset$, we have that $a^\top \hat{x} \geq \alpha$ and $a^\top w < \alpha$. Hence, from the equation $a^\top(\lambda\hat{x} + (1-\lambda)w) = \lambda a^\top \hat{x} + (1-\lambda)a^\top w$, there exists a value $0 < \lambda \leq 1$ such that $a^\top(\lambda\hat{x} + (1-\lambda)w) = \alpha$. Therefore, there exists a point $\bar{x} = \lambda\hat{x} + (1-\lambda)w$ for some $0 < \lambda \leq 1$, such that $\bar{x} \in \mathscr{C}_1 \cap \mathscr{A}^=$, but $\bar{x} \notin \mathscr{C}_2$, which is a contradiction. An analogous argument can be used when $\hat{x} \in \mathscr{B}$ and $\hat{x} \notin \mathscr{A}$. □

As mentioned at the beginning of Section 1, Propositions 1 and 2 are rather general in that they apply to any convex set $\mathscr{E}$. However, their hypotheses, (3) and (4), are hard to satisfy and hence limit their applicability. To explore the full potential of this result remains the subject of future research. In this paper we demonstrate the power of this tool by exploring a class of MICO, the class of MISOCO problems, for which the assumptions are satisfied under mild conditions.

In the general setting, cone $\mathscr{K}$ or cylinder $\mathscr{C}$ of Propositions 1 and 2 can be used as a conic cut in MICO problems. For example, in Branch-and-Cut algorithms if either $\mathscr{K}$ or $\mathscr{C}$ exists for a disjunctive set, then $\mathscr{K}$ or $\mathscr{C}$ can be used to help tightening the description of a MICO problem. For practical use of this methodology, one needs to prove that a cone $\mathscr{K}$ or cylinder $\mathscr{C}$ exists that satisfies Definitions 2 or 5, respectively, and one needs to provide an easy to compute algebraic representation of the cone or cylinder. In the following section we analyze MISOCO problems, where the feasible set $\mathscr{E}$ comes from the intersection of a second order cone and an affine space. Given that for this case we can prove the existence of the cone and we can give a method to compute its algebraic representation, the resulting conic cut can be embedded in Branch-and-Cut algorithms to solve MISOCO problems.

4 The Convex Hull of the Intersection of an Ellipsoid and a Disjunctive Set

In the remainder of the paper, we turn our attention to the convex hull $\text{conv}(\mathscr{E}\cap(\mathscr{A}\cup\mathscr{B}))$ in a special case of (1) where $\mathscr{K}$ is a Lorentz cone, i.e., $\mathscr{K}=\mathbb{L}^n$. Therefore $\mathscr{E}$ is an ellipsoid resulting from the intersection of a second order cone and an affine space. Consider, for example, the problem

$$\begin{array}{rl} \text{minimize:} & 3x_1 \; +2x_2 \; +2x_3 \; +x_4 \\ \text{subject to:} & 9x_1 \; +x_2 \; +x_3 \; +x_4 = 10 \\ & (x_1,x_2,x_3,x_4) \in \mathbb{L}^4 \\ & x_4 \in \mathbb{Z}. \end{array} \tag{5}$$

The feasible set of Problem (5) can be represented as an ellipsoid in $\mathbb{R}^3$ in terms of the variables x_2, x_3, x_4, as shown in Figure 5. In general, we consider the n-dimensional ellipsoid

$$\mathscr{E}=\{x\in\mathbb{R}^n | x^\top Qx+2q^\top x+\rho\leq 0\}, \tag{6}$$

where $Q\in\mathbb{R}^{n\times n}$ is a symmetric positive definite matrix, $x,q\in\mathbb{R}^n$, and $\rho\in\mathbb{R}$.

The main goal of this section is to show the existence of the cone $\mathscr{K}$ or cylinder $\mathscr{C}$, as defined in Definitions 2 or 5, in order to use Proposition 1 or 2 for finding $\text{conv}(\mathscr{E}\cap(\mathscr{A}\cup\mathscr{B}))$. We are interested in two cases. In the first case, the hyperplanes $\mathscr{A}^=$ and $\mathscr{B}^=$ are parallel (Section 4.1), while the two hyperplanes are in a general

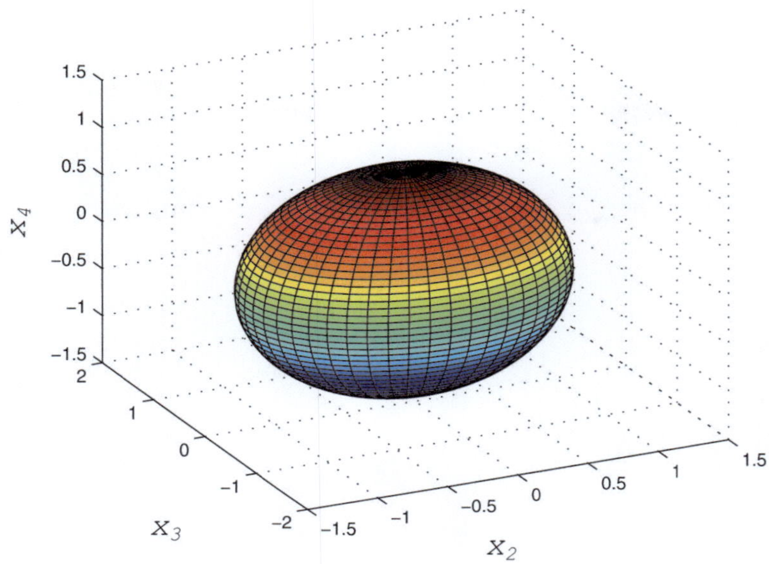

Fig. 5 The feasible region of Problem (5)

position in the second case (Section 4.2). In both cases, we are able to show that $\text{conv}(\mathscr{E} \cap (\mathscr{A} \cup \mathscr{B}))$ is obtained by intersecting $\mathscr{E}$ with a scaled second order cone $\mathscr{K}$ or a cylinder $\mathscr{C}$ and we show how to construct them.

4.1 *Parallel Disjunctions*

In this section, we consider a disjunctive set $\mathscr{A} \cup \mathscr{B}$ such that $\mathscr{A} = \{x \in \mathbb{R}^n | a^\top x \geq \alpha\}$ and $\mathscr{B} = \{x \in \mathbb{R}^n | a^\top x \leq \beta\}$, i.e., the hyperplanes $\mathscr{A}^=$ and $\mathscr{B}^=$ are parallel. We may assume w.l.o.g. that $\|a\| = 1$. We illustrate this case by using Problem (5), where one can use $\mathscr{A} = \{x \in \mathbb{R}^4 | x_4 \geq 0\}$ and $\mathscr{B} = \{x \in \mathbb{R}^4 | x_4 \leq -1\}$ to define a disjunctive set $\mathscr{A} \cup \mathscr{B}$. Figure 6a shows the hyperplanes defining this disjunctive set $\mathscr{A} \cup \mathscr{B}$ along with the feasible set of Problem (5).

4.1.1 Geometry of $\mathscr{E}$ and the Hyperplanes $\mathscr{A}^=$, and $\mathscr{B}^=$

We begin this analysis by recalling some results from [8], where the authors study several properties of quadrics. A quadric is defined as

$$\mathscr{Q} = \left\{x | x^\top Qx + 2q^\top x + \rho \leq 0\right\}, \tag{7}$$

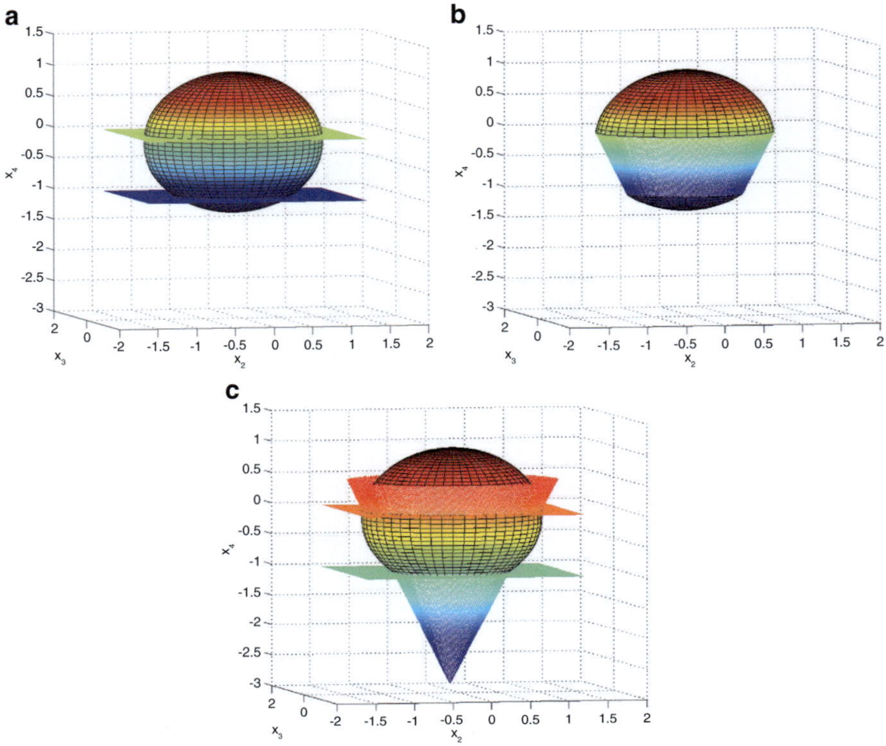

Fig. 6 The convex hull of the intersection of a parallel disjunction and an ellipsoid. (**a**) $\mathscr{A}^{=}$, $\mathscr{B}^{=}$, and $\mathscr{E}$. (**b**) $\text{conv}(\mathscr{E} \cap (\mathscr{A} \cup \mathscr{B}))$. (**c**) The cone yielding $\text{conv}(\mathscr{E} \cap (\mathscr{A} \cup \mathscr{B}))$

where $Q \in \mathbb{R}^{n \times n}$ is symmetric, $q, x \in \mathbb{R}^n$ and $\rho \in \mathbb{R}$, and is denoted by the triplet $\mathscr{Q} = (Q, q, \rho)$. Note that under this definition, $\mathscr{E}$ is a quadric with a positive definite matrix Q. We first recall Theorem 3.2 of [8], which defines a uniparametric family of quadrics $\mathscr{Q}(\tau)$ parametrized by $\tau \in \mathbb{R}$ having the same intersection with two fixed parallel hyperplanes. This result is stated here as Theorem 1.

Theorem 1 ([8]). *Consider an ellipsoid $\mathscr{E} = (Q, q, \rho)$ and two parallel hyperplanes $\mathscr{A}^{=} = (a, \alpha)$ and $\mathscr{B}^{=} = (a, \beta)$. The uniparametric family of quadrics $\mathscr{Q}(\tau)$ parametrized by $\tau \in \mathbb{R}$ and having the same intersection with $\mathscr{A}^{=}$ and $\mathscr{B}^{=}$ as ellipsoid $\mathscr{E}$ is given by*

$$Q(\tau) = Q + \tau a a^{\top}$$

$$q(\tau) = q - \tau \frac{\alpha + \beta}{2} a$$

$$\rho(\tau) = \rho + \tau \alpha \beta.$$

From Theorem 1, for any $\tau \in \mathbb{R}$ the quadric $\mathscr{Q}(\tau)$ is such that $\mathscr{Q}(\tau) \cap \mathscr{A}^= = \mathscr{E} \cap \mathscr{A}^=$ and $\mathscr{Q}(\tau) \cap \mathscr{B}^= = \mathscr{E} \cap \mathscr{B}^=$. Hence, from Lemma 3 we need to investigate if there exists a value $\bar{\tau}$ such that $Q(\bar{\tau})$ is a two-sided cone one side of which, denoted by $\mathscr{K}$, satisfies the conditions of Proposition 1 with a vertex $x^* \notin \bar{\mathscr{A}} \cap \bar{\mathscr{B}}$ or a convex cylinder $\mathscr{C}$ that satisfies the conditions of Proposition 2. As a result, we obtain that the intersections $\mathscr{K} \cap \mathscr{E}$ or $\mathscr{C} \cap \mathscr{E}$ would be the convex hull for $\mathscr{E} \cap (\mathscr{A} \cup \mathscr{B})$. Figures 6b, c illustrate the convex hull of $\mathscr{E} \cap (\mathscr{A} \cup \mathscr{B})$ and the four sets $\mathscr{E}, \mathscr{A}^=, \mathscr{B}^=, \mathscr{K}$ for Problem (5).

Now, let us assume first that we are given a τ such that $Q(\tau)$ is non-singular. Under this assumption one can rewrite the quadric set $\mathscr{Q}(\tau)$ in (7) as

$$\left\{x \middle| (x+Q(\tau)^{-1}q(\tau))^\top Q(\tau)(x+Q(\tau)^{-1}q(\tau)) \leq q(\tau)^\top Q(\tau)^{-1}q(\tau) - \rho(\tau)\right\}. \tag{8}$$

From (8), one can easily verify that the quadric $\mathscr{Q}(\tau)$ is empty if the matrix $Q(\tau)$ is positive definite and $q(\tau)^\top Q(\tau)^{-1}q(\tau) - \rho(\tau) < 0$. Belotti et al. [8] prove that the quadric $\mathscr{Q}(\tau)$ defines a cone if $Q(\tau)$ is a non-singular symmetric matrix with exactly one negative eigenvalue and $q(\tau)^\top Q(\tau)^{-1}q(\tau) - \rho(\tau) = 0$. They also prove that for any $\tau \in \mathbb{R}$, the matrix $Q(\tau)$ has at most one negative eigenvalue and at least $n-1$ positive eigenvalues. Therefore, we need to focus on $Q(\tau)$ to explore those τ values for which $q(\tau)^\top Q(\tau)^{-1}q(\tau) - \rho(\tau) = 0$.

Let us define the vectors $u_a = Q^{-1/2}a$ and $u_q = Q^{-1/2}q$, where $Q^{-1/2}$ is the unique symmetric square root of Q^{-1}. Then, from Theorem 1 we can get the following expression, which is derived in Section 3.2.1 in [8]:

$$\begin{aligned} &q(\tau)^\top Q(\tau)^{-1}q(\tau) - \rho(\tau) \\ &= \frac{(\alpha-\beta)^2\|u_a\|^2}{4(1+\tau\|u_a\|^2)}\tau^2 + \frac{\left(4\|u_a\|^2(\|u_q\|^2-\rho) - (\alpha+\beta+2u_a^\top u_q)^2 + (\alpha-\beta)^2\right)}{4(1+\tau\|u_a\|^2)}\tau \\ &\quad + \frac{4(\|u_q\|^2-\rho)}{4(1+\tau\|u_a\|^2)}. \end{aligned} \tag{9}$$

Hence $q(\tau)^\top Q(\tau)^{-1}q(\tau) - \rho(\tau)$ is the ratio of two polynomials in τ. Two remarks are in order: first, note that at value $\hat{\tau} = -1/\|u_a\|^2$, the denominator of (9) becomes zero. Additionally, at $\hat{\tau}$, the matrix $Q(\tau)$ is positive semidefinite with one zero eigenvalue. Lemma 3.3 in [8] characterizes the behavior of $Q(\tau)$ at $\hat{\tau}$. There are two main ranges in this characterization. On the one hand, for $\tau > \hat{\tau}$, the matrix $Q(\tau)$ is positive definite. On the other hand, for $\tau < \hat{\tau}$, the matrix $Q(\tau)$ is indefinite with one negative eigenvalue.

Second, for any $\tau \neq \hat{\tau}$, $q(\tau)^\top Q(\tau)^{-1}q(\tau) - \rho(\tau)$ becomes zero only at the roots $\bar{\tau}_1, \bar{\tau}_2$ of the numerator of (9). Let f be a function of τ that denotes the quadratic function in the numerator of (9). Hence, both roots $\bar{\tau}_1, \bar{\tau}_2$ of f are less than or equal to $\hat{\tau}$ [8]. Then, from Lemma 3.3 in [8] the two roots $\bar{\tau}_1, \bar{\tau}_2$ correspond to the cones or the cylinders in the family of Theorem 1. A characterization of the family $\mathscr{Q}(\tau)$ for $\tau \in \mathbb{R}$ depending on the geometry of $\mathscr{E}$ and the hyperplanes $\mathscr{A}^=$, and $\mathscr{B}^=$ is presented in Theorem 3.4 of [8], which we recall here.

Theorem 2 ([8]). *Depending on the geometry of $\mathscr{E}$, $\mathscr{A}$, and $\mathscr{B}$, $\mathscr{Q}(\tau)$ can have the following shapes for $\tau \in \mathbb{R}$:*

- *$f(\tau)$ has two distinct roots $\bar{\tau}_1 < \bar{\tau}_2$ and $\bar{\tau}_2 < \hat{\tau}$: this is the general case, $\mathscr{Q}(\hat{\tau})$ is a paraboloid, and $\mathscr{Q}(\bar{\tau}_1), \mathscr{Q}(\bar{\tau}_2)$ are two cones.*
- *$f(\tau)$ has two distinct roots $\bar{\tau}_1 < \bar{\tau}_2$, and $\bar{\tau}_2 = \hat{\tau}$: the two hyperplanes are symmetric about the center of $\mathscr{E}$. $\mathscr{Q}(\bar{\tau}_1)$ is cone and $\mathscr{Q}(\bar{\tau}_2)$ is a cylinder.*
- *The two roots $\bar{\tau}_1, \bar{\tau}_2$ of $f(\tau)$ are equal, and $\bar{\tau}_2 < \hat{\tau}$: the discriminant of $f(\tau)$ is zero, which means that one of the hyperplanes intersects $\mathscr{E}$ in only one point. $\mathscr{Q}(\hat{\tau})$ is a paraboloid and $\mathscr{Q}(\bar{\tau}_2)$ is a cone.*
- *The two roots $\bar{\tau}_1, \bar{\tau}_2$ of $f(\tau)$ coincide with $\hat{\tau}$: this is the most degenerate case as both hyperplanes intersect $\mathscr{E}$ in only one point, and as such they are symmetric about the center of $\mathscr{E}$. In this case $\mathscr{Q}(\hat{\tau})$ is a line.*

4.1.2 Building a Disjunctive Conic Cut

We can use the geometrical analysis of Section 4.1.1 to build a conic cut to convexify the intersection of a MISOCO problem with a parallel disjunction. To simplify the analysis, we separate the cylinder and conic cases.

Cylinders: First, we study the families $\mathscr{Q}(\tau)$, $\tau \in \mathbb{R}$ described in the second and fourth cases in Theorem 2, where there is a cylinder $\mathscr{C}$ at $\mathscr{Q}(\hat{\tau})$. In particular, $\mathscr{C}$ is given by $\mathscr{Q}(\bar{\tau}_2)$ in these cases. From Eq. (7), we have that

$$\mathscr{Q}(\bar{\tau}_2) = \left\{ x \in \mathbb{R}^n \;\middle|\; x^\top Q(\bar{\tau}_2) x + 2q(\bar{\tau}_2)^\top x + \rho(\bar{\tau}_2) \le 0 \right\}, \tag{10}$$

where $Q(\bar{\tau}_2)$ is a positive semidefinite matrix. Hence, it follows from (10) that the quadric $\mathscr{Q}(\bar{\tau}_2)$ is a convex set and Proposition 2 proves that $\mathscr{C} \cap \mathscr{E}$ is the convex hull of $\mathscr{E} \cap (\mathscr{A} \cup \mathscr{B})$. Finally, notice that the cylinder $\mathscr{C}$ described by (10) can be represented in terms of a second order cone, for that reason we classify $\mathscr{C}$ as a conic cut in this section.

Cones: Now we focus on the cones described in the first and third cases of Theorem 2. Our strategy is to show that the quadrics $\mathscr{Q}(\bar{\tau}_1)$ and $\mathscr{Q}(\bar{\tau}_2)$ can be written as the union of two convex cones. Then, we derive a criterion to identify which cone gives the convex hull of $\mathscr{E} \cap (\mathscr{A} \cup \mathscr{B})$.

Consider the roots $\bar{\tau}_i \neq \hat{\tau}$, $i = 1,2$, and let $x(\bar{\tau}_i) = -Q(\bar{\tau}_i)^{-1} q(\bar{\tau}_i)$. Recall from Section 4.1.1 that $Q(\bar{\tau}_i)$ is a symmetric and non-singular matrix that has exactly one negative eigenvalue. Then, $Q(\bar{\tau}_i)$ can be diagonalized as $U(\bar{\tau}_i) D(\bar{\tau}_i) U(\bar{\tau}_i)^\top$, where $U(\bar{\tau}_i)$ is an orthogonal matrix and $D(\bar{\tau}_i)$ is a diagonal matrix having the eigenvalues of $Q(\bar{\tau}_i)$ in its diagonal. Let the index j_i be such that $D(\bar{\tau}_i)_{j_i j_i} < 0$, and let $W(\bar{\tau}_i) = U(\bar{\tau}_i) \bar{D}(\bar{\tau}_i)^{1/2}$, where $\bar{D}(\bar{\tau}_i)_{l,k} = |D(\bar{\tau}_i)_{l,k}|$. Thus, we may write $\mathscr{Q}(\bar{\tau}_i)$ in terms of $W(\bar{\tau}_i)$ as follows:

$$\left\{ x \in \mathbb{R}^n \,\middle|\, (x - x(\bar{\tau}_i))^\top W(\bar{\tau}_i)_{i \neq j_i} W(\bar{\tau}_i)^\top_{i \neq j_i} (x - x(\bar{\tau}_i)) \leq \left(W(\bar{\tau}_i)^\top_{j_i} (x - x(\bar{\tau}_i)) \right)^2 \right\},$$

where $W(\bar{\tau}_i)_{i \neq j_i}$ has the columns of $W(\bar{\tau}_i)$ that are different from j_i. Now, let us define the sets $\mathscr{Q}(\bar{\tau}_i)^+$, $\mathscr{Q}(\bar{\tau}_i)^-$ as follows:

$$\mathscr{Q}(\bar{\tau}_i)^+ \equiv \left\{ x \in \mathbb{R}^n \;\middle|\; \left\| W(\bar{\tau}_i)^\top_{i \neq j_i} (x - x(\bar{\tau}_i)) \right\| \leq W(\bar{\tau}_i)^\top_{j_i} (x - x(\bar{\tau}_i)) \right\}, \tag{11}$$

$$\mathscr{Q}(\bar{\tau}_i)^- \equiv \left\{ x \in \mathbb{R}^n \;\middle|\; \left\| W(\bar{\tau}_i)^\top_{i \neq j_i} (x - x(\bar{\tau}_i)) \right\| \leq -W(\bar{\tau}_i)^\top_{j_i} (x - x(\bar{\tau}_i)) \right\}, \tag{12}$$

which are two second order cones. These two cones satisfy the general definition of a cone with the vertex at $x(\bar{\tau}_i)$ presented in Remark 1. It is easy to verify that $\mathscr{Q}(\bar{\tau}_i) = \mathscr{Q}(\bar{\tau}_i)^+ \cup \mathscr{Q}(\bar{\tau}_i)^-$. Also, it is clear from (11) and (12) that $\mathscr{Q}(\bar{\tau}_i)^+$ and $\mathscr{Q}(\bar{\tau}_i)^-$ are two convex sets. This shows that the quadrics $\mathscr{Q}(\bar{\tau}_1)$ and $\mathscr{Q}(\bar{\tau}_2)$ can be written as the union of two convex cones.

Given the convex cones, we need a criterion to identify which cone gives the convex hull of $\mathscr{E} \cap (\mathscr{A} \cup \mathscr{B})$. First, we choose one of the two quadrics $\mathscr{Q}(\bar{\tau}_i)$, $i = 1,2$. For this purpose we can use Lemma 3, thus we need to verify if at least one of $\mathscr{Q}(\bar{\tau}_i)$, $i = 1,2$ contains a cone with a vertex $x(\bar{\tau}_i) \notin \mathscr{A} \cap \mathscr{B}$. This criterion is presented in Lemma 9. The interested reader can review the proof in section "Proof of Lemma 9" in Appendix 1.

Lemma 9. *The quadric $\mathscr{Q}(\bar{\tau}_2)$ found at the larger root $\bar{\tau}_2$ of $f(\tau)$ in the family $\mathscr{Q}(\tau)$ of the first and third case of Theorem 2 contains a cone that satisfies Definition 2.*

From Lemma 9, we reduce the choices to the cones $\mathscr{Q}(\bar{\tau}_2)^+$ and $\mathscr{Q}(\bar{\tau}_2)^-$. We now decide between the two cones using the sign of $W(\bar{\tau}_2)_1^\top(-Q^{-1}q - x(\bar{\tau}_2))$. Thus, we choose $\mathscr{Q}(\bar{\tau}_2)^+$ if $W(\bar{\tau}_2)_1^\top(-Q^{-1}q - x(\bar{\tau}_2)) > 0$, and we choose $\mathscr{Q}(\bar{\tau}_2)^-$ when $W(\bar{\tau}_2)_1^\top(-Q^{-1}q - x(\bar{\tau}_2)) < 0$. Finally, it follows from Proposition 1 that the selected cone gives the convex hull for $\mathscr{E} \cap (\mathscr{A} \cup \mathscr{B})$. Note that if $W(\bar{\tau}_2)_1^\top(-Q^{-1}q - x(\bar{\tau}_2)) = 0$ the center of the ellipsoid $\mathscr{E}$ coincides with the vertex of the selected cone. In this case the feasible set is a single point, so by identifying this unique solution the problem is solved. This completes the procedure.

We have shown that for all the cases in Theorem 2, we can find a cone $\mathscr{K}$ or a cylinder $\mathscr{C}$ that satisfies Definitions 2 or 5, respectively. Hence, by combining Theorem 2 with Propositions 1 and 2 we can provide a procedure to find the convex hull of $\mathscr{E} \cap (\mathscr{A} \cup \mathscr{B})$, where the disjunctive set $\mathscr{A} \cup \mathscr{B}$ is such that the hyperplanes $\mathscr{A}^=$ and $\mathscr{B}^=$ are parallel. Thus, we have given easy to compute procedures to identify disjunctive conic cuts, and disjunctive cylindrical cuts in the respective cases of Theorem 2.

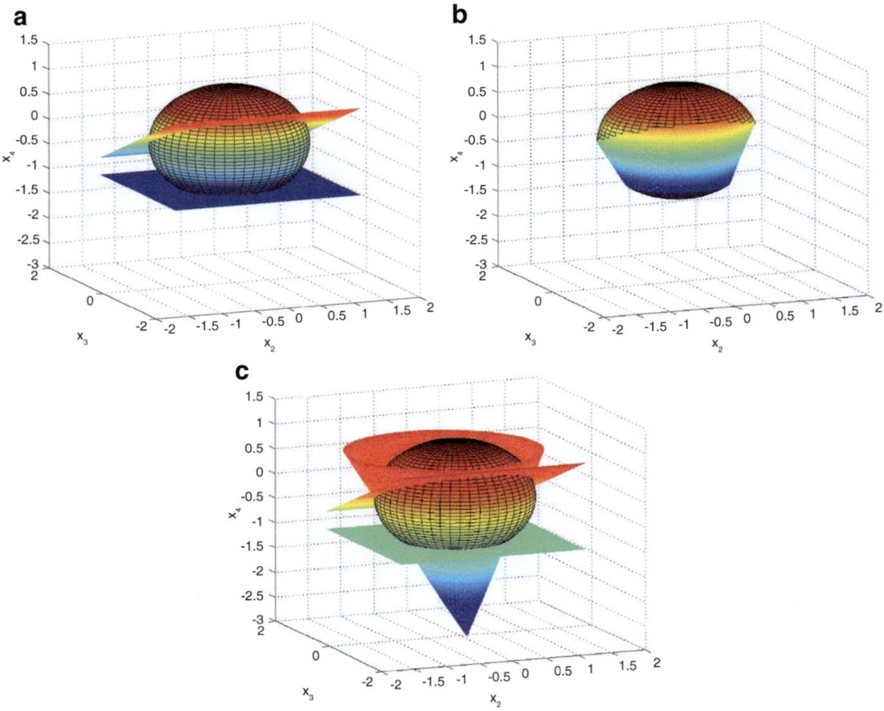

Fig. 7 Convex hull of the intersection between a non-parallel disjunction and an ellipsoid. (**a**) $\mathscr{A}^{=}$, $\mathscr{B}^{=}$, and $\mathscr{E}$. (**b**) $\mathrm{conv}(\mathscr{E} \cap (\mathscr{A} \cup \mathscr{B}))$. (**c**) The cone yielding $\mathrm{conv}(\mathscr{E} \cap (\mathscr{A} \cup \mathscr{B}))$

4.2 General Disjunctions

Some of the results in Section 4.1 can be extended to general disjunctive sets $\mathscr{A} \cup \mathscr{B}$, where $\mathscr{A} = \{x \in \mathbb{R}^n | a^\top x \geq \alpha\}$ and $\mathscr{B} = \{x \in \mathbb{R}^n | b^\top x \leq \beta\}$ are defined such that there exists no $\kappa \in \mathbb{R}$ such that $b = \kappa a$.

An important example of general disjunction is given by *complementarity constraints*, usually described in the form $x_i x_j = 0$ and hence equivalent to the disjunction $x_i = 0 \vee x_j = 0$. An example of disjunctive cuts separated for problems with complementarity constraints is given by Júdice et al. [18], who study a problem where complementarity constraints are the only nonlinear ones, and whose relaxation yields an LP. Disjunctive cuts are separated using violated complementarity constraints by observing that both variables are basic and then applying a disjunctive procedure to the corresponding tableau rows.

We may assume w.l.o.g. that $\|a\| = \|b\| = 1$. These disjunctive sets are illustrated in Figure 7a for Problem (5) using $\mathscr{A} = \{x \in \mathbb{R}^4 \mid 0.45x_3 + 0.89x_4 \geq 0\}$ and $\mathscr{B} = \{x \in \mathbb{R}^4 \mid x_4 \leq -1\}$ to define the disjunctive set $\mathscr{A} \cup \mathscr{B}$.

4.2.1 Geometry of $\mathscr{E}$ and the Hyperplanes $\mathscr{A}^=$ and $\mathscr{B}^=$

We begin this analysis recalling Theorem 4.1 in [8]. This theorem defines a family of quadrics $\mathscr{Q}(\tau)$ for $\tau \in \mathbb{R}$ such that $\mathscr{Q}(\tau) \cap \mathscr{A}^= = \mathscr{Q} \cap \mathscr{A}^=$ and $\mathscr{Q}(\tau) \cap \mathscr{B}^= = \mathscr{Q} \cap \mathscr{B}^=$.

Theorem 3 ([8]). *Consider an ellipsoid $\mathscr{E} = (Q, q, q_0)$ and two nonparallel hyperplanes $\mathscr{A}^=$ and $\mathscr{B}^=$. The uniparametric family of quadrics $\mathscr{Q}(\tau)$ parametrized by $\tau \in \mathbb{R}$ and having the same intersection with $\mathscr{A}^=$ and $\mathscr{B}^=$ as the ellipsoid $\mathscr{E}$ is given by*

$$\begin{aligned} Q(\tau) &= Q + \tau \frac{ab^\top + ba^\top}{2} \\ q(\tau) &= q - \tau \frac{\beta a + \alpha b}{2} \\ \rho(\tau) &= \rho + \tau\alpha\beta. \end{aligned} \tag{13}$$

We need to investigate if there is a value $\bar{\tau}$ in the family of Theorem 3 for which $\mathscr{Q}(\bar{\tau})$ is a cone $\mathscr{K}$ or a cylinder $\mathscr{C}$. Thus, either $\mathscr{K} \cap \mathscr{E}$ or $\mathscr{C} \cap \mathscr{E}$ will give the convex hull for $\mathscr{E} \cap (\mathscr{A} \cup \mathscr{B})$. Figure 7b, c illustrate this for the example in Problem (5).

Note that in Theorem 3, $Q(\tau)$ has a rank-2 update. This opens the possibility of having a matrix with two negative eigenvalues. However, it can be verified that under the assumption of Q being positive definite, $Q(\tau)$ can have at most one non-positive eigenvalue [8, § 4]. This property reduces the case of general disjunctive sets to the same set of geometrical objects that were considered in Section 4.1.1.

For any vector d, define $u_d = Q^{-1/2} d$. Using this notation, we get from Theorem 3 the following [8, § 4]:

$$q(\tau)^\top Q(\tau)^{-1} q(\tau) - \rho(\tau) = \frac{f(\tau)}{g(\tau)}, \tag{14}$$

where

$$g(\tau) = \tau^2 \left(\left(u_a^\top u_b \right)^2 - \|u_a\|^2 \|u_b\|^2 \right) + 4 u_a^\top u_b \tau + 4$$

and

$$\begin{aligned} f(\tau) =& \tau^2 \Big[\|u_a\|^2 (\beta + u_b^\top u_q)^2 + \|u_b\|^2 (\alpha + u_a^\top u_q)^2 \\ &+ ((u_a^\top u_b)^2 - \|u_a\|^2 \|u_b\|^2)(\|u_q\|^2 - \rho) - 2 u_a^\top u_b (u_a^\top u_q + \alpha)(u_b^\top u_q + \beta) \Big] \\ &+ 4\tau \Big[u_a^\top u_b (\|u_q\|^2 - \rho) - (\alpha + u_a^\top u_q)(\beta + u_b^\top u_q) \Big] + 4 \Big[\|u_q\|^2 - \rho \Big], \end{aligned} \tag{15}$$

which are two quadratic functions in τ. Let the two roots of $g(\tau)$ be denoted as $\hat{\tau}_1$ and $\hat{\tau}_2$, and we may assume w.l.o.g. that $\hat{\tau}_1 \leq \hat{\tau}_2$. It is proven in [8, § 4] that at these two values $Q(\tau)$ is a positive semidefinite matrix with one zero eigenvalue. Now, let the roots of $f(\tau)$ be denoted as $\bar{\tau}_1$ and $\bar{\tau}_2$, and we may also assume w.l.o.g. that $\bar{\tau}_1 \leq \bar{\tau}_2$. It is easy to verify that (14) becomes zero for these two values when $Q(\bar{\tau}_i)$ is non-singular, $i = 1, 2$. Additionally, in [8, §4] it is shown that the situations $\hat{\tau}_1 < \bar{\tau}_1 < \hat{\tau}_2$, or $\hat{\tau}_1 < \bar{\tau}_2 < \hat{\tau}_2$, are only possible when the quadric $\mathcal{Q}$ is a single point, which is a trivial case. We use these observations in Theorem 4 to summarize the behavior of the family $\mathcal{Q}(\tau)$ when the quadric $\mathcal{Q}$ is not a single point, based on the values $\hat{\tau}_1, \hat{\tau}_2, \bar{\tau}_1, \bar{\tau}_2$. The interested reader can review the details of this theorem in [8, §4.2].

Theorem 4 ([8]). *Depending on the geometry of $\mathcal{E}$, $\mathcal{A}$, and $\mathcal{B}$, $\mathcal{Q}(\tau)$ can have the following shapes for $\tau \in \mathbb{R}$:*

- *$f(\tau)$ has two distinct roots $\bar{\tau}_1 < \bar{\tau}_2$ such that $\hat{\tau}_2 < \bar{\tau}_1$, or $\bar{\tau}_2 < \hat{\tau}_1$, or $\bar{\tau}_1 < \hat{\tau}_1 \leq \hat{\tau}_2 < \bar{\tau}_2$: this is the general case, $\mathcal{Q}(\hat{\tau}_1)$ and $\mathcal{Q}(\hat{\tau}_2)$ are paraboloids, and $\mathcal{Q}(\bar{\tau}_1)$ and $\mathcal{Q}(\bar{\tau}_2)$ are cones.*
- *$f(\tau)$ has two distinct roots $\bar{\tau}_1 < \bar{\tau}_2$, and exactly one of them coincides with either $\hat{\tau}_1$ or $\hat{\tau}_2$: this case has two possibilities. First, $\mathcal{Q}(\hat{\tau}_1)$ is a cylinder and $\mathcal{Q}(\hat{\tau}_2)$ is a paraboloid. Second, $\mathcal{Q}(\hat{\tau}_2)$ is a cylinder and $\mathcal{Q}(\hat{\tau}_1)$ is a paraboloid. In both situations we have that either $\mathcal{Q}(\bar{\tau}_1)$ is a cylinder and $\mathcal{Q}(\bar{\tau}_2)$ is a cone or that $\mathcal{Q}(\bar{\tau}_2)$ is a cylinder and $\mathcal{Q}(\bar{\tau}_1)$ is a cone.*
- *$f(\tau)$ has two distinct roots $\bar{\tau}_1 < \bar{\tau}_2$ such that $\bar{\tau}_1 = \hat{\tau}_1$ and $\bar{\tau}_2 = \hat{\tau}_2$: in this case both hyperplanes contain the center $-Q^{-1}q$ of the ellipsoid $\mathcal{E}$. Both quadrics $\mathcal{Q}(\hat{\tau}_1)$ and $\mathcal{Q}(\hat{\tau}_2)$ are cylinders.*
- *The two roots of $f(\tau)$ coincide, and either $\bar{\tau}_1 = \bar{\tau}_2 < \hat{\tau}_1$ or $\hat{\tau}_2 < \bar{\tau}_1 = \bar{\tau}_2$: in this case the discriminant of $f(\tau)$ is zero, which implies that one of the hyperplanes intersects $\mathcal{E}$ in only one point. We have that $\mathcal{Q}(\bar{\tau}_1)$ is a cone and the quadrics $\mathcal{Q}(\hat{\tau}_1)$, $\mathcal{Q}(\hat{\tau}_2)$ are two paraboloids.*
- *The two roots of $f(\tau)$ coincide and either $\bar{\tau}_1 = \bar{\tau}_2 = \hat{\tau}_1$ or $\hat{\tau}_2 = \bar{\tau}_1 = \bar{\tau}_2$: in this case both hyperplanes intersect $\mathcal{E}$ in only one point. Then, either $\mathcal{Q}(\hat{\tau}_1)$ is a line and $\mathcal{Q}(\hat{\tau}_2)$ is a paraboloid or $\mathcal{Q}(\hat{\tau}_2)$ is a line and $\mathcal{Q}(\hat{\tau}_1)$ is a paraboloid.*

4.2.2 Building a Disjunctive Conic Cut

Using the results of the geometrical analysis of Section 4.2.1 we give now the guidelines to build a conic cut to convexify the intersection of a MISOCO feasible set with a general disjunction.

First of all, observe that from Assumption 1 the third case in Theorem 4 cannot occur. Hence, this case is not considered for building a cut for general disjunctions. We classify the remaining cases as cylinders and cones.

Cylinders: We look at the cylinders $\mathcal{C}$ in the families $\mathcal{Q}(\tau)$ described in the second and fifth cases of Theorem 4 of [8]. Observe that in general, $\mathcal{C}$ can be found at either $\hat{\tau}_2$ or $\hat{\tau}_1$. This can be decided by comparing $\hat{\tau}_2$ or $\hat{\tau}_1$ with the roots $\bar{\tau}_1$ and

$\bar{\tau}_2$ using the criteria described in Theorem 4. Let $\hat{\tau}$ be a value such that $\mathcal{Q}(\hat{\tau})$ is a cylinder. From Eq. (7) it is easy to verify that $\mathcal{Q}(\hat{\tau})$ is a convex set. Consequently, from Proposition 2 we get that $\mathscr{C} \cap \mathscr{E}$ is the convex hull of $\mathscr{E} \cap (\mathscr{A} \cup \mathscr{B})$. Finally, note that the cylinder $\mathscr{C}$ can be represented in terms of a second order cone. For that reason, we classify $\mathscr{C}$ as a conic cut in this section too.

Cones: We need to focus now on the cones described in the first and fourth cases of Theorem 4. Let $\bar{\tau}_i \neq \hat{\tau}_1, \hat{\tau}_2$, $i = 1,2$. In these two cases $Q(\bar{\tau}_i)$ is symmetric and non-singular matrix with exactly one negative eigenvalue. This is a similar situation as the first and third cases of Theorem 2. From the analysis in Section 4.1.2 it follows that $\mathcal{Q}(\bar{\tau}_i) = \mathcal{Q}(\bar{\tau}_i)^+ \cup \mathcal{Q}(\bar{\tau}_i)^-$, where $\mathcal{Q}(\bar{\tau}_i)^+, \mathcal{Q}(\bar{\tau}_i)^-$ are the second order cones (11) and (12). Observe that $x(\bar{\tau}_i) = -q(\bar{\tau}_i)^\top Q(\bar{\tau}_i)$ is the vertex of $\mathcal{Q}(\bar{\tau}_i)^+$ and $\mathcal{Q}(\bar{\tau}_i)^-$. Then, using Lemma 3 we can verify if there is a cone in $\mathcal{Q}(\bar{\tau}_i)^+, \mathcal{Q}(\bar{\tau}_i)^-$, $i = 1,2$, that satisfies Definition 2. In particular, we need to prove that there is one $x(\bar{\tau}_i)$, $i = 1,2$, that is either in $\mathscr{A}$ or $\mathscr{B}$. This criteria is stated in Lemma 10.

Lemma 10. *Let the two roots $\bar{\tau}_i$, $i = 1,2$ of $f(\tau)$ be different from $\hat{\tau}_1$, and $\hat{\tau}_2$. Then, in the first and fourth cases of Theorem 4, the cone $\mathcal{Q}(\bar{\tau}_2)$ contains a convex cone that satisfies Definition 2.*

The proof of Lemma 10 is presented in section "Proof of Lemma 10" in Appendix 1. Now we can define a procedure to identify a conic cut. We need to identify which of the cones $\mathcal{Q}(\bar{\tau}_2)^+, \mathcal{Q}(\bar{\tau}_2)^-$ gives the conic cut. For this purpose we use the sign of $W(\bar{\tau}_2)_1^\top(-Q^{-1}q - x(\bar{\tau}_2))$. Hence, we choose $\mathcal{Q}(\bar{\tau}_2)^+$ if $W(\bar{\tau}_2)_1^\top(-Q^{-1}q - x(\bar{\tau}_2)) > 0$, and we choose $\mathcal{Q}(\bar{\tau}_2)^-$ when $W(\bar{\tau}_2)_1^\top(-Q^{-1}q - x(\bar{\tau}_2)) < 0$. This completes the procedure.

In summary, excluding the third case in Theorem 4, we have shown that it is possible to find a cone $\mathscr{K}$ or cylinder $\mathscr{C}$ satisfying Definitions 2 or 5 for all the relevant cases in Theorem 4. Hence, combining Theorem 4 with Propositions 1 and 2 we provided a procedure to find a disjunctive conic cut for $\mathscr{E} \cap (\mathscr{A} \cup \mathscr{B})$ for a general disjunctive set $\mathscr{A} \cup \mathscr{B}$.

5 Disjunctive Conic Cut vs Nonlinear Conic Mixed Integer Rounding Inequality

Atamtürk and Narayanan [1] present a procedure for generating a *nonlinear conic mixed integer rounding* cut. Since this is a conic cut, we examine how it compares to the disjunctive conic cut introduced here. For this purpose, let us consider the following example

$$\begin{aligned} \text{minimize: } & -x \quad -y \\ \text{subject to: } & \left\| \begin{matrix} x - \frac{4}{3} \\ y - 1 \end{matrix} \right\| \leq \frac{4}{3} - \frac{x}{2} - \frac{y}{2} \\ & x \in \mathbb{Z},\ y \in \mathbb{R}. \end{aligned} \tag{16}$$

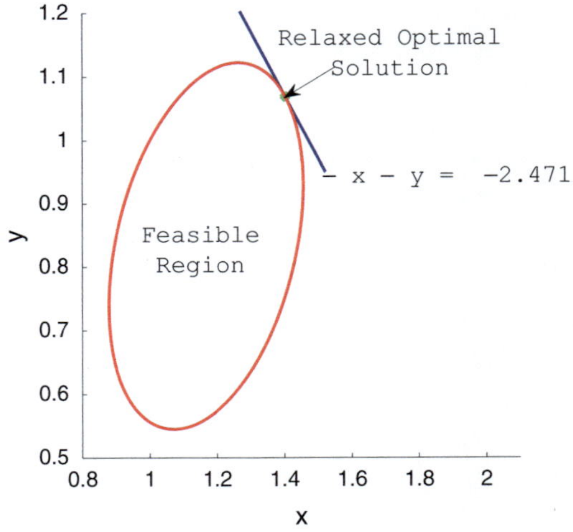

Fig. 8 An optimal solution of problem (17)

First, notice that the example in (16) is in the format used in [1], which is different from the one in (1). The main difference is the way we write the conic constraint. Despite this difference we can still construct a disjunctive conic cut, because the feasible region of this problem is an ellipsoid in the (x,y) space.

Relaxing the integrality constraint, the resulting relaxation from problem (16) can be solved easily (the KKT conditions give a 2×2 linear system). First, notice that this relaxation is just a problem of minimizing a linear function over an ellipsoid. Particularly, we can rewrite the relaxation of problem (16) as follows:

$$\begin{aligned} \text{minimize: } & -x-y \\ \text{subject to: } & \tfrac{3}{4}x^2+\tfrac{3}{4}y^2-\tfrac{1}{2}xy-\tfrac{4}{3}x-\tfrac{2}{3}y+1\le 0 \\ & x,\ y\in\mathbb{R}. \end{aligned} \tag{17}$$

The feasible set of this problem, which is an ellipsoid, is presented in Figure 8. The optimal objective function value is -2.471, and the relaxed optimal solution for the example in problem (16) is $(x^*,y^*)=(1.402,1.069)$.

We can rewrite problem (16) in the following form:

$$\begin{aligned} \text{minimize: } & -x-y \\ \text{subject to: } & x+y+2t=\tfrac{8}{3} \\ & \sqrt{(x-\tfrac{4}{3})^2+(y-1)^2}\le t \\ & x\in\mathbb{Z},\ y\in\mathbb{R},\ t\in\mathbb{R}. \end{aligned} \tag{18}$$

Figure 9 presents the feasible region of this equivalent problem. Using a branch-and-bound procedure one can easily solve the mixed integer problem in (18), and get that the optimal solution is $(t^*,x^*,y^*)=(1/3,1,1)$ with the optimal cost of -2.

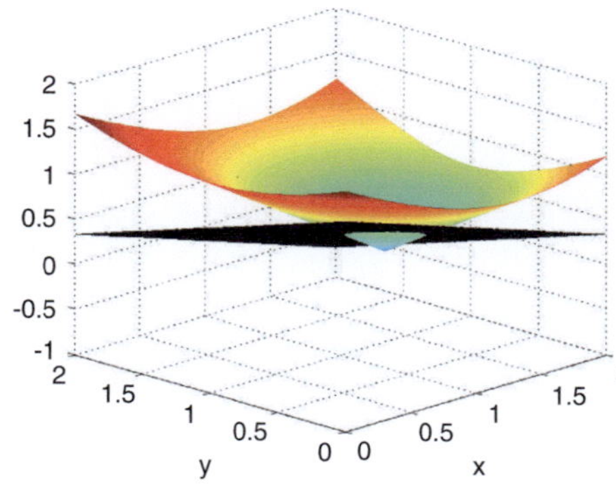

Fig. 9 The feasible region of the continuous relaxation of problem (18)

The problem reformulation (18) presents a case similar to the one studied in Example 1 in [1], which shows how to obtain a nonlinear conic mixed integer rounding inequality for the set

$$T_0 = \left\{ (x,y,t) \in \mathbb{Z} \times \mathbb{R} \times \mathbb{R} : \sqrt{(x - \frac{4}{3})^2 + (y-1)^2} \le t \right\}, \tag{19}$$

which is the set of solutions satisfying the last constraint in (18). In general, the procedure discussed by Atamtürk and Narayanan [1] focuses on generating the convex hull for each *polyhedral second-order conic constraint* in the problem. Then, by adding those new cuts they tighten the original formulation. In particular, applying that procedure to the set in (19) they obtain the cut

$$\sqrt{\left(\frac{x}{3}\right)^2 + (y-1)^2} \le t, \tag{20}$$

which is a valid cut for the problem in (18).

Analyzing the relaxed solution showed in Figure 8, we can see that the solution is not feasible for the integer problem. First, observe that if we use the disjunction $x \le 1 \vee x \ge 2$ it is not possible to apply the disjunctive conic cut here, because the line $x = 2$ does not intersect the set of feasible solutions that is an ellipsoid, violating one of the assumptions in Section 3. However, we can still use the nonlinear conic mixed integer rounding inequality procedure. Figure 10 shows the result of applying the nonlinear conic cut (19) to the problem in (18). The point $(t^*, x^*, y^*) = (1/3, 1, 1)$ is the new optimal solution for the continuous relaxation of the resulting problem with the cut added, which turns out to be optimal for the mixed integer problem. The optimal objective value is -2.

Now, let us modify the first constraint in (18) as follows

$$x + y + 2t = \frac{14}{3}.$$

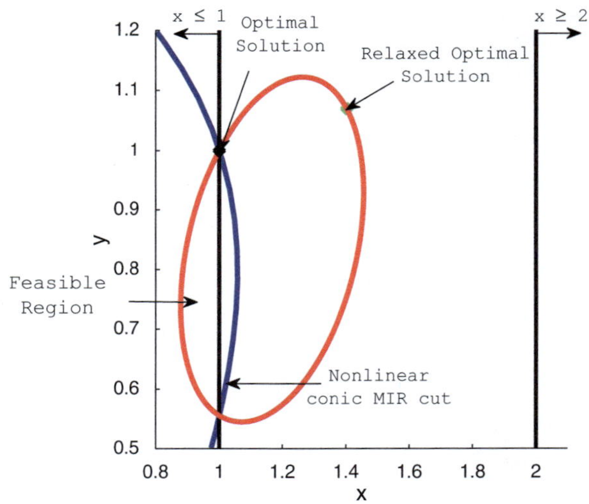

Fig. 10 Nonlinear conic mixed integer rounding inequality

Figure 11 shows the new feasible region. With this modification, the relaxed optimal solution is $(t^*, x^*, y^*) = (0.68, 1.81, 1.48)$, which is not feasible for the integer problem. Now, for this example we can use the disjunction $x \leq 1 \vee x \geq 2$ and obtain a disjunctive conic cut that can be represented in the (x, y) space as follows:

$$\sqrt{(y - 0.33x + 0.22)^2} \leq 2.67 - 0.93x. \tag{21}$$

Observe that the nonlinear conic mixed integer rounding inequality (20) stays the same, since we have not modified the conic constraint. Figure 11 shows these two cuts and highlights the difference between applying the nonlinear conic mixed integer rounding inequality and the disjunctive conic cut to the modified problem. More specifically, the disjunctive conic cut gives the convex hull of the intersection between the disjunction $x \leq 1 \vee x \geq 2$ and the feasible set of problem (18). This is not the case for the nonlinear conic mixed integer rounding inequality (20). The new optimal solution for the relaxed problem when either of the cuts is applied is $(t^*, x^*, y^*) = (0.71, 2.0, 1.25)$. In particular, we can see that any of the cuts is enough to find the optimal solution. The optimal value for the objective function is -3.25.

Finally, we perform an additional test modifying the first constraint in (18) as follows:

$$x + y + 2t = 8.$$

In this case we use the disjunction $x \leq 2 \vee x \geq 3$. Then, we can obtain a disjunctive conic cut that can be represented in the (x, y) space as follows:

$$\sqrt{(y - 0.33x + 1.33)^2} \leq 6.04 - 1.21x.$$

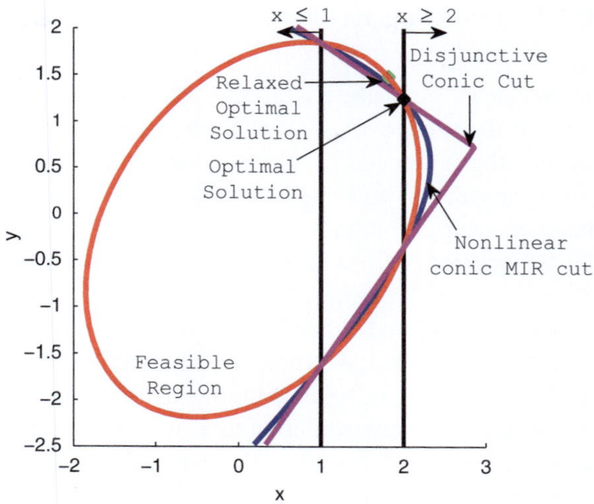

Fig. 11 The Disjunctive conic cut and the Nonlinear conic mixed integer rounding inequality cutting off the relaxed optimal solution

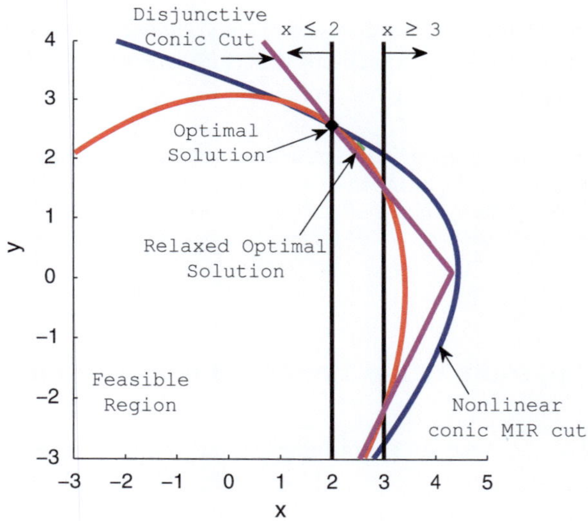

Fig. 12 The nonlinear conic mixed integer rounding inequality fails to cut off the optimal solution for the relaxed problem

For this example the nonlinear conic mixed integer rounding inequality (20) fails to eliminate the continuous optimal solution found for the relaxed problem, as illustrated in Figure 12. Thus, there is no gain in adding this cut to the problem. However, the disjunctive conic cut is violated by the current fractional solution, and the addition of the disjunctive conic cut is enough to find the integer solution for the problem.

6 Concluding Remarks

In this paper, we analyzed the convex hull of the intersection of a convex set $\mathscr{E}$ and a linear disjunctive set $\mathscr{A} \cup \mathscr{B}$. This analysis is done for general convex sets. We assume the existence of a convex cone $\mathscr{K}$ (resp. a convex cylinder $\mathscr{C}$) that has the same intersection with the boundary $\mathscr{A}^=$, $\mathscr{B}^=$ of the disjunction as $\mathscr{E}$. Given the cone $\mathscr{K}$ (resp. cylinder $\mathscr{C}$), we proved that the convex hull $\text{conv}(\mathscr{E} \cap (\mathscr{A} \cup \mathscr{B}))$ is $\mathscr{E} \cap \mathscr{K}$ (resp. $\mathscr{E} \cap \mathscr{C}$). Additionally, we were able to prove that if $\mathscr{K}$ (resp. a cylinder $\mathscr{C}$) exists, then it is unique.

We then showed the existence of such a cone $\mathscr{K}$ (resp. a cylinder $\mathscr{C}$) for MISOCO problems. We consider the feasible set of the continuous relaxation of a MISOCO problem, assumed to be an ellipsoid, intersected with a general linear disjunction. We showed that in this case $\mathscr{K}$ is a second order cone, and provided a closed formula to describe $\mathscr{K}$ (resp. a cylinder $\mathscr{C}$) for MISOCO problems. This cone provides a novel conic cut for MISOCO and because it gives the convex hull of the disjunction, it is the strongest possible cut for MISOCO problems. Having a closed form for this disjunctive conic cut makes them ready to use. The development of an efficient Branch-and-Cut software package for MISOCO problems is the subject of ongoing research.

Acknowledgements The authors Pietro Belotti, Imre Pólik and Tamás Terlaky acknowledge the support of Lehigh University with a start up package for the development of this research. The authors Julio C. Góez and Tamás Terlaky acknowledge the support of the Airforce Research Office grant # FA9550-10-1-0404 for the development of this research.

Appendix 1: The Proofs of Lemmas 9 and 10

For the sake of simplifying the algebra of these proofs we use the following observation. If $Q \succ 0$ and the quadric $\mathscr{Q}$ is not single point, $\mathscr{Q}$ can be transformed to a unit hypersphere $\{y \in \mathbb{R}^n \mid \|y\|^2 \leq 1\}$ using the affine transformation

$$y = \frac{Q^{1/2}(x + Q^{-1}q)}{\sqrt{\|u_q\|^2 - \rho}}. \tag{22}$$

Observe that this transformation preserves the inertia of Q, hence the classification of the quadric is not changed. Additionally, observe that if we apply the same transformation to two parallel hyperplanes, the resulting hyperplanes are still parallel. Hence, throughout this proof, if $Q \succ 0$, we assume w.l.o.g. that the quadric $\mathscr{Q}$ is a unit hypersphere centered at the origin. In this case, we have that the positive definite matrix Q of Section 4 is the identity matrix, the vector q is the zero vector, $\rho = -1$. Additionally, given Assumption 2 and the assumption that $\|a\| = \|b\| = 1$, we have that $|\alpha| \leq 1$, and $|\beta| \leq 1$. Finally, recall Assumption 1, then we may assume w.l.o.g. that $\alpha > \beta$.

Proof of Lemma 9

From Section 4.1 we have that $\hat{\tau} = -1$, and the numerator of the right-hand side of (9) reduces to

$$f(\tau) = \tau^2 \frac{(\alpha-\beta)^2}{4} + \tau(1-\alpha\beta) + 1.$$

Recall from Section 4.1.1 that the quadrics $\mathscr{Q}(\bar{\tau}_1)$ and $\mathscr{Q}(\bar{\tau}_2)$ in the family $\{\mathscr{Q}(\tau) \mid \tau \in \mathbb{R}\}$, are computed using the roots $\bar{\tau}_1$ and $\bar{\tau}_2$ of the function $f(\tau)$. Particularly, we have that

$$\bar{\tau}_1 = 2\left(\frac{\alpha\beta - 1 - \sqrt{(1-\alpha^2)(1-\beta^2)}}{(\alpha-\beta)^2}\right),$$

$$\bar{\tau}_2 = 2\left(\frac{\alpha\beta - 1 + \sqrt{(1-\alpha^2)(1-\beta^2)}}{(\alpha-\beta)^2}\right),$$

where $\bar{\tau}_1 \le \bar{\tau}_2$. Note that if $\alpha = \beta$, then $f(\tau)$ is a linear function. In this case we would have that $\mathscr{A} \cup \mathscr{B} = \mathbb{R}^n$ and it is easy to verify that $\operatorname{conv}(\mathscr{Q} \cap (\mathscr{A} \cup \mathscr{B})) = \mathscr{Q}$. However, recall that our assumption is $\beta \neq \alpha$. Hence, for the rest of this proof we assume w.l.o.g. that $\alpha > \beta$, which results from Assumption 1.

The vertices of the cones $\mathscr{Q}(\bar{\tau}_1)$ and $\mathscr{Q}(\bar{\tau}_2)$ are $x(\bar{\tau}_i) = -Q(\bar{\tau}_i)^{-1}q(\bar{\tau}_i)$, $i = 1,2$. We can express $x(\bar{\tau}_i)$ in terms of a, α, and β as follows:

$$\begin{aligned} x(\bar{\tau}_i) = -Q(\bar{\tau}_i)^{-1}q(\bar{\tau}_i) &= -\left(I - \frac{\bar{\tau}_i}{(1+\bar{\tau}_i)} aa^\top\right)\left(-\bar{\tau}_i \frac{(\alpha+\beta)}{2} a\right) \\ &= \bar{\tau}_i \frac{(\alpha+\beta)}{2}\left(1 - \frac{\bar{\tau}_i}{(1+\bar{\tau}_i)}\right) a \\ &= \bar{\tau}_i \frac{(\alpha+\beta)}{2(1+\bar{\tau}_i)} a. \end{aligned}$$

Consider the inner product

$$a^\top x(\bar{\tau}_i) = -a^\top Q(\bar{\tau}_i)^{-1} q(\bar{\tau}_i) = \bar{\tau}_i \frac{(\alpha+\beta)}{2(1+\bar{\tau}_i)} a^\top a = \bar{\tau}_i \frac{(\alpha+\beta)}{2(1+\bar{\tau}_i)}.$$

Note that if $\alpha = -\beta$ then $a^\top x(\bar{\tau}_i) = 0$. Recall from Theorem 2 that in that case $Q(\bar{\tau}_1)$ is a cylinder. For that reason, we assume that $\alpha \neq -\beta$ for the rest of this proof.

Next, note that since $\mathscr{A}^=$ and $\mathscr{B}^=$ are parallel, then $\mathscr{A} \cap \mathscr{B} = \emptyset$. Then, we need to show that in the first and third cases of Theorem 2 the vertex $x(\bar{\tau}_2)$ cannot be in the set $\overline{\mathscr{A}} \cap \overline{\mathscr{B}}$. Assume to the contrary that $x(\bar{\tau}_2) \in \overline{\mathscr{A}} \cap \overline{\mathscr{B}}$. Now, since we are analyzing the first and third cases of Theorem 2 we know that $\bar{\tau}_2 < -1$. Thus, if

$a^\top x(\bar{\tau}_2) < \alpha$ and $a^\top x(\bar{\tau}_2) > \beta$, then

$$\bar{\tau}_2(\beta - \alpha) > 2\alpha \quad \text{and} \quad \bar{\tau}_2(\alpha - \beta) < 2\beta. \tag{23}$$

Substituting $\bar{\tau}_2$ in (23) we obtain that $\sqrt{\frac{(1-\alpha^2)}{(1-\beta^2)}} = 1$. The last inequality is possible only if either $\alpha = -\beta$ or $\alpha = \beta$. Hence, in the first and third cases of Theorem 2 the vertex $x(\bar{\tau}_2)$ cannot be in the set $\overline{\mathscr{A}} \cap \overline{\mathscr{B}}$.

Thus, since the intersections $\mathscr{Q}(\bar{\tau}_2) \cap \mathscr{A}^=$ and $\mathscr{Q}(\bar{\tau}_2) \cap \mathscr{B}^=$ are bounded, then one of the following two cases holds:

- Case 1: $\mathscr{Q}^+(\bar{\tau}_2) \cap \mathscr{A}^= = \mathscr{E} \cap \mathscr{A}^=$ and $\mathscr{Q}^+(\bar{\tau}_2) \cap \mathscr{B}^= = \mathscr{E} \cap \mathscr{B}^=$;
- Case 2: $\mathscr{Q}^-(\bar{\tau}_2) \cap \mathscr{A}^= = \mathscr{E} \cap \mathscr{A}^=$ and $\mathscr{Q}^-(\bar{\tau}_2) \cap \mathscr{B}^= = \mathscr{E} \cap \mathscr{B}^=$.

Consequently, we have that one of the cones $\mathscr{Q}^+(\bar{\tau}_2)$ and $\mathscr{Q}^-(\bar{\tau}_2)$ found at the root $\bar{\tau}_2$ satisfy Proposition 1.

□

Proof of Lemma 10

Recall from Section 4.2.1 that the quadrics $\mathscr{Q}(\bar{\tau}_1)$ and $\mathscr{Q}(\bar{\tau}_2)$ in the family $\{\mathscr{Q}(\tau) \mid \tau \in \mathbb{R}\}$ of Theorem 4 are computed using the roots of the function (15), which in this case simplifies to

$$f(\tau) = \left((\alpha\beta - a^\top b)^2 - (1-\alpha^2)(1-\beta^2)\right)\tau^2 + 4(a^\top b - \alpha\beta)\tau + 4.$$

The roots of $f(\tau)$ are

$$\bar{\tau}_1 = 2\left(\frac{\alpha\beta - a^\top b - \sqrt{(1-\alpha^2)(1-\beta^2)}}{(\alpha\beta - a^\top b)^2 - (1-\alpha^2)(1-\beta^2)}\right) = \frac{2}{\alpha\beta - a^\top b + \sqrt{(1-\alpha^2)(1-\beta^2)}},$$

$$\bar{\tau}_2 = 2\left(\frac{\alpha\beta - a^\top b + \sqrt{(1-\alpha^2)(1-\beta^2)}}{(\alpha\beta - a^\top b)^2 - (1-\alpha^2)(1-\beta^2)}\right) = \frac{2}{\alpha\beta - a^\top b - \sqrt{(1-\alpha^2)(1-\beta^2)}},$$

where $\bar{\tau}_1 \leq \bar{\tau}_2$.

Also, recall that the classification of the quadrics $\mathscr{Q}(\bar{\tau}_1)$ and $\mathscr{Q}(\bar{\tau}_2)$ is done based on the ratio $f(\tau)/g(\tau)$, where $g(\tau)$ simplifies in this case to

$$g(\tau) = ((a^\top b)^2 - 1)\tau^2 + 4a^\top b\tau + 4.$$

Note that if $(a^\top b)^2 - 1 = 0$, then we obtain that g is an affine function with a zero at -1. However, since $\|a\| = \|b\| = 1$, in this case we either obtain that $a^\top b =$

$\cos(0)$, which implies that $a = b$, or we obtain that $a^\top b = -\cos(0)$, which implies that $a = -b$. This is the case when we have parallel hyperplanes, which was already analyzed in section "Proof of Lemma 9" in Appendix 1 and will not be considered in the rest of this proof. Now, the roots of $g(\tau)$ are

$$\hat{\tau}_1 = -\frac{2}{a^\top b + 1} < 0 \quad \text{and} \quad \hat{\tau}_2 = -\frac{2}{a^\top b - 1} > 0.$$

The vertex of the cone $\mathscr{Q}(\bar{\tau}_2)$ is $x(\bar{\tau}_2) = -Q(\bar{\tau}_2)^{-1}q(\bar{\tau}_2)$. We can express $x(\bar{\tau}_2)$ in terms of a, b, α, and β as follows:

$$\begin{aligned} x(\bar{\tau}_2) &= -Q(\bar{\tau}_2)^{-1}q(\bar{\tau}_2) \\ &= -\left(I - \frac{(aa^\top + bb^\top)\bar{\tau}_2^2 - (a^\top b\bar{\tau}_2^2 + 2\bar{\tau}_2)(ba^\top + ab^\top)}{(1-(a^\top b)^2)\bar{\tau}_2^2 - 4a^\top b\bar{\tau}_2 - 4}\right)\left(-\bar{\tau}_2\frac{\beta a + \alpha b}{2}\right) \\ &= \frac{\bar{\tau}_2\left(((\alpha - a^\top b\beta)\bar{\tau}_2 - 2\beta)\,a + ((\beta - a^\top b\alpha)\bar{\tau}_2 - 2\alpha)\,b\right)}{(1-(a^\top b)^2)\bar{\tau}_2^2 - 4a^\top b\bar{\tau}_2 - 4}. \end{aligned}$$

Consider the inner products

$$a^\top x(\bar{\tau}_2) = \frac{\bar{\tau}_2\left((1-(a^\top b)^2)\alpha\bar{\tau}_2 - 2(a^\top b\alpha + \beta)\right)}{(1-(a^\top b)^2)\bar{\tau}_2^2 - 4a^\top b\bar{\tau}_2 - 4}$$

and

$$b^\top x(\bar{\tau}_2) = \frac{\bar{\tau}_2\left((1-(a^\top b)^2)\beta\bar{\tau}_2 - 2(a^\top b\beta + \alpha)\right)}{(1-(a^\top b)^2)\bar{\tau}_2^2 - 4a^\top b\bar{\tau}_2 - 4}.$$

Next, we show that in the first and fourth cases of Theorem 4 the vertex $x(\bar{\tau}_2)$ cannot be in the set $\overline{\mathscr{A}} \cap \overline{\mathscr{B}}$. Assume to the contrary that $x(\bar{\tau}_2) \in \overline{\mathscr{A}} \cap \overline{\mathscr{B}}$. Note that $\hat{\tau}_1$ and $\hat{\tau}_2$ are the roots of $(1-(a^\top b)^2)\tau^2 - 4a^\top b\tau - 4 = -g(\tau)$. Now, since we are analyzing the first and fourth cases of Theorem 4 we know that $\hat{\tau}_2 < \bar{\tau}_1$, or $\bar{\tau}_2 < \hat{\tau}_1$, or $\bar{\tau}_1 < \hat{\tau}_1 < \hat{\tau}_2 < \bar{\tau}_2$. Even more, since $1-(a^\top b)^2 \geq 0$ we have that $(1-(a^\top b)^2)\bar{\tau}_2^2 - 4a^\top b\bar{\tau}_2 - 4 \geq 0$. Thus, if $a^\top x(\bar{\tau}_2) < \alpha$ and $b^\top x(\bar{\tau}_2) > \beta$, then

$$(a^\top b\alpha - \beta)\bar{\tau}_2 < -2\alpha \quad \text{and} \quad (a^\top b\beta - \alpha)\bar{\tau}_2 > -2\beta. \tag{24}$$

Substituting $\bar{\tau}_2$ in (24) we obtain that $\frac{\alpha}{\sqrt{1-\alpha^2}} = -\frac{\beta}{\sqrt{1-\beta^2}}$, which implies that $\alpha = -\beta$. This is possible if $\bar{\tau}_2 = \hat{\tau}_1$, which is not in the cases being considered. Hence, in the first and fourth cases of Theorem 4 $x(\bar{\tau}_2)$ cannot be in the set $\overline{\mathscr{A}} \cap \overline{\mathscr{B}}$.

Similarly, we can show that in the first and fourth cases of Theorem 4 the vertex $x(\bar{\tau}_2)$ cannot be in the set $\mathscr{A} \cap \mathscr{B}$. In particular, if $a^\top x(\bar{\tau}_2) > \alpha$ and $b^\top x(\bar{\tau}_2) < \beta$, then

$$(a^\top b\alpha-\beta)\bar{\tau}_2 > -2\alpha \quad \text{and} \quad (a^\top b\beta-\alpha)\bar{\tau}_2 < -2\beta. \tag{25}$$

Substituting $\bar{\tau}_2$ in (25) we obtain that $\frac{\alpha}{\sqrt{1-\alpha^2}} = -\frac{\beta}{\sqrt{1-\beta^2}}$. This implies that $\bar{\tau}_2 = \hat{\tau}_1$, which is not in the cases being considered. Hence, the vertex $x(\bar{\tau}_2)$ cannot be in the set $\mathscr{A} \cap \mathscr{B}$.

Thus, since the intersections $\mathscr{Q}(\bar{\tau}_2) \cap \mathscr{A}^=$ and $\mathscr{Q}(\bar{\tau}_2) \cap \mathscr{B}^=$ are bounded, then one of the following two cases is true:

- Case 1: $\mathscr{Q}^+(\bar{\tau}_2) \cap \mathscr{A}^= = \mathscr{E} \cap \mathscr{A}^=$ and $\mathscr{Q}^+(\bar{\tau}_2) \cap \mathscr{B}^= = \mathscr{E} \cap \mathscr{B}^=$.
- Case 2: $\mathscr{Q}^-(\bar{\tau}_2) \cap \mathscr{A}^= = \mathscr{E} \cap \mathscr{A}^=$ and $\mathscr{Q}^-(\bar{\tau}_2) \cap \mathscr{B}^= = \mathscr{E} \cap \mathscr{B}^=$.

Consequently, we have that one of the cones $\mathscr{Q}^+(\bar{\tau}_2)$, $\mathscr{Q}^-(\bar{\tau}_2)$ found at the root $\bar{\tau}_2$ satisfies Proposition 1. □

Appendix 2: Additional Lemma

Lemma 11. *Let $\mathscr{C} \subset \mathbb{R}^n$ be a cylinder with a compact base. Then $\mathscr{C}$ is closed.*

Proof. Let $\mathscr{D}$ be a compact base for $\mathscr{C} = \{x \in \mathbb{R}^n | x = d + \sigma d_0, d \in \mathscr{D}, \sigma \in \mathbb{R}\}$ and let $u \in \mathbb{R}^n$ be a vector such that $u \notin \mathscr{C}$. Our goal is to show that there is a neighborhood $\mathscr{U}$ of u such that $\mathscr{U} \cap \mathscr{C} = \emptyset$.

Let $\delta = \max\{\|u-x\| \mid x \in \mathscr{D}\} > 0$ be the maximum distance from a point $x \in \mathscr{D}$ to u. Let us choose $\sigma_o = (\delta+1)/\|d_0\|$ and let $\mathscr{B}$ be the open ball of radius 1 centered at u. Define the set $\mathscr{C}_1 = \{x \in \mathbb{R}^n \mid x = d + \sigma d_o, d \in \mathscr{D}, \sigma \leq -\sigma_o\} \cup \{x \in \mathbb{R}^n \mid x = d + \sigma d_0, d \in \mathscr{D}, \sigma \geq \sigma_o\}$. Then, we have that $\mathscr{B} \cap \mathscr{C}_1 = \emptyset$.

Let $\mathscr{X} = \mathscr{D} \times [-\sigma_o, \sigma_o]$, and consider the map $h : \mathscr{X} \mapsto \mathbb{R}^n$, defined by $h(\sigma, x) = x + \sigma d_o$. Since $\mathscr{D}$ and $[-\sigma_o, \sigma_o]$ are compact we have that $\mathscr{X}$ is compact. Since h is continuous in $\mathscr{X}$ we have that the image $h(\mathscr{X})$ is a compact set as well, and hence closed in $\mathbb{R}^n$. Furthermore, note that $h(\mathscr{X}) \subset \mathscr{C}$, thus $u \notin h(\mathscr{X})$. Hence, there is a neighborhood $\mathscr{N}$ of u such that $\mathscr{N} \cap h(\mathscr{X}) = \emptyset$. Let $\mathscr{U} = \mathscr{B} \cap \mathscr{N}$, then for any $\sigma \in \mathbb{R}$ we have that $\mathscr{U} \cap (\sigma d_0 + \mathscr{D}) = \emptyset$. This proves that the complement of $\mathscr{C}$ is open, thus $\mathscr{C}$ is closed. □

References

1. Atamtürk, A., Narayanan, V.: Conic mixed-integer rounding cuts. Math. Program. **122**(1), 1–20 (2010)
2. Atamtürk, A., Narayanan, V.: Lifting for conic mixed-integer programming. Math. Program. A **126**, 351–363 (2011)
3. Atamtürk, A., Berenguer, G., Shen, Z.J.: A conic integer programming approach to stochastic joint location-inventory problems. Oper. Res. **60**(2), 366–381 (2012)

4. Balas, E.: Disjunctive programming. In: Hammer, P.L., Johnson, E.L., Korte, B.H. (eds.) Annals of Discrete Mathematics 5: Discrete Optimization, pp. 3–51. North Holland, Amsterdam, The Netherlands (1979)
5. Balas, E.: Disjunctive programming: properties of the convex hull of feasible points. Discret. Appl. Math. **89**(1–3), 3–44 (1998)
6. Balas, E., Ceria, S., Cornuéjols, G.: A lift-and-project cutting plane algorithm for mixed 0–1 programs. Math. Program. **58**, 295–324 (1993)
7. Barvinok, A.: A Course in Convexity. American Mathematical Society, Providence, RI (2002)
8. Belotti, P., Góez, J., Pólik, I., Ralphs, T., Terlaky, T.: On families of quadratic surfaces having fixed intersections with two hyperplanes. Discret. Appl. Math. **161**(16–17), 2778–2793 (2013)
9. Ben-Tal, A., Nemirovski, A.: On polyhedral approximations of the second-order cone. Math. Oper. Res. **26**(2), 193–205 (2001)
10. Bertsimas, D., Shioda, R.: Algorithm for cardinality-constrained quadratic optimization. Comput. Optim. Appl. **43**(1), 1–22 (2009)
11. Çezik, M., Iyengar, G.: Cuts for mixed 0–1 conic programming. Math. Program. **104**(1), 179–202 (2005)
12. Cornuéjols, G.: Valid inequalities for mixed integer linear programs. Math. Program. **112**(1), 3–44 (2008)
13. Cornuéjols, G., Lemaréchal, C.: A convex-analysis perspective on disjunctive cuts. Math. Program. **106**(2), 567–586 (2006)
14. Dadush, D., Dey, S., Vielma, J.: The split closure of a strictly convex body. Oper. Res. Lett. **39**(2), 121–126 (2011)
15. Drewes, S.: Mixed integer second order cone programming. Ph.D. thesis, Technische Universität Darmstadt, Germany (2009)
16. Fampa, M., Maculan, N.: Using a conic formulation for finding Steiner minimal trees. Numer. Algorithms **35**(2), 315–330 (2004)
17. Grossmann, I.E.: Review of nonlinear mixed-integer and disjunctive programming techniques. Optim. Eng. **3**, 227–252 (2002)
18. Júdice, J.J., Sherali, H.D., Ribeiro, I.M., Faustino, A.M.: A complementarity-based partitioning and disjunctive cut algorithm for mathematical programming problems with equilibrium constraints. J. Glob. Optim. **36**, 89–114 (2006)
19. Krokhmal, P.A., Soberanis, P.: Risk optimization with p-order conic constraints: a linear programming approach. Eur. J. Oper. Res. **201**(3), 653–671 (2010)
20. Kumar, M., Torr, P., Zisserman, A.: Solving Markov random fields using second order cone programming relaxations. In: 2006 IEEE Computer Society Conference on Computer Vision and Pattern Recognition, vol. 1, pp. 1045–1052 (2006)
21. Masihabadi, S., Sanjeevi, S., Kianfar, K.: n-Step conic mixed integer rounding inequalities. Optimization Online (2011). http://www.optimization-online.org/DB_HTML/2011/11/3251.html
22. Nemhauser, G.L., Wolsey, L.A.: Integer and Combinatorial Optimization. Wiley-Interscience, New York (1999)
23. Rockafellar, R.T.: Convex Analysis. Princeton University Press, Princeton (1970)
24. Stubbs, R.A., Mehrotra, S.: A branch-and-cut method for 0–1 mixed convex programming. Math. Program. **86**(3), 515–532 (1999)
25. Vielma, J., Ahmed, S., Nemhauser, G.: A lifted linear programming branch-and-bound algorithm for mixed-integer conic quadratic programs. INFORMS J. Comput. **20**(3), 438–450 (2008)

Runge–Kutta Methods for Ordinary Differential Equations

J.C. Butcher

Abstract Since their first discovery by Runge (Math Ann 46:167–178, 1895), Heun (Z Math Phys 45:23–38, 1900) and Kutta (Z Math Phys 46:435–453, 1901), Runge–Kutta methods have been one of the most important procedures for the numerical solution of ordinary differential equation systems. This survey paper ranges over many aspects of Runge–Kutta methods, including order conditions, order barriers, the efficient implementation of implicit methods, effective order methods and strong stability-preserving methods. Finally, applications to the analysis and implementation of G-symplectic methods will be discussed.

Keywords Runge–Kutta methods • Order conditions • Taylor series • Rooted trees • Elementary differentials • Low order methods • Order barriers • B-series • Effective order • Implicit methods • Singly-implicit methods • Efficient implementation • Strong stability preserving methods • G-symplectic methods

1 Introduction to Runge–Kutta Methods

Differential equations, especially initial value problems, are a vital component in mathematical modelling. However, since the majority of the differential equations arising in physics, engineering and other areas of application do not have analytical solutions, numerical methods become necessary.

One of the most important classes of methods for obtaining numerical approximations is the class of Runge–Kutta methods, which dates from the work of Runge in 1895 [18]. The idea is to study the Taylor series for the solution to a generic problem and to compare this series with the series produced by a particular numerical scheme which contains unspecified parameters. The parameters are then determined to force the two series to agree for as many terms in the expansions as possible.

J.C. Butcher (✉)
The University of Auckland, Auckland, New Zealand
e-mail: butcher@math.auckland.ac.nz

M. Al-Baali et al. (eds.), *Numerical Analysis and Optimization*, Springer Proceedings in Mathematics & Statistics 134, DOI 10.1007/978-3-319-17689-5_2

The generic initial value problem generally takes one of three forms. These are

$$y'(x) = f(x, y(x)), \quad y(x_0) = y_0, \quad f : \mathbb{R} \times \mathbb{R} \to \mathbb{R} \tag{1}$$

$$y'(x) = f(x, y(x)), \quad y(x_0) = y_0, \quad f : \mathbb{R} \times \mathbb{R}^N \to \mathbb{R}^N \tag{2}$$

$$y'(x) = f(y(x)), \quad y(x_0) = y_0, \quad f : \mathbb{R}^N \to \mathbb{R}^N \tag{3}$$

The standard problem (1) was used in the derivation of methods in [18] and other pioneering papers. But, derivations using this Ansatz cannot be trusted above order 4 because some of the order conditions found from an analysis based on (3) are absent in a scalar analysis. Hence, we must revert to an N dimensional problem, such as (2) or (3). The first correctly derived fifth order method was due to Nyström [17]. The order conditions based on these model problems are identical, but it will be simpler to use the autonomous formulation (3), even though it is convenient in practical applications to use (2).

The classical Euler method can be written as a formula for progressing from an approximation $y_{n-1} \approx y(x_{n-1}) = y(x_0 + (n-1)h)$ to the next step value y_n:

$$y_n = y_{n-1} + hf(y_{n-1}), \quad h = x_n - x_{n-1}. \tag{4}$$

The method given by (4) can be made more accurate by using either the mid-point or the trapezoidal rule quadrature formula:

$$y_n = y_{n-1} + hf\big(y_{n-1} + \tfrac{1}{2}hf(y_{n-1})\big), \tag{5}$$

$$y_n = y_{n-1} + \tfrac{1}{2}hf(y_{n-1}) + \tfrac{1}{2}hf\big(y_{n-1} + hf(y_{n-1})\big). \tag{6}$$

These methods, from Runge's 1895 paper [18], are "second order" because the error in a single step behaves like $O(h^3)$. At a specific output point the accumulated error is $O(h^2)$. In 1900, Heun [14] took these ideas further and gave a full explanation of order 3 methods and in 1900, Kutta [15] gave a detailed analysis of order 4 methods.

In the early days of Runge–Kutta methods the aim was to find explicit methods of higher and higher order. Later the aim shifted to finding methods that seemed to be optimal in terms of local truncation error and to finding built-in error estimators.

With the emergence of stiff problems as an important application area, attention moved to implicit methods, such as those based on Gaussian quadrature, which have superior stability properties compared with explicit methods. In contrast to the use of computationally demanding implicit methods, strong stability-preserving (SSP) *explicit* methods may sometimes be more efficient.

The structure of this paper is as follows. Section 2 describes the formulation of Runge–Kutta methods and their representation using tableaux. This is followed in Section 3 by a review of Runge–Kutta order theory and in Section 4 by a description of how methods up to order 4 are constructed. In Section 5 the order barrier result on the non-existence of methods with $s = p > 4$ is established; in Section 6, the theory is given an algebraic, or B-series, emphasis and, in Section 7, implicit methods and their efficient implementation are introduced. This is followed in Section 8 by an introduction to SSP methods. Finally we will consider the role of Runge–Kutta methods and their associated algebraic structures, in the analysis of G-symplectic general linear methods. This will include, in Section 9, a B-series analysis of the order conditions for a sample method and, in Section 10, Runge–Kutta methods will be derived for the practical implementation of starting methods.

2 Formulation of Methods

In carrying out a single step, we evaluate s stage values $Y_1, Y_2, \ldots, Y_s$, together with s stage derivatives $F_1, F_2, \ldots, F_s$, using the formula $F_i = f(Y_i)$.

Each Y_i is found as y_0 plus a linear combination of the F_j:

$$Y_i = y_0 + h\sum_{j=1}^{s} a_{ij}F_j \approx y(x_0 + c_i h)$$

and the approximation at $x_1 = x_0 + h$ is found from

$$y_1 = y_0 + h\sum_{i=1}^{s} b_i F_i \approx y(x_0 + h).$$

This procedure for approximating y_1 is repeated to obtain $y_2, y_3, \ldots, y_n$.

We represent the method by a tableau:

$$\begin{array}{c|cccc} c_1 & a_{11} & a_{12} & \cdots & a_{1s} \\ c_2 & a_{21} & a_{22} & \cdots & a_{2s} \\ \vdots & \vdots & \vdots & & \vdots \\ c_s & a_{s1} & a_{s2} & \cdots & a_{ss} \\ \hline & b_1 & b_2 & \cdots & b_s \end{array},$$

or, if the method is explicit, by the simplified tableau

$$\begin{array}{c|ccccc} 0 & & & & & \\ c_2 & a_{21} & & & & \\ \vdots & \vdots & \vdots & \ddots & & \\ c_s & a_{s1} & a_{s2} & \cdots & a_{s,s-1} & \\ \hline & b_1 & b_2 & \cdots & b_{s-1} & b_s \end{array}.$$

As examples, the Runge methods (5) and (6) have the tableaux

$$\begin{array}{c|cc} 0 & & \\ \frac{1}{2} & \frac{1}{2} & \\ \hline & 0 & 1 \end{array}\;, \qquad \begin{array}{c|cc} 0 & & \\ 1 & 1 & \\ \hline & \frac{1}{2} & \frac{1}{2} \end{array}\;.$$

3 The Order Conditions

3.1 Taylor Expansion of the Exact Solution

The analysis presented here is from [2], following the work of Gill [11] and Merson [16]. To obtain a series for the exact solution, we need formulae for the second, third, …, derivatives. These are easily found using the chain rule:

$$\begin{aligned} y'(x) &= f(y(x)), \\ y''(x) &= f'(y(x))y'(x) \\ &= f'(y(x))f(y(x))), \\ y'''(x) &= f''(y(x))(f(y(x)),y'(x)) + f'(y(x))f'(y(x))y'(x) \\ &= f''(y(x))(f(y(x)),f(y(x))) + f'(y(x))f'(y(x))f(y(x)). \end{aligned}$$

This process will become increasingly complicated as we evaluate higher derivatives and we look for a systematic pattern.

Write $\mathbf{f} = f(y(x))$, $\mathbf{f}' = f'(y(x))$, $\mathbf{f}'' = f''(y(x))$, …. We can now write the terms in $y'(x), \ldots, y'''(x)$ in a compact style. At the end of each line is a diagram showing the tree-like structure of each term.

$$y'(x) = \mathbf{f},$$

$$y''(x) = \mathbf{f}'\mathbf{f},$$

$$y'''(x) = \mathbf{f}''(\mathbf{f},\mathbf{f}) \quad + \quad \mathbf{f}'\mathbf{f}'\mathbf{f}.$$

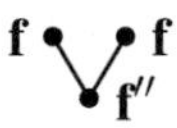

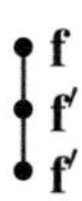

The various terms have a structure related to rooted trees. Because of this connection, we will introduce the set of all rooted trees and some functions on this set.

Let T denote the set of rooted trees:

$T = \Big\{$ 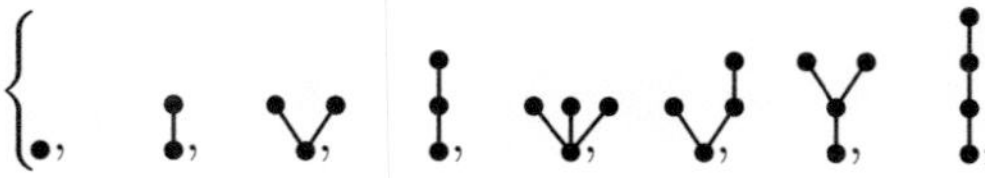 $\dots \Big\}$

We identify the following functions on T. Write t as a typical tree

$\|t\|$	order of t = number of vertices
$\sigma(t)$	symmetry of t = order of automorphism group
$t!$	density or factorial of t; sometimes written $\gamma(t)$
$\alpha(t)$	number of ways of labelling with an ordered set
$\beta(t)$	number of ways of labelling with an unordered set
$F(t)(y_0)$	elementary differential

We will give examples of these functions based on a specific tree

$t =$ 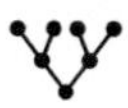

$|t| = 7$

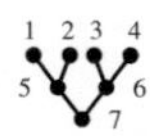

$\sigma(t) = 8$

$t! = 63$

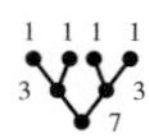

$$\alpha(t) = \frac{|t|!}{\sigma(t)t!} = 10$$

$$\beta(t) = \frac{|t|!}{\sigma(t)} = 630$$

$$F(t) = \mathbf{f}''\big(\mathbf{f}''(\mathbf{f},\mathbf{f}),\mathbf{f}''(\mathbf{f},\mathbf{f})\big)$$

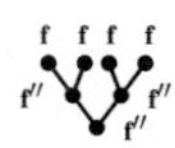

In Table 1, these functions are given for trees up to order 4. Note that the function $\Phi(t)$, with values given in the final row of Table 1, will be explained in Section 3.2.

Table 1 Functions on trees to order 4

t								
$\|t\|$	1	2	3	3	4	4	4	4
$\sigma(t)$	1	1	2	1	6	1	2	1
$t!$	1	2	3	6	4	8	12	24
$\alpha(t)$	1	1	1	1	1	3	1	1
$\beta(t)$	1	2	3	6	4	24	12	24
$F(t)$	$\mathbf{f}$	$\mathbf{f}'\mathbf{f}$	$\mathbf{f}''(\mathbf{f},\mathbf{f})$	$\mathbf{f}'\mathbf{f}'\mathbf{f}$	$\mathbf{f}'''(\mathbf{f},\mathbf{f},\mathbf{f})$	$\mathbf{f}''(\mathbf{f},\mathbf{f}'\mathbf{f})$	$\mathbf{f}'\mathbf{f}''(\mathbf{f},\mathbf{f})$	$\mathbf{f}'\mathbf{f}'\mathbf{f}'\mathbf{f}$
$\Phi(t)$	$\sum b_i$	$\sum b_i c_i$	$\sum b_i c_i^2$	$\sum b_i a_{ij} c_j$	$\sum b_i c_i^3$	$\sum b_i c_i a_{ij} c_j$	$\sum b_i a_{ij} c_j^2$	$\sum b_i a_{ij} a_{jk} c_k$

The formal Taylor expansion of the solution at x_0+h is

$$y(x_0+h) = y_0 + \sum_{t\in T} \frac{\alpha(t)h^{|t|}}{|t|!} F(t)(y_0)$$

and, using the known formula for $\alpha(t)$, we can write this as

$$y(x_0+h) = y_0 + \sum_{t\in T} \frac{h^{|t|}}{\sigma(t)t!} F(t)(y_0). \tag{7}$$

Our aim will now be to find a corresponding formula for the result computed by one step of a Runge–Kutta method. By comparing these formulae term by term, we will obtain conditions for a specific order of accuracy.

3.2 *Taylor Expansion for Numerical Approximation*

We need to evaluate various expressions which depend on the tableau for a particular method. These are known as "elementary weights".

We use the example tree we have already considered, to illustrate the construction of the elementary weight $\Phi(t)$. First label the vertices $i, j, k, \ldots, o$ as shown:

$t =$ 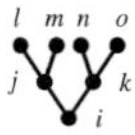

Now form the sum

$$\Phi(t) = \sum_{i,j,k,l,m,n,o=1}^{s} b_i a_{ij} a_{jl} a_{jm} a_{ik} a_{kn} a_{ko},$$

where the factor b_i indicates that i is the label attached to the root. Furthermore, factors such as a_{ij} indicate that the rooted tree contains an edge from i to j.

Simplify by summing over l,m,n,o, noting, for example, that $\sum_{l=1}^{s} a_{jl} = c_j$. This gives

$$\Phi(t) = \sum_{i,j,k=1}^{s} b_i a_{ij} c_j^2 a_{ik} c_k^2$$

Expressions for $\Phi(t)$ up to order 4 trees are given in Table 1. The formal Taylor expansion of the *computed* solution at x_0+h is

$$y_1 = y_0 + \sum_{t\in T} \frac{\beta(t)h^{|t|}}{|t|!} \Phi(t)F(t)(y_0)$$

and, using the known formula for $\beta(t)$, we can write this as

$$y_1 = y_0 + \sum_{t\in T} \frac{h^{|t|}}{\sigma(t)} \Phi(t)F(t)(y_0). \tag{8}$$

3.3 *Order Conditions*

To match the Taylor series (7) and (8) up to h^p terms we need to ensure that

$$\Phi(t) = \frac{1}{t!},$$

for all trees such that

$$|t| \le p.$$

These are the "order conditions".

4 Construction of Low Order Explicit Methods

We will attempt to construct methods of order $p = s$ with s stages for $s = 1,2,\ldots$. We will find that this is possible up to order 4 but not for $p \ge 5$.

The usual approach will be to first choose $c_2, c_3, \ldots, c_s$ and then solve for $b_1, b_2, \ldots, b_s$. (Recall that $c_1 = 0$.) After this, solve for those of the a_{ij} which can be found as solutions to linear equations.

The order equations for specific orders can now be given, together with some sample solutions for $p = 2$ and $p = 3$. In the case of $p = 4$, further details are given.

Order 2

$$b_1 + b_2 = 1,$$
$$b_2 c_2 = \tfrac{1}{2}.$$

$$\begin{array}{c|cc} 0 & & \\ c_2 & c_2 & \\ \hline & 1-\frac{1}{2c_2} & \frac{1}{2c_2} \end{array} \qquad \begin{array}{c|cc} 0 & & \\ \frac{1}{2} & \frac{1}{2} & \\ \hline & 0 & 1 \end{array} \qquad \begin{array}{c|cc} 0 & & \\ 1 & 1 & \\ \hline & \frac{1}{2} & \frac{1}{2} \end{array}$$

Order 3

$$
\begin{aligned}
b_1 + b_2 + b_3 &= 1,\\
b_2 c_2 + b_3 c_3 &= \tfrac{1}{2},\\
b_2 c_2^2 + b_3 c_3^2 &= \tfrac{1}{3},\\
b_3 a_{32} c_2 &= \tfrac{1}{6}.
\end{aligned}
$$

$$
\begin{array}{c|ccc}
0 & & & \\
\frac{1}{2} & \frac{1}{2} & & \\
1 & -1 & 2 & \\
\hline
 & \frac{1}{6} & \frac{2}{3} & \frac{1}{6}
\end{array}
\qquad
\begin{array}{c|ccc}
0 & & & \\
\frac{2}{3} & \frac{2}{3} & & \\
\frac{2}{3} & 0 & \frac{2}{3} & \\
\hline
 & \frac{1}{4} & \frac{3}{8} & \frac{3}{8}
\end{array}
\qquad
\begin{array}{c|ccc}
0 & & & \\
\frac{2}{3} & \frac{2}{3} & & \\
0 & -1 & 1 & \\
\hline
 & 0 & \frac{3}{4} & \frac{1}{4}
\end{array}
$$

Order 4

$$b_1 + b_2 + b_3 + b_4 = 1, \tag{9}$$

$$b_2 c_2 + b_3 c_3 + b_4 c_4 = \tfrac{1}{2}, \tag{10}$$

$$b_2 c_2^2 + b_3 c_3^2 + b_4 c_4^2 = \tfrac{1}{3}, \tag{11}$$

$$b_3 a_{32} c_2 + b_4 a_{42} c_2 + b_4 a_{43} c_3 = \tfrac{1}{6}, \tag{12}$$

$$b_2 c_2^3 + b_3 c_3^3 + b_4 c_4^3 = \tfrac{1}{4}, \tag{13}$$

$$b_3 c_3 a_{32} c_2 + b_4 c_4 a_{42} c_2 + b_4 c_4 a_{43} c_3 = \tfrac{1}{8}, \tag{14}$$

$$b_3 a_{32} c_2^2 + b_4 a_{42} c_2^2 + b_4 a_{43} c_3^2 = \tfrac{1}{12}, \tag{15}$$

$$b_4 a_{43} a_{32} c_2 = \tfrac{1}{24}. \tag{16}$$

To solve these equations, treat c_2, c_3, c_4 as parameters, and solve for b_1, b_2, b_3, b_4 from (9), (10), (11), (13). Now solve for a_{32}, a_{42}, a_{43} from (12), (14), (15). Finally use (16) to obtain a consistency condition on c_2, c_3, c_4. The outcome of this analysis is that $c_4 = 1$.

We will prove a stronger result in another way.

Lemma 1. *Let U and V be 3×3 matrices such that*

$$UV = \begin{bmatrix} w_{11} & w_{12} & 0 \\ w_{21} & w_{22} & 0 \\ 0 & 0 & 0 \end{bmatrix}, \quad \textit{where} \quad \begin{bmatrix} w_{11} & w_{12} \\ w_{21} & w_{22} \end{bmatrix} \textit{ is non-singular,}$$

then either the last row of U is zero or the last column of V is zero.

Proof. Let $W = UV$. Either U or V is singular. If U is singular, let x be a non-zero vector such that $x^T U = 0$. Therefore $x^T W = 0$ and it follows that the first two components of x are zero. Hence, the last row of U is zero. The second case follows similarly if V is singular.

We will apply this result with specific U and V. Let

$$U = \begin{bmatrix} b_2 & b_3 & b_4 \\ b_2 c_2 & b_3 c_3 & b_4 c_4 \\ \sum_i b_i a_{i2} - b_2(1-c_2) & \sum_i b_i a_{i3} - b_3(1-c_3) & \sum_i b_i a_{i4} - b_4(1-c_4) \end{bmatrix},$$

$$V = \begin{bmatrix} c_2 & c_2^2 & \sum_j a_{2j}c_j - \frac{1}{2}c_2^2 \\ c_3 & c_3^2 & \sum_j a_{3j}c_j - \frac{1}{2}c_3^2 \\ c_4 & c_4^2 & \sum_j a_{4j}c_j - \frac{1}{2}c_4^2 \end{bmatrix}.$$

It is found that

$$UV = \begin{bmatrix} \frac{1}{2} & \frac{1}{3} & 0 \\ \frac{1}{3} & \frac{1}{4} & 0 \\ 0 & 0 & 0 \end{bmatrix}.$$

Using Lemma 1, we conclude that $b_4 = 0$, $c_2 = 0$ or $c_4 = 1$. However, the first two options are impossible because they contradict (16) and we conclude that $c_4 = 1$ and the last row of U is zero. The construction of fourth order Runge–Kutta methods now becomes straightforward.

In his famous 1901 paper, Kutta classified all solutions to the fourth order conditions, for $s = 4$. In particular we have his famous method:

$$\begin{array}{c|cccc} 0 & & & & \\ \frac{1}{2} & \frac{1}{2} & & & \\ \frac{1}{2} & 0 & \frac{1}{2} & & \\ 1 & 0 & 0 & 1 & \\ \hline & \frac{1}{6} & \frac{1}{3} & \frac{1}{3} & \frac{1}{6} \end{array}$$

5 Order Barriers

We will review what is achievable up to order 8. In Table 2, N_p is the number of order conditions to achieve this order. Furthermore, $M_s = s(s+1)/2$ is the number of free parameters to satisfy the order conditions for the required s stages.

Table 2 Minimum number of stages for a given order

p	N_p	s	M_s
1	1	1	1
2	2	2	3
3	4	3	6
4	8	4	10
5	17	6	21
6	37	7	28
7	115	9	45
8	200	11	66

We will now present the first order barrier result, [3].

Theorem 1. *There does not exist an explicit Runge–Kutta method with s stages and order p, where $s = p = 5$.*

Proof. Let $\hat{b}_j = \sum_{i=1}^{5} b_i a_{ij}$, $j = 1,2,3,4$ and let

$$U = \begin{bmatrix} \hat{b}_2 & \hat{b}_3 & \hat{b}_4 \\ \hat{b}_2 c_2 & \hat{b}_3 c_3 & \hat{b}_4 c_4 \\ \sum_i \hat{b}_i a_{i2} - \frac{1}{2}\hat{b}_2(1-c_2) & \sum_i \hat{b}_i a_{i3} - \frac{1}{2}\hat{b}_3(1-c_3) & \sum_i \hat{b}_i a_{i4} - \frac{1}{2}\hat{b}_4(1-c_4) \end{bmatrix},$$

$$V = \begin{bmatrix} c_2 & c_2^2 & \sum_j a_{2j}c_j - \frac{1}{2}c_2^2 \\ c_3 & c_3^2 & \sum_j a_{3j}c_j - \frac{1}{2}c_3^2 \\ c_4 & c_4^2 & \sum_j a_{4j}c_j - \frac{1}{2}c_4^2 \end{bmatrix},$$

then

$$UV = \begin{bmatrix} \frac{1}{6} & \frac{1}{12} & 0 \\ \frac{1}{12} & \frac{1}{20} & 0 \\ 0 & 0 & 0 \end{bmatrix}.$$

Using Lemma 1, we deduce that $c_4 = 1$. Now use the lemma again with

$$U = \begin{bmatrix} b_2(1-c_2) & b_3(1-c_3) & b_5(1-c_5) \\ b_2c_2(1-c_2) & b_3c_3(1-c_3) & b_5c_5(1-c_5) \\ \begin{array}{c}\sum_i b_i a_{i2}(1-c_2) \\ -b_2(1-c_2)^2\end{array} & \begin{array}{c}\sum_i b_i a_{i3}(1-c_3) \\ -b_3(1-c_3)^2\end{array} & \begin{array}{c}\sum_i b_i a_{i5}(1-c_5) \\ -b_5(1-c_5)^2\end{array} \end{bmatrix},$$

$$V = \begin{bmatrix} c_2 & c_2^2 & \sum_j a_{2j}c_j - \frac{1}{2}c_2^2 \\ c_3 & c_3^2 & \sum_j a_{3j}c_j - \frac{1}{2}c_3^2 \\ c_5 & c_5^2 & \sum_j a_{5j}c_j - \frac{1}{2}c_5^2 \end{bmatrix},$$

then

$$UV = \begin{bmatrix} \frac{1}{6} & \frac{1}{12} & 0 \\ \frac{1}{12} & \frac{1}{20} & 0 \\ 0 & 0 & 0 \end{bmatrix}.$$

It follows that $c_5 = 1$. Since we already know that $c_4 = 1$, we obtain a contradiction from

$$0 = \sum b_i(1-c_i)a_{ij}a_{jk}c_k = \tfrac{1}{120}.$$

By modifying the details slightly, we can prove that $s = p \geq 5$ is never possible. The proof that $s = p+1$ is impossible when $p \geq 7$ is more complicated. The proof that $s = p+2$ is impossible when $p \geq 8$ is much more complicated.

6 Algebraic Interpretation

We will introduce an algebraic system [4] which represents individual Runge–Kutta methods and also compositions of methods. This centres on the meaning of order for Runge–Kutta methods and leads to a possible generalisation to "effective order".

Denote by G the group consisting of mappings of (rooted) trees to real numbers where the group operation is defined according to the algebraic theory of Runge–Kutta methods or to the (equivalent) theory of B-series.

We will illustrate this operation in Table 3, where we also introduce the special member $E \in G$.

G_p will denote the normal subgroup defined by $t \mapsto 0$ for $|t| \leq p$. If $\alpha \in G$, then this maps canonically to $\alpha G_p \in G/G_p$.

Table 3 The group operation $(\alpha\beta)(t)$ and the values of $E(t)$

$r(t_i)$	i	t_i	$\alpha(t_i)$	$\beta(t_i)$	$(\alpha\beta)(t_i)$	$E(t_i)$
1	1		α_1	β_1	$\alpha_1+\beta_1$	1
2	2		α_2	β_2	$\alpha_2+\alpha_1\beta_1+\beta_2$	$\frac{1}{2}$
3	3		α_3	β_3	$\alpha_3+\alpha_1^2\beta_1+2\alpha_1\beta_2+\beta_3$	$\frac{1}{3}$
3	4		α_4	β_4	$\alpha_4+\alpha_2\beta_1+\alpha_1\beta_2+\beta_4$	$\frac{1}{6}$
4	5	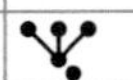	α_5	β_5	$\alpha_5+\alpha_1^3\beta_1+3\alpha_1^2\beta_2+3\alpha_1\beta_3+\beta_5$	$\frac{1}{4}$
4	6	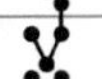	α_6	β_6	$\alpha_6+\alpha_1\alpha_2\beta_1+(\alpha_1^2+\alpha_2)\beta_2+\alpha_1(\beta_3+\beta_4)+\beta_6$	$\frac{1}{8}$
4	7		α_7	β_7	$\alpha_7+\alpha_3\beta_1+\alpha_1^2\beta_2+2\alpha_1\beta_4+\beta_7$	$\frac{1}{12}$
4	8		α_8	β_8	$\alpha_8+\alpha_4\beta_1+\alpha_2\beta_2+\alpha_1\beta_4+\beta_8$	$\frac{1}{24}$

If α is defined from the elementary weights for a Runge–Kutta method, then order p can be written as

$$\alpha G_p = E G_p.$$

Effective order p is defined by the existence of β such that

$$\beta\alpha G_p = E\beta G_p.$$

The computational interpretation of effective order is that the sequence of steps, corresponding to α, is preceded by a starting step corresponding to β, with a finishing step corresponding to β^{-1} inserted at the completion of the calculation. This is equivalent to many steps all corresponding to $\beta\alpha\beta^{-1}$. Thus, the benefits of high order can be enjoyed by high effective order.

To analyse the conditions for effective order 4 we can, without loss of generality, assume that $\beta(t_1)=0$. The details are

i	$(\beta\alpha)(t_i)$	$(E\beta)(t_i)$
1	α_1	1
2	$\beta_2+\alpha_2$	$\frac{1}{2}+\beta_2$
3	$\beta_3+\alpha_3$	$\frac{1}{3}+2\beta_2+\beta_3$
4	$\beta_4+\beta_2\alpha_1+\alpha_4$	$\frac{1}{6}+\beta_2+\beta_4$
5	$\beta_5+\alpha_5$	$\frac{1}{4}+3\beta_2+3\beta_3+\beta_5$
6	$\beta_6+\beta_2\alpha_2+\alpha_6$	$\frac{1}{8}+\frac{3}{2}\beta_2+\beta_3+\beta_4+\beta_6$
7	$\beta_7+\beta_3\alpha_1+\alpha_7$	$\frac{1}{12}+\beta_2+2\beta_4+\beta_7$
8	$\beta_8+\beta_4\alpha_1+\beta_2\alpha_2+\alpha_8$	$\frac{1}{24}+\frac{1}{2}\beta_2+\beta_4+\beta_8$

Of these eight conditions, only five are conditions on α. Once α is known, there remain three conditions on β.

The five order conditions, written in terms of the Runge–Kutta tableau, are

$$\begin{aligned}
\sum b_i &= 1,\\
\sum b_i c_i &= \tfrac{1}{2},\\
\sum b_i a_{ij} c_j &= \tfrac{1}{6},\\
\sum b_i a_{ij} a_{jk} c_k &= \tfrac{1}{24},\\
\sum b_i c_i^2(1-c_i) + \sum b_i a_{ij} c_j (2c_i - c_j) &= \tfrac{1}{4}.
\end{aligned}$$

7 Implicit Runge–Kutta Methods

Given the existence of order barriers, it is natural to ask whether these barriers also apply to implicit methods. Even though explicit methods, and the solution of the order conditions, become increasingly complicated as the order increases, everything becomes simpler for implicit methods.

For example, the following method has order 5:

$$\begin{array}{c|cccc}
0 & & & & \\
\frac{1}{4} & \frac{1}{8} & \frac{1}{8} & & \\
\frac{7}{10} & -\frac{1}{100} & \frac{14}{25} & \frac{3}{20} & \\
1 & \frac{2}{7} & 0 & \frac{5}{7} & \\
\hline
 & \frac{1}{14} & \frac{32}{81} & \frac{250}{567} & \frac{5}{54}
\end{array}$$

This method has limited applications and we will consider instead methods where A is a full lower triangular matrix.

If all the diagonal elements are equal, we get the DIRK methods of R. Alexander [1] and others. The following third order L-stable method illustrates what is possible for DIRK methods

$$\begin{array}{c|ccc}
\lambda & \lambda & & \\
\frac{1}{2}(1+\lambda) & \frac{1}{2}(1-\lambda) & \lambda & \\
1 & \frac{1}{4}(-6\lambda^2+16\lambda-1) & \frac{1}{4}(6\lambda^2-20\lambda+5) & \lambda \\
\hline
 & \frac{1}{4}(-6\lambda^2+16\lambda-1) & \frac{1}{4}(6\lambda^2-20\lambda+5) & \lambda
\end{array},$$

where $\lambda \approx 0.4358665215$ satisfies $\frac{1}{6}-\frac{3}{2}\lambda+3\lambda^2-\lambda^3=0$. Methods of this type have a limited value in practical computation and instead we will consider a more general family of methods.

7.1 Singly Implicit Runge–Kutta Methods

The main advantage of the DIRK methods is that the stages can be computed independently and sequentially from equations of the form

$$Y_i - h\lambda f(Y_i) = \text{a known quantity.}$$

Each stage requires the same factorised matrix $I - h\lambda \mathscr{J}$, where $\mathscr{J} \approx \partial f/\partial y$, to permit solution by a modified Newton iteration process.

A SIRK method is characterised by the equation $\sigma(A) = \{\lambda\}$. That is, A has a one-point spectrum. We will consider the possible implementation of SIRK methods in an efficient manner. The secret lies in the inclusion of a transformation to Jordan canonical form in the computation.

Suppose the matrix T transforms A to canonical form as follows:

$$T^{-1}AT = \overline{A},$$

where

$$\overline{A} = \lambda(I - J) = \begin{bmatrix} \lambda & 0 & 0 & \cdots & 0 & 0 \\ -\lambda & \lambda & 0 & \cdots & 0 & 0 \\ 0 & -\lambda & \lambda & \cdots & 0 & 0 \\ \vdots & \vdots & \vdots & & \vdots & \vdots \\ 0 & 0 & 0 & \cdots & \lambda & 0 \\ 0 & 0 & 0 & \cdots & -\lambda & \lambda \end{bmatrix}.$$

Consider a single modified Newton iteration, in which the same approximate Jacobian $\mathscr{J}$ is used for each stage. Assume the incoming approximation is y_0 and that we are attempting to evaluate

$$y_1 = y_0 + h(b^T \otimes I)F,$$

where F is made up from the s subvectors $F_i = f(Y_i)$, $i = 1, 2, \ldots, s$.

The implicit equations to be solved are

$$Y = 1 \otimes y_0 + h(A \otimes I)F,$$

where 1 is the vector in $\mathbb{R}^n$ with every component equal to 1 and Y has subvectors Y_i, $i = 1, 2, \ldots, s$.

The modified Newton process consists, in each iteration, of solving the linear system

$$(I_s \otimes I - hA \otimes \mathscr{J})D = Y - e \otimes y_0 - h(A \otimes I)F,$$

Table 4 Components of implementation costs

	Without transformations	With transformations
LU factorisation	s^3N^3	N^3
Transformation		s^2N
Backsolves	s^2N^2	sN^2
Transformation		s^2N

and then updating $Y \to Y - D$. To benefit from the SI property, write

$$\overline{Y} = (T^{-1} \otimes I)Y, \quad \overline{F} = (T^{-1} \otimes I)F, \quad \overline{D} = (T^{-1} \otimes I)D,$$

so that

$$(I_s \otimes I - h\overline{A} \otimes \mathcal{J})\overline{D} = \overline{Y} - \overline{e} \otimes y_0 - h(\overline{A} \otimes I)\overline{F}.$$

Table 4 summarises the costs of this process. We see that the use of transformations enables the very high LU factorisation cost to be reduced to a level comparable to BDF methods. Also the back substitution costs are also reduced to the same work per stage as for DIRK or BDF methods. By comparison, the additional transformation costs are insignificant for large problems.

To gain the full benefit of SI methods, we will show how stage order s can be achieved. This would mean that

$$\sum_{j=1}^{s} a_{ij}\phi(c_i) = \int_0^{c_i} \phi(t)dt,$$

for ϕ any polynomial of degree $s-1$. This implies in turn that

$$Ac^{k-1} = \frac{1}{k}c^k, \qquad k = 1, 2, \ldots, s, \tag{17}$$

where the vector powers are interpreted component by component.

It can be verified that (17) is equivalent to

$$A^k c^0 = \frac{1}{k!}c^k, \qquad k = 1, 2, \ldots, s. \tag{18}$$

From the Cayley–Hamilton theorem

$$(A - \lambda I)^s c^0 = 0,$$

and hence

$$\sum_{i=0}^{s} \binom{s}{i} (-\lambda)^{s-i} A^i c^0 = 0.$$

Substitute from (18) and it is found that

$$\sum_{i=0}^{s} \frac{1}{i!} \binom{s}{i} (-\lambda)^{s-i} c^i = 0.$$

Hence each component of c satisfies

$$\sum_{i=0}^{s} \frac{1}{i!} \binom{s}{i} \left(-\frac{x}{\lambda}\right)^i = 0,$$

which can be written

$$L_s\left(\frac{x}{\lambda}\right) = 0,$$

where L_S denotes the Laguerre polynomial of degree s.

Let $\xi_1, \xi_2, \ldots, \xi_s$ denote the zeros of L_s so that

$$c_i = \lambda \xi_i, \qquad i = 1, 2, \ldots, s.$$

For methods defined in this way, a suitable choice of the transformation T is known. This is

$$T = \begin{bmatrix} L_0(\xi_1) & L_1(\xi_1) & L_2(\xi_1) & \cdots & L_{s-1}(\xi_1) \\ L_0(\xi_2) & L_1(\xi_2) & L_2(\xi_2) & \cdots & L_{s-1}(\xi_2) \\ L_0(\xi_3) & L_1(\xi_3) & L_2(\xi_3) & \cdots & L_{s-1}(\xi_3) \\ \vdots & \vdots & \vdots & & \vdots \\ L_0(\xi_s) & L_1(\xi_s) & L_2(\xi_s) & \cdots & L_{s-1}(\xi_s) \end{bmatrix}.$$

The question now is, how should λ be chosen?

Unfortunately, to obtain A-stability, at least for orders $p > 2$, λ has to be assigned a value requiring that some of the c_i lie outside the interval $[0,1]$. This effect becomes more severe for increasingly high orders and can be seen as a major disadvantage of these methods. However, there are two ways in which SIRK methods can be generalised to overcome this disadvantage.

The first generalisation is to add additional diagonally implicit stages [8], so that the coefficient matrix becomes

$$\begin{bmatrix} \hat{A} & 0 \\ W & \lambda I \end{bmatrix},$$

where the spectrum of the $p \times p$ submatrix $\hat{A}$ is

$$\sigma(\hat{A}) = \{\lambda\}$$

For $s - p = 1, 2, 3, \ldots$, we get improvements to the behaviour of these methods.

A second generalisation is to replace "order" by "effective order", [9]. This allows us to locate the abscissae where we wish.

8 Strong Stability-Preserving Methods

Ordinary differential equation systems, formed by semi-discretisation applied to time-dependent partial differential equations, can impose heavy computational demands. Because these systems are at least mildly stiff, the use of implicit methods is often appropriate. However, it might be more economical to use explicit methods, but with a severe restriction on the stepsize.

Assume that for sufficiently small h

$$\|Y + hf(Y)\| \leq \|Y\|. \tag{19}$$

If (19) holds for $h \leq H$, then for the Euler method,

$$\|y_n\| \leq \|y_{n-1}\|, \qquad h \leq CH, \tag{20}$$

where $C = 1$.

A Runge–Kutta is said to be "strong stability-preserving" (SSP; also known as "total variation diminishing" or TVD) [20] if (19) implies that (20) holds for some $C > 0$.

For the second order method, (6), the stages and the output are given by

$$\begin{aligned}
Y_1 &= y_{n-1}, \\
Y_2 &= y_{n-1} + hf(Y_1) &&= Y_1 + hf(Y_1), &\quad (21)\\
y_n &= y_{n-1} + \tfrac{1}{2}hf(Y_1) + \tfrac{1}{2}hf(Y_2) &&= \tfrac{1}{2}y_{n-1} + \tfrac{1}{2}(Y_2 + hf(Y_2)). &\quad (22)
\end{aligned}$$

Given that $h \leq H$, (21) implies that $\|Y_2\| \leq \|y_{n-1}\|$ and from (22) we have

$$\|y_n\| \leq \tfrac{1}{2}\|y_{n-1}\| + \tfrac{1}{2}\|Y_2 + hf(Y_2)\| \leq \tfrac{1}{2}\|y_{n-1}\| + \tfrac{1}{2}\|Y_2\| \leq \|y_{n-1}\|.$$

Note that if this analysis is attempted for the second order method (5) based on the midpoint rule, so that

$$\begin{aligned}
Y_1 &= y_{n-1},\\
Y_2 &= y_{n-1} + \tfrac{1}{2}hf(Y_1) = Y_1 + \tfrac{1}{2}hf(Y_1),\\
y_n &= y_{n-1} + hf(Y_2) \;\; = (1-\theta)(Y_1 - \tfrac{\theta}{2(1-\theta)}hf(Y_1)) + \theta(Y_2 + \theta^{-1}hf(Y_2)),
\end{aligned} \tag{23}$$

it is not possible to obtain a similar result. No choice of the parameter θ is possible in the interval $(0,1)$ (so that the addition of the two terms in (23) would be a convex combination) which also satisfies $-\frac{\theta}{2(1-\theta)} \geq 0$. Hence, this method is not SSP.

The Shu–Osher transformation used in the systematic analysis of the SSP property was introduced in [21]. See also [12] for recent developments of methods possessing SSP.

9 Order Analysis for G-Symplectic Methods

Symplectic Runge–Kutta methods have become important in recent years because of their ability to preserve symplectic behaviour of Hamiltonian systems and to conserve quadratic invariants; refer to [19]. If a method satisfies the condition

$$b_i a_{ij} + b_j a_{ji} = b_i b_j, \qquad i,j = 1,2,\ldots,s,$$

then for a problem (3) such that $\langle Y, Qf(Y)\rangle$, where Q is symmetric, the value of $\langle y_n, Qy_n\rangle$ is constant, just as, for the exact solution, $\langle y(x), Qy(x)\rangle$ is conserved.

Although multi-value methods are incapable of preserving quadratic invariants or symplectic behaviour, a more general conservation law is satisfied in G-symplectic methods; refer to [5, 6, 10, 13].

If instead of a Runge–Kutta method, we were to use a two value method, such as

$$\begin{aligned}
Y_1 &= y_{n-1} + z_{n-1}, & F_1 &= f(Y_1),\\
Y_2 &= \tfrac{2}{3}hF_1 + y_{n-1} - z_{n-1}, & F_2 &= f(Y_2),\\
Y_3 &= \tfrac{2}{5}hF_1 - \tfrac{3}{10}hF_2 + \tfrac{1}{2}hF_3 + y_{n-1} - \tfrac{1}{5}z_{n-1}, & F_3 &= f(Y_3),\\
y_n &= \tfrac{1}{3}hF_1 - \tfrac{3}{8}hF_2 + \tfrac{25}{24}hF_3 + y_{n-1},\\
z_n &= \tfrac{1}{3}hF_1 + \tfrac{3}{8}hF_2 - \tfrac{5}{24}hF_3 - z_{n-1},
\end{aligned} \tag{24}$$

it is possible to obtain many of the geometric benefits of symplectic Runge–Kutta methods but at a lower computational cost. For this particular method, the conserved quantity is $\langle y_n, Qy_n\rangle - \langle z_n, Qz_n\rangle$.

To investigate the order of (24) and show that it is equal to 4, it will be sufficient to analyse only the first step. We need to choose suitable Taylor series for y_0 and z_0 expanded about $y(x_0)$ in the form

$$y_0 = \varphi_h(y(x_0)),$$
$$z_0 = \psi_h(y(x_0)),$$

so that, to within $O(h^5)$,

$$y_1 = \varphi_h(y(x_1)),$$
$$z_1 = \psi_h(y(x_1)).$$

It will be verified that the choices of y_0 and z_0 given in Table 5 achieve this purpose. Shown in this table are the coefficients of y_0 and of $h^{|t_i|}F(t_i)(y_0)/\sigma(t_i)$, $i = 1,2,\ldots,8$. as well as the coefficients in the case of the other quantities used in the calculation. In each case the elementary differentials **f**, **f′f**, ... are evaluated at $y(x_0)$.

A calculation of $\varphi_h(y(x_1))$ and $\psi_h(y(x_1))$, respectively, in each case expanded in Taylor series about x_0, yields coefficients identical with the series for y_1 and z_1 respectively. This verifies that the order of the method (24) is 4.

10 Implementation of G-Symplectic Methods

In the implementation of (24) and similar methods it is necessary to construct a starting method with a single input and two outputs which match the B-series coefficients given for y_0 and z_0 in Table 5. It is convenient to do this in two steps. First a Runge–Kutta, with tableau given by (25), to produce the series for $y_0 = \varphi_h y(x_0)$ and secondly a mapping $\chi_h = \psi_h \circ \varphi_h^{-1}$ to be used to calculate $z_0 = \chi_h y_0$. Because the coefficient of $y(x_0)$ in the Taylor expansion of χ_h is zero and not one, as for a standard Runge–Kutta method, the tableau representing the second step in the process has to be interpreted appropriately. This modification is indicated by the additional 0 in the final row of the tableau for χ_h shown in (26).

$$\begin{array}{c|ccc}
0 & & & \\
-\frac{7}{135} & -\frac{7}{135} & & \\
\frac{7}{135} & -\frac{1}{2} & \frac{149}{270} & \\
\hline
 & \frac{135}{224} & 0 & -\frac{135}{224}
\end{array} \tag{25}$$

$$\begin{array}{c|cccc}
0 & & & & \\
-\frac{13}{30} & -\frac{13}{30} & & & \\
-\frac{1}{10} & -\frac{67}{1170} & -\frac{5}{117} & & \\
-\frac{51}{100} & -\frac{1330607}{2366000} & -\frac{626773}{3380000} & \frac{432837}{1820000} & \\
\hline
0 & -\frac{1863}{1768} & \frac{4245}{4784} & \frac{855}{656} & -\frac{56875}{64124}
\end{array} \tag{26}$$

Table 5 Taylor coefficients for a two value four stage method

	$y(x_0)$	$h\mathbf{f}$	$h^2\mathbf{f}'\mathbf{f}$	$\frac{1}{2}h^3\mathbf{f}''(\mathbf{f},\mathbf{f})$	$h^3\mathbf{f}'\mathbf{f}'\mathbf{f}$	$\frac{1}{6}h^4\mathbf{f}'''(\mathbf{f},\mathbf{f},\mathbf{f})$	$h^4\mathbf{f}''(\mathbf{f},\mathbf{f}'\mathbf{f})$	$\frac{1}{2}h^4\mathbf{f}'\mathbf{f}''(\mathbf{f},\mathbf{f})$	$h^4\mathbf{f}'\mathbf{f}'\mathbf{f}'\mathbf{f}$
y_0	1	0	$-\frac{1}{32}$	$-\frac{7}{4320}$	$\frac{149}{8640}$	$-\frac{49}{583200}$	$\frac{1043}{1166400}$	$-\frac{1043}{1166400}$	0
z_0	0	$\frac{1}{4}$	$-\frac{1}{16}$	$-\frac{49}{960}$	$-\frac{13}{384}$	$\frac{2543}{57600}$	$\frac{193}{7680}$	$\frac{619}{34560}$	$\frac{163}{69120}$
Y_1	1	$\frac{1}{4}$	$-\frac{3}{32}$	$-\frac{91}{1728}$	$-\frac{287}{17280}$	$\frac{205591}{4665600}$	$\frac{242839}{9331200}$	$\frac{79393}{4665600}$	$\frac{163}{69120}$
Y_2	1	$\frac{5}{12}$	$\frac{19}{96}$	$\frac{787}{8640}$	$-\frac{197}{17280}$	$-\frac{6311}{186624}$	$-\frac{371951}{9331200}$	$-\frac{251537}{4665600}$	$-\frac{557}{41472}$
Y_3	1	$\frac{11}{20}$	$\frac{37}{160}$	$\frac{1147}{8640}$	$\frac{739}{17280}$	$\frac{274441}{4665600}$	$\frac{236521}{9331200}$	$\frac{63031}{4665600}$	$\frac{6113}{345600}$
y_1	1	1	$\frac{15}{32}$	$\frac{1163}{4320}$	$\frac{1319}{8640}$	$\frac{88241}{583200}$	$\frac{110393}{1166400}$	$\frac{99937}{1166400}$	$\frac{187}{4320}$
z_1	0	$\frac{1}{4}$	$\frac{3}{16}$	$\frac{71}{960}$	$\frac{11}{384}$	$-\frac{2677}{57600}$	$-\frac{73}{2560}$	$-\frac{1001}{34560}$	$-\frac{1457}{69120}$

The details are easily confirmed using Table 6, where the entries for φ_h and χ_h are calculated as the elementary weights of (25) and (26), respectively, and $\psi_h = \chi_h \circ \varphi_h$ is found from the composition formula shown to order 4 in Table 3.

Table 6 Taylor coefficients for components of starting method

	y	$h\mathbf{f}$	$h^2\mathbf{f}'\mathbf{f}$	$\frac{1}{2}h^3\mathbf{f}''(\mathbf{f},\mathbf{f})$	$h^3\mathbf{f}'\mathbf{f}'\mathbf{f}$	$\frac{1}{6}h^4\mathbf{f}'''(\mathbf{f},\mathbf{f},\mathbf{f})$	$h^4\mathbf{f}''(\mathbf{f},\mathbf{f}'\mathbf{f})$	$\frac{1}{2}h^4\mathbf{f}'\mathbf{f}''(\mathbf{f},\mathbf{f})$	$h^4\mathbf{f}'\mathbf{f}'\mathbf{f}'\mathbf{f}$
$\varphi_h(y)$	1	0	$-\frac{1}{32}$	$-\frac{7}{4320}$	$\frac{149}{8640}$	$-\frac{49}{583200}$	$\frac{1043}{1166400}$	$-\frac{1043}{1166400}$	0
$\chi_h(y)$	0	$\frac{1}{4}$	$-\frac{1}{16}$	$-\frac{49}{960}$	$-\frac{5}{192}$	$\frac{2543}{57600}$	$\frac{89}{3840}$	$\frac{211}{11520}$	$-\frac{1}{256}$
$\psi_h(y)$	0	$\frac{1}{4}$	$-\frac{1}{16}$	$-\frac{49}{960}$	$-\frac{13}{384}$	$\frac{2543}{57600}$	$\frac{193}{7680}$	$\frac{619}{34560}$	$\frac{163}{69120}$

In simulations reported in [7], the method (24), equipped with the starting process described in this section, was shown to act in a similar way to a symplectic Runge–Kutta for millions of time steps.

Acknowledgements The author expresses his thanks for support from the Marsden Fund and for helpful comments from an anonymous referee.

References

1. Alexander, R.: Diagonally implicit Runge–Kutta methods for stiff ODEs. SIAM J. Numer. Anal. **14**, 1006–1021 (1977)
2. Butcher, J.C.: Coefficients for the study of Runge–Kutta integration processes. J. Austral. Math. Soc. **3**, 185–201 (1963)
3. Butcher, J.C.: On the attainable order of Runge–Kutta methods. Math. Comput. **19**, 408–417 (1965)
4. Butcher, J.C.: An algebraic theory of integration methods. Math. Comput. **26**, 79–106 (1972)
5. Butcher, J.C.: Numerical Methods for Ordinary Differential Equations, 2nd edn. Wiley, New York (2008)
6. Butcher, J.C.: Dealing with parasitic behaviour in G-symplectic integrators. In: Ansorge, R. (ed.) Recent Developments in the Numerics of Nonlinear Hyperbolic Conservation Laws, pp. 105–123. Springer, Heidelberg (2013)
7. Butcher, J.C.: The cohesiveness of G-symplectic methods. Numer. Algorithms (2015) DOI:10.1007/s11075-015-9964-y
8. Butcher, J.C., Cash, J.R.: Towards efficient Runge–Kutta methods for stiff systems. SIAM J. Numer. Anal. **27**, 753–761 (1990)
9. Butcher, J.C., Chartier, P.: A generalization of singly-implicit Runge–Kutta methods. Appl. Numer. Math. **24**, 343–350 (1997)
10. Butcher, J.C., Habib, Y., Hill, A.T., Norton, T.J.T.: The control of parasitism in G-symplectic methods. SIAM J. Numer. Anal. **52**, 2440–2465 (2014)
11. Gill, S.: A process for the step-by-step integration of differential equations in an automatic computing machine. Proc. Camb. Philos. Soc. **47**, 96–108 (1951)
12. Gottlieb, S., Ketcheson, D.I., Shu, C.-W.: Strong Stability Preserving Runge–Kutta and Multistep Time Discretizations. World Scientific Press, Singapore (2011)
13. Hairer, E., Lubich, C., Wanner, G.: Geometric Numerical Integration. Structure-Preserving Algorithms for Ordinary Differential Equations, 1st edn. Springer, New York (2003)

14. Heun, K.: Neue Methoden zur approximativen Integration der Differentialgleichungen einer unabhängigen Veränderlichen. Z. Math. Phys. **45**, 23–38 (1900)
15. Kutta, W.: Beitrag zur näherungsweisen Integration totaler Differentialgleichungen. Z. Math. Phys. **46**, 435–453 (1901)
16. Merson, R.H.: An operational method for the study of integration processes. In: Proceedings of Symposium on Data Processing. Weapons Research Establishment, Salisbury, S. Australia (1957)
17. Nyström, E.J.: Über die numerische integration von differentialgleichungen. Acta Soc. Sci. Fennicae **50**(13), 55pp. (1925)
18. Runge, C.: Über die numerische auflösung von differentialgleichungen. Math. Ann. **46**, 167–178 (1895)
19. Sanz-Serna, J.M.: Runge–Kutta schemes for Hamiltonian systems. BIT **28**, 877–883 (1988)
20. Shu, C.-W.: Total-variation diminishing time discretizations. SIAM J. Sci. Stat. Comput. **9**, 1073–1084 (1988)
21. Shu, C.-W., Osher, S.: Efficient implementation of essentially non-oscillatory shockcapturing schemes. J. Comput. Phys. **77**, 439–471 (1988)

A Positive Barzilai–Borwein-Like Stepsize and an Extension for Symmetric Linear Systems

Yu-Hong Dai, Mehiddin Al-Baali, and Xiaoqi Yang

Abstract The Barzilai and Borwein (BB) gradient method has achieved a lot of attention since it performs much more better than the classical steepest descent method. In this paper, we analyze a positive BB-like gradient stepsize and discuss its possible uses. Specifically, we present an analysis of the positive stepsize for two-dimensional strictly convex quadratic functions and prove the R-superlinear convergence under some assumption. Meanwhile, we extend BB-like methods for solving symmetric linear systems and find that a variant of the positive stepsize is very useful in the context. Some useful discussions on the positive stepsize are also given.

Keywords Unconstrained optimization • Barzilai and Borwein gradient method • Quadratic function • R-superlinear convergence • Condition number

1 Introduction

Consider the unconstrained quadratic optimization problem,

$$\min f(\mathbf{x}) = \frac{1}{2}\mathbf{x}^T A\mathbf{x} - \mathbf{b}^T\mathbf{x}, \tag{1}$$

Y.-H. Dai (✉)
State Key Laboratory of Scientific and Engineering Computing, Chinese Academy of Sciences, No. 55, ZhongGuanCun Donglu, P.O. Box 2719, Beijing 100190, P.R. China
e-mail: dyh@lsec.cc.ac.cn

M. Al-Baali
Department of Mathematics and Statistics, Sultan Qaboos University, Muscat, Oman
e-mail: albaali@squ.edu.om

X. Yang
Department of Applied Mathematics, The Hong Kong Polytechnic University, Kowloon, Hong Kong
e-mail: mayangxq@polyu.edu.hk

M. Al-Baali et al. (eds.), *Numerical Analysis and Optimization*, Springer Proceedings in Mathematics & Statistics 134, DOI 10.1007/978-3-319-17689-5_3

where $A \in R^{n\times n}$ is a real symmetric positive definite matrix and $\mathbf{b} \in R^n$. The (negative) gradient method for solving (1) takes the negative gradient as its search direction and updates the solution approximation iteratively by

$$\mathbf{x}_{k+1} = \mathbf{x}_k - \alpha_k \mathbf{g}_k, \tag{2}$$

where $\mathbf{g}_k = \nabla f(\mathbf{x}_k)$ and α_k is some stepsize. Denote $\mathbf{s}_{k-1} = \mathbf{x}_k - \mathbf{x}_{k-1}$ and $\mathbf{y}_{k-1} = \mathbf{g}_k - \mathbf{g}_{k-1}$. Since the matrix $B_k = \alpha_k^{-1} I$, where I is the identity matrix, can be regarded as an approximation to the Hessian of f at $\mathbf{x}_k$, Barzilai and Borwein [2] choose the stepsize α_k such that B_k has a certain quasi-Newton property:

$$B_k = \arg\min_{B=\alpha^{-1}I} \|B\mathbf{s}_{k-1} - \mathbf{y}_{k-1}\|, \tag{3}$$

where $\|\cdot\|$ means the two norm, yielding the long stepsize

$$\alpha_k^{BB1} = \frac{\mathbf{s}_{k-1}^T \mathbf{s}_{k-1}}{\mathbf{s}_{k-1}^T \mathbf{y}_{k-1}}. \tag{4}$$

An alternative way is to approximate the inverse Hessian by the matrix $H_k = \alpha_k I$ and solve

$$H_k = \arg\min_{H=\alpha I} \|\mathbf{s}_{k-1} - H\mathbf{y}_{k-1}\|, \tag{5}$$

which gives the short stepsize

$$\alpha_k^{BB2} = \frac{\mathbf{s}_{k-1}^T \mathbf{y}_{k-1}}{\mathbf{y}_{k-1}^T \mathbf{y}_{k-1}}. \tag{6}$$

Comparing with the steepest descent (SD) method, which was due to Cauchy [4], the Barzilai–Borwein (BB) method often requires less computational work and speeds up the convergence greatly. Due to its simplicity and efficiency, the BB method has been extended or generalized in many occasions or applications. For example, Raydan [17] designed an efficient global Barzilai and Borwein algorithm for unconstrained optimization by incorporating the nonmonotone line search by Grippo et al. [15]. In the context of neural network, Dai and Liao [12] considered the one-delay method, that consists in the model

$$\frac{d\mathbf{x}(t)}{dt} = -P\nabla f(\mathbf{x}(t)), \quad t \geq 0, \tag{7}$$

where

$$P = I + \frac{\mathbf{s}\mathbf{s}^T}{\mathbf{s}^T\mathbf{y}}. \tag{8}$$

Here, $\mathbf{s} = \mathbf{x}(t-\Delta t) - \mathbf{x}(t-2\Delta t)$, $\mathbf{y} = \nabla f(\mathbf{x}(t-\Delta t)) - \nabla f(\mathbf{x}(t-2\Delta t))$, and Δt is the time delay. One advantage of the above model is that, if some modification is made so that the denominator in (8) is greater than zero, each eigenvalue of P will be not less than one, which makes the model not slower than the gradient neural network. The algorithm of Raydan was further generalized by Birgin et al. (2000) for minimizing a differentiable function on a closed convex set, yielding an efficient projected gradient methods. Efficient projected algorithms based on BB-like methods have also been designed (see Serafini et al. [18] and Dai and Fletcher [10]) for special quadratic programs arising from training support vector machine. The BB method has also received much attention in finding sparse approximation solutions to large underdetermined linear systems of equations from signal/image processing and statics (for example, see Wright et al. [20]).

Several attention have been paid to theoretical properties of the BB method in spite of the potential difficulties due to its heavy nonmonotone behavior. These analyses proceed in the unconstrained quadratic case (this is also the case in this paper). Specifically, Barzilai and Borwein [2] present an interesting R-superlinear convergence result for their method when the dimension is only two. For the general n-dimensional strong convex quadratic function, the BB method is also convergent (see Raydan 1993) and the convergence rate is R-linear (see Dai and Liao 2002). Further analysis on the asymptotic behavior of BB-like methods can be found in [8, 9].

One disadvantage of the BB stepsize, however, is that it may become negative for non-convex objective functions. In this case, one remedy used in [17] is to restrict the BB stepsize into some interval like $[10^{-30}, 10^{30}]$. The setting of such interval seems very artificial. The main purpose of this paper is to analyze the following positive stepsize

$$\alpha_k = \frac{\|\mathbf{s}_{k-1}\|}{\|\mathbf{y}_{k-1}\|}. \tag{9}$$

This stepsize is exactly the geometrical mean of the long BB stepsize and the short BB stepsize. Here we should remark that the stepsize (9) has been noticed by the authors for several times (see (4.28) in [7], an unpublished preprint [9] therein, Dai and Yang [13], or Cheng and Dai [5]). This stepsize has also been noticed by Al-Baali [1]. Vrahatis et al. [19] directly replaced the Lipschitz constant L in the constant stepsize $\frac{1}{2L}$ by the estimate $\frac{\|\mathbf{y}_{k-1}\|}{\|\mathbf{s}_{k-1}\|}$, yielding a stepsize similar but not identical to (9). Nevertheless, there is no any theoretical analysis for the stepsize (9) yet.

For simplicity, we refer to the gradient method (2) with the stepsize formula (9) as method (9). In the quadratic case, since $\mathbf{s}_{k-1} = -\alpha_{k-1}\mathbf{g}_{k-1}$ and $\mathbf{y}_{k-1} = A\mathbf{s}_{k-1}$, an equivalent expression of formula (9) is

$$\alpha_k = \frac{\|\mathbf{g}_{k-1}\|}{\|A\mathbf{g}_{k-1}\|}. \tag{10}$$

Therefore formula (10) can be regarded with the one-retard extension of the stepsize considered in [13],

$$\alpha_k^{DY} = \frac{\|\mathbf{g}_k\|}{\|A\mathbf{g}_k\|}. \tag{11}$$

Interestingly enough, for the gradient method with the stepsize formula (11), it was shown in [13] that the stepsize (11) will eventually tend to the stepsize that minimizes the modulus $\|I - \alpha A\|$ (this stepsize is called the optimal stepsize in [14]). More exactly,

$$\liminf_{k \to \infty} \alpha_k^{DY} = \frac{2}{\lambda_1 + \lambda_n}, \tag{12}$$

where λ_1 and λ_n are the minimal and maximal eigenvalues of the matrix A, respectively. Simultaneously, the eigenvectors corresponding to λ_1 and λ_n can be recovered from

$$\frac{\mathbf{g}_k}{\|\mathbf{g}_k\|} + \frac{\mathbf{g}_{k+1}}{\|\mathbf{g}_{k+1}\|} \quad \text{and} \quad \frac{\mathbf{g}_k}{\|\mathbf{g}_k\|} - \frac{\mathbf{g}_{k+1}}{\|\mathbf{g}_{k+1}\|},$$

respectively.

Though simple, the two-dimensional analysis has a special meaning to the BB method. As was just mentioned, the BB method is significantly faster than the SD method in practical computations, but there is still lack of theoretical evidences that explain why the BB method is better than the SD method in the n-dimensional case. Nevertheless, the notorious zigzagging phenomenon of the SD method is well known to us; namely, the search directions in the SD method usually tend to two orthogonal directions when applied to any-dimensional quadratic functions. Unlike the SD method, however, the BB method will not produce zigzags due to its R-superlinear convergence in the two-dimensional case. This explains to some extent the efficiency of the BB method over the SD method. In this paper, we shall also analyze the convergence properties of method (9) for two-dimensional quadratic functions.

The rest of this paper is organized as follows. In the next section, we devote ourselves into the analysis of method (9) in the two-dimensional case. After giving some basic analysis in Section 2.1, we will establish the R-superlinear convergence of method (9) under some assumptions in Section 2.2. Then we make some discussions in Section 2.3. In the third section, we provide the use of the BB-like methods for solving symmetric linear systems. A typical numerical example is presented in Section 3.1, which shows that BB-like gradient methods are still very useful for solving symmetric systems. Specifically, we will see that formula (9) has a stronger ability to approximate the eigenvalues (except the signs) of a symmetric (but not necessarily positive definite) matrix A than the BB stepsizes, since formula (53) is more efficient. Some related discussions on the topic are made in Section 3.2. Finally, concluding remarks are given in the last section.

2 Analysis of Method (9) for Solving (1)

2.1 Some Basic Analysis on Method (9)

We focus on method (9) for minimizing the quadratic function (1) with $n = 2$. In this case, since the method is invariant under translations and rotations, we assume without loss of generality that

$$A = \begin{bmatrix} 1 & 0 \\ 0 & \lambda \end{bmatrix}, \quad \mathbf{b} = \mathbf{0}, \tag{13}$$

where $\lambda \geq 1$. Assume that $\mathbf{x}_1$ and $\mathbf{x}_2$ are given with

$$g_1^{(i)} \neq 0, \qquad g_2^{(i)} \neq 0, \quad \text{for } i = 1 \text{ and } 2. \tag{14}$$

To analyze $\|\mathbf{g}_k\|$ for all $k \geq 3$, we denote $\mathbf{g}_k = (g_k^{(1)}, g_k^{(2)})^T$ and define

$$q_k = \frac{\left(g_k^{(1)}\right)^2}{\left(g_k^{(2)}\right)^2}. \tag{15}$$

Then it follows that

$$\|\mathbf{g}_k\|^2 = \left(g_k^{(2)}\right)^2 (1 + q_k),$$

$$\alpha_k = \frac{\|\mathbf{s}_{k-1}\|}{\|\mathbf{y}_{k-1}\|} = \frac{\|\mathbf{g}_{k-1}\|}{\|A\mathbf{g}_{k-1}\|} = \frac{\sqrt{1+q_{k-1}}}{\sqrt{\lambda^2 + q_{k-1}}}.$$

Noticing that $\mathbf{x}_{k+1} = \mathbf{x}_k - \alpha_k \mathbf{g}_k$ and $\mathbf{g}_k = A\mathbf{x}_k$, we have that

$$\mathbf{g}_{k+1} = (I - \alpha_k A)\mathbf{g}_k. \tag{16}$$

Writing the above relation in componentwise form,

$$\begin{pmatrix} g_{k+1}^{(1)} \\ g_{k+1}^{(2)} \end{pmatrix} = \left(\begin{bmatrix} 1 & \\ & 1 \end{bmatrix} - \frac{\sqrt{1+q_{k-1}}}{\sqrt{\lambda^2+q_{k-1}}} \begin{bmatrix} 1 & \\ & \lambda \end{bmatrix} \right) \begin{pmatrix} g_k^{(1)} \\ g_k^{(2)} \end{pmatrix}$$

$$= \begin{bmatrix} \frac{\sqrt{\lambda^2+q_{k-1}}-\sqrt{1+q_{k-1}}}{\sqrt{\lambda^2+q_{k-1}}} & \\ & \frac{\sqrt{\lambda^2+q_{k-1}}-\lambda\sqrt{1+q_{k-1}}}{\sqrt{\lambda^2+q_{k-1}}} \end{bmatrix} \begin{pmatrix} g_k^{(1)} \\ g_k^{(2)} \end{pmatrix}.$$

Therefore we get for all $k \geq 2$,

$$\begin{cases} \left(g_{k+1}^{(1)}\right)^2 = \dfrac{(\sqrt{\lambda^2+q_{k-1}} - \sqrt{1+q_{k-1}})^2}{\lambda^2+q_{k-1}} \left(g_k^{(1)}\right)^2, \\ \left(g_{k+1}^{(2)}\right)^2 = \dfrac{(\sqrt{\lambda^2+q_{k-1}} - \lambda\sqrt{1+q_{k-1}})^2}{\lambda^2+q_{k-1}} \left(g_k^{(2)}\right)^2. \end{cases} \tag{17}$$

In the case that $\lambda = 1$, which means that the object function has sphere contours, the method will take a unit stepsize $\alpha_2 = 1$ and give the exact solution at the third iteration. If $g_2^{(1)} = 0$ but $g_2^{(2)} \neq 0$, we have that $q_2 = 0$ and hence by (17) that $g_k^{(1)} = 0$ for $k \geq 3$ and $g_4^{(2)} = 0$, which means that the method gives the exact solution in at most four iterations. This is also true if $g_2^{(2)} = 0$ but $g_2^{(1)} \neq 0$ due to symmetry of the first and second components. If $g_1^{(1)} = 0$ but $g_1^{(2)} \neq 0$, we have that $q_1 = 0$ and $g_3^{(2)} = 0$. Then by considering $\mathbf{x}_2$ and $\mathbf{x}_3$ as two starting points, we must have $\mathbf{g}_k = 0$ for some $k \leq 5$. The symmetry works for the case that $g_1^{(2)} = 0$ but $g_1^{(1)} \neq 0$. Thus, similarly to the analysis for the BB method in [8], we may assume that $\lambda > 1$ and the assumption (14) holds, for otherwise the method has the finite termination property. On the other hand, if (14) holds, then we will have $g_k^{(1)} \neq 0$ and $g_k^{(2)} \neq 0$ for all $k \geq 1$ and hence q_k is always well defined.

Now, substituting (17) into the definition of q_{k+1}, we can obtain the following recurrence relation

$$\begin{aligned} q_{k+1} &= \left(\frac{\sqrt{\lambda^2+q_{k-1}} - \sqrt{1+q_{k-1}}}{\sqrt{\lambda^2+q_{k-1}} - \lambda\sqrt{1+q_{k-1}}}\right)^2 q_k \\ &= \left(\frac{(\sqrt{\lambda^2+q_{k-1}} - \sqrt{1+q_{k-1}})(\sqrt{\lambda^2+q_{k-1}} + \lambda\sqrt{1+q_{k-1}})}{(\lambda^2-1)q_{k-1}}\right)^2 q_k \\ &= \left(\frac{\lambda - q_{k-1} + \sqrt{\tau(q_{k-1})}}{\lambda+1}\right)^2 \frac{q_k}{q_{k-1}^2}, \end{aligned} \tag{18}$$

where τ is the following quadratic function

$$\tau(w) = (1+w)(\lambda^2+w), \quad \text{where } w \geq 0. \tag{19}$$

To proceed with our analysis, we denote $M_k = \log q_k$ and

$$h(w) = \frac{\lambda - w + \sqrt{\tau(w)}}{\lambda+1}, \quad \text{where } w \geq 0. \tag{20}$$

It follows from the recurrence relation (18) that

$$M_{k+1} = M_k - 2M_{k-1} + 2\log h(q_{k-1}). \tag{21}$$

2.2 R-Superlinear Convergence of Method (9)

Lemma 1.2.1 *Assume that* $\lambda > 1$. *The function* $h(w)$ *in (20) is monotonically increasing for* $w \in [0, +\infty)$. *Further, we have that*

$$h(w) \in \left[\frac{2\lambda}{\lambda+1}, \frac{\lambda+1}{2}\right), \quad \textit{for any } w \geq 0. \tag{22}$$

Proof. By the definition of τ in (19), we have that

$$(\tau')^2 - 4\tau = (\lambda^2 - 1)^2. \tag{23}$$

Then by direct calculations, we get

$$\begin{aligned} h'(w) &= \frac{-1 + \frac{1}{2}\tau^{-\frac{1}{2}}\tau'}{\lambda+1} \\ &= \frac{(\tau')^2 - 4\tau}{2(\lambda+1)\tau^{\frac{1}{2}}(\tau' + 2\tau^{\frac{1}{2}})} \\ &= \frac{(\lambda^2-1)^2}{2(\lambda+1)(\tau^{\frac{1}{2}}\tau' + 2\tau)}. \end{aligned} \tag{24}$$

Thus we see that $h'(w) > 0$ for $w \geq 0$, which indicates that $h(w)$ is monotonically increasing. Noticing that

$$h(0) = \frac{2\lambda}{\lambda+1} \quad \text{and} \quad \lim_{w \to \infty} h(w) = \frac{\lambda+1}{2},$$

we know that (22) holds. This completes our proof. □

Lemma 1.2.2 *Assume that* $\lambda > 1$. *Consider the function*

$$\psi(w) = \frac{wh'(w)}{h(w)}, \quad \textit{where } w \geq 0, \tag{25}$$

where $h(w)$ *is given in (20). Then* $\psi(w) \geq 0$ *for all* $w \geq 0$. *Further, it reaches its maximal value at* $w_{\max} = \lambda$ *and*

$$\psi_{\max} := \psi(w_{\max}) = \frac{1}{2} - \frac{\sqrt{\lambda}}{\lambda+1}. \tag{26}$$

Proof. The nonnegativity of $\psi(w)$ over $[0, +\infty)$ is obvious due to Lemma 1.2.1. To analyze the maximal value of $\psi(w)$ for $w > 0$, by setting $\psi'(w) = 0$ and noting that

$h'(w) \neq 0$ for $\lambda \neq 1$, we can get that

$$\frac{1}{w} - \frac{h'(w)}{h(w)} = -\frac{h''(w)}{h'(w)}. \tag{27}$$

Direct calculations show that

$$\begin{aligned}\frac{1}{w} - \frac{h'(w)}{h(w)} &= \frac{1}{w} - \frac{-1 + \frac{1}{2}\tau^{-\frac{1}{2}}\tau'}{\lambda - w + \tau^{\frac{1}{2}}} = \frac{\lambda + \tau^{\frac{1}{2}} - \frac{1}{2}w\tau^{-\frac{1}{2}}\tau'}{w(\lambda - w + \tau^{\frac{1}{2}})} \\ &= \frac{\lambda\tau + \tau^{\frac{1}{2}}(\tau - \frac{1}{2}w\tau')}{w\tau(\lambda - w + \tau^{\frac{1}{2}})} = \frac{2\lambda\tau + \tau^{\frac{1}{2}}\left[2\lambda^2 + (\lambda^2+1)w\right]}{2w\tau(\lambda - w + \tau^{\frac{1}{2}})}. \end{aligned} \tag{28}$$

On the other hand, noticing that $\tau'' = 2$, we have by (23) and direct calculations that

$$(\lambda+1)h''(w) = -\frac{1}{4}\tau^{-\frac{3}{2}}\left[(\tau')^2 - 2\tau\tau''\right] = -\frac{1}{4}(\lambda^2-1)^2\tau^{-\frac{3}{2}}.$$

The above relation indicates that $h(w)$ is a concave function. It follows from this relation and (24) that

$$-\frac{h''(w)}{h'(w)} = \tau^{-\frac{1}{2}}(1 + \frac{1}{2}\tau^{-\frac{1}{2}}\tau'). \tag{29}$$

Substituting (28) and (29) into the Equation (27) and noticing that $\tau' = 1 + \lambda^2 + 2w$, we can get

$$2\lambda\tau + \tau^{\frac{1}{2}}\left[2\lambda^2 + (\lambda^2+1)w\right] = w\left(\lambda - w + \tau^{\frac{1}{2}}\right)\left(1 + \lambda^2 + 2w + 2\tau^{\frac{1}{2}}\right). \tag{30}$$

The relation (30) is equivalent to

$$(\lambda - w)\left[-w(1+\lambda^2+2w) + 2\lambda\tau^{\frac{1}{2}} + 2\tau\right] = 0.$$

Substituting $\tau = (1+w)(\lambda^2+w)$ into the above relation yields

$$(\lambda - w)\left[2\lambda^2 + (1+\lambda^2)w + 2\lambda\tau^{\frac{1}{2}}\right] = 0. \tag{31}$$

Thus, to meet (27), which is equivalent to (31), we must have that $w = \lambda$. Since $\psi(0) = 0$ and $\psi(w) > 0$ for $w > 0$, we know that $\psi(w)$ must reach its maximal value at its unique stationary point $w = \lambda$. Therefore $w_{\max} = \lambda$. Noticing that

$$h(\lambda) = \sqrt{\lambda} \quad \text{and} \quad h'(\lambda) = \frac{(\sqrt{\lambda}-1)^2}{2(\lambda+1)\sqrt{\lambda}},$$

we can deduce (26). □

In addition to the function $\psi(w)$ in (25), we consider the function

$$\phi(w) = \begin{cases} \dfrac{\log h(w) - \log h(1)}{\log w}, & \text{if } w > 0 \text{ but } w \neq 1; \\ \dfrac{h'(1)}{h(1)}, & \text{if } w = 1. \end{cases} \tag{32}$$

Lemma 1.2.3 *For the function $\phi(w)$ defined in (32), we have that*

$$0 < \phi(w) \leq \psi_{\max}, \quad \textit{for all } w > 0,$$

where $\psi_{\max}$ is given by (26).

Proof. It is obvious that ϕ is continuous in $(0,+\infty)$ and continuously differentiable over $(0,1) \cup (1,+\infty)$. Due to Lemma 1.2.1, we can also see that $\phi(w)$ tends to zero when w tends to zero or when w tends to $+\infty$. Further, by setting the derivative of $\phi(w)$ to be zero, we know that the optimal w^* that maximize $\phi(w)$ over $(0,1) \cup (1,+\infty)$ must satisfy

$$\phi(w^*) = \frac{w^* h'(w^*)}{h(w^*)}.$$

Consequently, by Lemma 1.2.2, we have for $w > 0$,

$$0 < \phi(w) \leq \max\{\phi(w^*), \phi(1)\} = \max\{\psi(w^*), \psi(1)\} \leq \psi_{\max}.$$

This completes our proof. □

Now, noticing the relation (21) and using the definition of ϕ, we can get that

$$M_{k+1} = M_k - 2(1 - \phi(q_{k-1}))M_{k-1} + 2\log h(1). \tag{33}$$

By Lemma 1.2.3, we know that the coefficient of M_{k-1} belongs to the interval

$$\left(-2, -1 - \frac{2\sqrt{\lambda}}{\lambda+1}\right]. \tag{34}$$

This interval, however, cannot enable us to find some suitable parameter γ such that the sequence of $|M_k + \gamma M_{k-1}|$ is monotonically increasing with k. To do so, we have to strengthen the upper bound of $\phi(w)$ in Lemma 1.2.3. Meanwhile, we still need some suitable assumption on the initial value of M_1 and M_2 similarly to Lemma 1.2.4. Based on this reason, we directly work with the recursive relation (21). Pick γ to be any root of the equation $\gamma^2 - \gamma + 2 = 0$; namely,

$$\gamma = \frac{1 \pm \sqrt{7}i}{2}, \tag{35}$$

where i is the imaginary unit (sometimes i is also used as an index, but it is easy for the reader to tell). We have the following lemma.

Lemma 1.2.4 *Consider the sequence $\{M_k\}$ satisfying (21). Denote $\xi_k = M_k + (\gamma - 1)M_{k-1}$, where γ is given in (35). If*

$$|\xi_2| > 2\log\frac{\lambda+1}{2}, \tag{36}$$

there exist some positive constants c_1 and c_2 such that

$$|\xi_k| \geq c_1 2^{k-2} + c_2, \quad \textit{for all } k \geq 2. \tag{37}$$

Proof. It follows from the definition of ξ_k, the relation (21) and the choice of γ that

$$\xi_{k+1} = \gamma M_k - 2M_{k-1} + 2\log h(q_{k-1}) = \gamma\xi_k + 2\log h(q_{k-1}).$$

Noticing that $|\gamma| = 2$ and by Lemma 1.2.1, $|\log h(q_{k-1})| < \log\frac{\lambda+1}{2}$, we have from the above relation that

$$|\xi_{k+1}| \geq 2|\xi_k| - c_2, \tag{38}$$

where $c_2 = 2\log\frac{\lambda+1}{2}$. The relation (38) is equivalent to

$$|\xi_{k+1}| - c_2 \geq 2\left(|\xi_k| - c_2\right). \tag{39}$$

Denoting $c_1 = |\xi_2| - c_2$, which is strictly greater than zero by assumption, we can know from the repeated use of (39) that (37) holds. □

Notice that $|\gamma - 1| = 2$ and hence

$$|\xi_k| \leq |M_k| + 2|M_{k-1}| \leq 3\max\{|M_k|, |M_{k-1}|\}.$$

This with (37) gives that

$$\max\{|M_k|, |M_{k-1}|\} \geq \frac{1}{3}\left(c_1 2^{k-2} + c_2\right), \quad \text{for all } k \geq 2. \tag{40}$$

Lemma 1.2.5 *Consider the sequence $\{M_k\}$ satisfying (21). Under the same assumption in Lemma 1.2.4, we have for all $k \geq 2$ that*

$$\max_{-1\leq i\leq 3} M_{k+i} \geq \frac{1}{3}c_1 2^{k-2} - 4\log\frac{\lambda+1}{2} \tag{41}$$

and

$$\min_{-1\leq i\leq 3} M_{k+i} \leq -\frac{1}{3}c_1 2^{k-2} + 4\log\frac{\lambda+1}{2}. \tag{42}$$

Proof. It follows from the recursive relation (21) that

$$M_{k+2} = -M_k - 2M_{k-1} + 2\log h(q_k) + 2\log h(q_{k-1}). \tag{43}$$

We focus on the relation (40). If there exists some $i = 0$ or 1 such that

$$M_{k-i} \geq \frac{1}{3}\left(c_1 2^{k-2} + c_2\right),$$

then it is obvious that (41) holds. Otherwise, we must have that

$$M_{k-i} \leq -\frac{1}{3}\left(c_1 2^{k-2} + c_2\right)$$

holds for some $i = 0$ or 1. In this case, noticing Lemma 1.2.1, we can see from (21) and (43) (with $k-1$ replaced with $k-i$) that the following relation

$$M_{k-i+j} \geq \frac{2}{3}\left(c_1 2^{k-2} + c_2\right) - 4\log\frac{\lambda+1}{2}$$

holds for $j = 2$ or 3. As a matter of fact, we can use the relation (21) if $M_{k-i+1} \geq 0$ or the relation (43) otherwise. Therefore (41) must be true. The proof of (42) is similar. □

The above lemma indicates that there must exist two subsequences of $\{M_k\}$ which tend to $+\infty$ and $-\infty$, respectively, at a geometrical rate. Then we are able to show that both the components of the gradient tend to zero R-superlinearly and hence the whole gradient norm is R-superlinearly convergent.

Theorem 1.2.6 *Consider method (9). Assume that (14) and (36) hold. Then the sequence of the gradient norm $\{\|\mathbf{g}_k\|\}$ converges to zero and the convergence is R-superlinear.*

Proof. First, noticing that $\alpha_k \in (\lambda^{-1}, 1)$ for any k, we know from (16) that

$$|g_{k+1}^{(i)}| \leq (\lambda - 1)|g_k^{(i)}| \tag{44}$$

holds for $i = 1$ and 2 and all $k \geq 1$. Let us focus on the second component of $\mathbf{g}_k$. By the second relation in (17), it is not difficult to prove that

$$\begin{aligned} |g_{k+1}^{(2)}| &\leq \frac{(\lambda^2-1)q_{k-1}}{\sqrt{\lambda^2+q_{k-1}}(\sqrt{\lambda^2+q_{k-1}}+\lambda\sqrt{1+q_{k-1}})}|g_k^{(2)}| \\ &\leq \frac{(\lambda^2-1)q_{k-1}}{2\lambda^2}|g_k^{(2)}| \\ &< (\lambda-1)q_{k-1}|g_k^{(2)}|. \end{aligned} \tag{45}$$

Combining (44) and (45), we can get that

$$|g_{k+5}^{(2)}| \leq (\lambda-1)^5 \left(\min_{-1\leq i\leq 3} q_{k-1} \right) |g_k^{(2)}|,$$

which, with $M_k = \log q_k$ and the relation (42), yields

$$|g_{k+5}^{(2)}| \leq (\lambda-1)^5 \exp\left(-\frac{1}{3}c_1 2^{k-2} + 4\log\frac{\lambda+1}{2} \right) |g_k^{(2)}|.$$

Similarly, we can build

$$|g_{k+5}^{(1)}| \leq \frac{1}{2}(\lambda+1)(\lambda-1)^5 \exp\left(-\frac{1}{3}c_1 2^{k-2} + 4\log\frac{\lambda+1}{2} \right) |g_k^{(1)}|.$$

Thus we can obtain for all k,

$$||\mathbf{g}_{k+5}|| \leq \frac{1}{2}(\lambda+1)(\lambda-1)^5 \exp\left(-\frac{1}{3}c_1 2^{k-2} + 4\log\frac{\lambda+1}{2} \right) ||\mathbf{g}_k||. \tag{46}$$

Therefore we can see that $\|\mathbf{g}_k\|$ converges to zero and the convergence is R-superlinear. □

2.3 Some Discussions

Comparing the above two-dimensional analysis for method (9) with those for the BB method, we can see that the analysis in this paper is more difficult. The current analysis requires an assumption on the initial points, that is (36), so that we can prove the divergence of a subsequence of $\{M_k\}$ (see Lemma 1.2.4). Then we are able to show that there are two subsequences of $\{M_k\}$ which tend to $+\infty$ and $-\infty$, respectively (see Lemma 1.2.5). A direct implication of this result is that there are two subsequences of $\{\alpha_k\}$ which converges to the two eigenvalues of the matrix A. Finally, we can establish the R-superlinear convergence for method (9) in the two-dimensional case. Although our numerical observations show that the assumption (36) is not necessary, we do not know yet whether this assumption can be removed or not.

Since by Lemma 1.2.1, the last term in the recursive relation (21) is bounded above and below, we may think that the properties of the sequence $\{M_k\}$ are similar to the one satisfying the linear recursion relation $M_{k+1} = M_k - 2M_{k-1}$. The latter is exactly what Barzilai and Borwein [2] obtained for the BB method. Therefore, we might feel that method (9) itself performs not better than the BB method. An illustrative example is as following. Consider the 1000-dimensional example

$$A = diag(1:1000), \qquad \mathbf{b} = zeros(1000,1). \tag{47}$$

Here and below *diag* and *zeros* are standard matlab languages. The starting point and the stopping criterion are

$$\mathbf{x}_1 = ones(1000), \qquad \|\mathbf{g}_k\| \le 10^{-12}, \tag{48}$$

respectively. It was found that, to reach the stopping criterion, the BB1 method, the BB2 method, and method (9) require 590, 697, and 1,139, iterations, respectively.

Nevertheless, when applied the BB method for general nonconvex optimization, it is possible that $\mathbf{s}_{k-1}^T \mathbf{y}_{k-1} < 0$, in which case some truncations are often done. For example, by projecting the BB stepsizes onto the interval like $[10^{-30}, 10^{30}]$. With the help of the stepsize (9), we may now consider, for example, the following possibilities

$$\bar{\alpha}_k^{BB1} = \max\left\{ \frac{\mathbf{s}_{k-1}^T \mathbf{s}_{k-1}}{\mathbf{s}_{k-1}^T \mathbf{y}_{k-1}}, \frac{\|\mathbf{s}_{k-1}\|}{\|\mathbf{y}_{k-1}\|} \right\} \tag{49}$$

and

$$\bar{\alpha}_k^{BB2} = \max\left\{ \frac{\mathbf{s}_{k-1}^T \mathbf{y}_{k-1}}{\mathbf{y}_{k-1}^T \mathbf{y}_{k-1}}, \frac{\|\mathbf{s}_{k-1}\|}{\|\mathbf{y}_{k-1}\|} \right\}. \tag{50}$$

It is easy to see that if $\mathbf{s}_{k-1}^T \mathbf{y}_{k-1} > 0$, $\bar{\alpha}_k^{BB1} = \alpha_k^{BB1}$ and $\bar{\alpha}_k^{BB2}$ reduces to the stepsize (9). However, if $\mathbf{s}_{k-1}^T \mathbf{y}_{k-1} \le 0$, the stepsize (9) will be used instead. Theoretically, by the analysis in [7], it is not difficult to see that all the stepsizes (9), (49), and (50) possess the so-called Property (A) and hence the corresponding gradient methods are R-linearly convergent for any-dimensional strictly convex quadratic functions. More numerical experiments are still required to test the efficiency of the proposed variants.

3 Solving Symmetric Linear Systems

This section aims to expose another good property of the stepsize (9). More exactly, if the Hessian matrix A is only symmetric, but not necessarily positive definite, we will find that formula (9) has stronger ability to approximate the eigenvalues (except the signs) of A than the formulae (4) and (6).

3.1 A Typical Numerical Example

In this section, we shall consider the symmetric linear system

$$A\mathbf{x} = \mathbf{b}, \tag{51}$$

where $A \in R^{n\times n}$ is assumed to be symmetric and invertible and $\mathbf{b} \in R^n$. It is obvious that if A is symmetric positive definite, the unconstrained quadratic optimization problem (1) is equivalent to the linear system (51). In this subsection, however, we only assume A to be symmetric, but not necessarily positive definite. As BB-like gradient methods have achieved great success in various aspects, there seem not many studies on the methods for solving symmetric linear systems.

For easy illustration, for any dimension n, we define the n-dimensional vector $\mathbf{v}$ with $v(i) = (-1)^i i$ and consider the following example

$$A = diag(\mathbf{v}), \qquad \mathbf{b} = zeros(n,1).$$

Here again, *diag* and *zeros* are standard matlab languages. In the context of linear systems, we define $\mathbf{g}_k = A\mathbf{x}_k - \mathbf{b}$, which is exactly the derivative of the quadratic function in (1). The starting point and the stopping criterion are

$$\mathbf{x}_1 = ones(n) \quad \text{and} \quad \|\mathbf{g}_k\| \le 10^{-6}, \tag{52}$$

respectively. In practical computations, we consider the following five values of n: $n = 10, 20, 30, 40, 50$.

Firstly, we tried the naive use of the classical steepest descent method, that is, the method (2) with the stepsize

$$\alpha_k^{SD} = \frac{\mathbf{g}_k^T \mathbf{g}_k}{\mathbf{g}_k^T A \mathbf{g}_k},$$

and found that the norm $\|\mathbf{g}_k\|$ goes to infinity at a fast rate and cannot converge at all.

Secondly, we tested the two choices, (4) and (6), of the Barzilai–Borwein methods. They are denoted by BB1 and BB2, respectively. In this case, the steepest descent stepsize is used for the first iteration. In Table 1, we listed the number of

Table 1 Comparing different methods for symmetric linear systems

	n				
Method	10	20	30	40	50
BB1	1,117	2,806	2,568	2,948	4,685
BB2	238	499	1,138	2,104	2,345
(53)	147	426	607	687	847

iterations required by each method for each problem. It is remarkable to see that both BB1 and BB2 can provide a solution satisfying the stopping criterion in (52). Further, unlike the unconstrained optimization, where it is believed that BB1 is preferable to BB2, the BB2 method requires significantly fewer iterations than the BB1 method does.

Now, we think of how to make use of the stepsize (9) for solving the symmetric linear system (51). Due to its equivalent definition (10) of the stepsize, it is easy to see that α_k^2 is an approximation to some inverse eigenvalue of the matrix A^2. To decrease the components of the residual vector $\mathbf{g}_k$ corresponding to the negative eigenvalues of A, we need to design a mechanism how to choose the sign of the stepsize α_k. An easy way is to consider the function

$$sign(a) = \begin{cases} 1, & \text{if } a \geq 0; \\ -1, & \text{otherwise.} \end{cases}$$

Then we calculate the stepsize in the following way

$$\alpha_k = sign(\mathbf{s}_{k-1}^T \mathbf{y}_{k-1}) \frac{\|\mathbf{s}_{k-1}\|}{\|\mathbf{y}_{k-1}\|}. \tag{53}$$

In other words, the stepsize (53) aims to approximate the inverse eigenvalue of the matrix A based on the sign of the inner product $\mathbf{s}_{k-1}^T \mathbf{y}_{k-1}$. If $\mathbf{s}_{k-1}^T \mathbf{y}_{k-1}$ is greater than or equal to zero, it tends to estimate the inverse of the positive eigenvalues; otherwise, if $\mathbf{s}_{k-1}^T \mathbf{y}_{k-1}$ is less than zero, it goes to approximate the inverse of the negative eigenvalues. The iterations required by the method (53) are denoted in Table 1 in the row "(53)". Again, the steepest descent stepsize is used for the first iteration. From the table, we can see that the new method performs much efficient than the BB1 and BB2 methods.

3.2 *Some Discussions*

It is obvious that more numerical experiments with symmetric linear systems are needed to check the efficiency of the new method. Nevertheless, the above example is typical, which explains that the new method performs much better than the BB1 and BB2 methods. The example also provides some reason of directly using the BB1 and BB2 stepsizes in the context of optimization, instead of truncating them to be some tiny positive numbers like $\alpha_{\min} = 10^{-30}$ as mentioned in the introduction. The first author once found that the direct use of the negative BB stepsizes can reduce the number of iterations.

As there have been a lot of Barzilai–Borwein-like gradient methods in the context of optimization, we do not know yet whether there exists more efficient stepsizes in the gradient method for solving symmetric linear systems. Another issue is the application of the new stepsize in nonlinear systems. Cruz et al. [6] built an efficient gradient algorithm for nonlinear systems based on the BB stepsizes. Can we improve their gradient algorithms by using our new stepsize?

Finally, an important theoretical question related to the new method (or the BB1 and BB2 methods) is, does the new method converge for general symmetric linear systems? Although our numerical experiments show that the answer might be yes, it seems very difficult for us to provide a proof.

4 Concluding Remarks

In this paper, we have analyzed a positive BB-like gradient stepsize and discussed its possible uses. We provide an analysis of the positive stepsize for two-dimensional strictly convex quadratic functions and prove the R-superlinear convergence under the assumption (36). It is not known yet whether the assumption (36) can be removed or not. At the same time, we have extended BB-like methods for solving symmetric linear systems and found that a variant of the positive stepsize, that is (53), is very useful in the context. More numerical experiments are required to examine the efficiency of the stepsize (53) for symmetric linear systems. The convergence of BB-like methods in this context is also not known to us in theory.

From the discussions in Sections 2.3 and 3.2, we have seen two possibilities to deal with the case where the BB stepsizes are negative. The first is by truncation [for example, see (49) and (50)]. The second is still to use the BB stepsize even when negative values of the BB stepsize have been detected due to the successful numerical example in Section 3.1. On the whole, the proposition of the positive stepsize (9) might provide much room in finding more efficient and reasonable BB-like gradient methods.

Acknowledgements The authors are grateful to Dr. Bo Jiang for checking an early version of this manuscript and to Ms. Liaoyuan Zeng for her editing of this paper. They also thank an anonymous referee for his/her useful suggestions and comments.

References

1. Al-Baali, M.: On alternate steps for gradient methods. Talk at 22-nd Biennial Conference on Numerical Analysis, University of Dundee, Scotland, 26–29 June 2007
2. Barzilai, J., Borwein, J.M.: Two-point step size gradient methods. IMA J. Numer. Anal. **8**, 141–148 (1988)

3. Birgin, E.G., Martínez, J.M., Raydan, M.: Nonmonotone spectral projected gradient methods on convex sets. SIAM J. Optim. **10**, 1196–1211 (2000)
4. Cauchy, A.: Méthode générale pour la résolution des systèms d'équations simultanées. Comput. Rend. Sci. Paris **25**, 536–538 (1847)
5. Cheng, M., Dai, Y.H.: Adaptive nonmonotone spectral residual method for large-scale nonlinear systems. Pac. J. Optim. **8**, 15–25 (2012)
6. Cruz, W.L., Martínez, J.M., Raydan, M.: Spectral residual method without gradient information for solving large-scale nonlinear systems of equations. Math. Comput. **75**, 1429–1448 (2006)
7. Dai, Y.H.: Alternate step gradient method. Optimization **52**, 395–415 (2003)
8. Dai, Y.H.: A new analysis on the Barzilai-Borwein gradient method. J. Oper. Res. Soc. China **1**, 187–198 (2013)
9. Dai, Y.H., Fletcher, R.: On the asymptotic behaviour of some new gradient methods. Math. Program. Ser. A **103**, 541–559 (2005)
10. Dai, Y.H., Fletcher, R.: New algorithms for singly linearly constrained quadratic programs subject to lower and upper bounds. Math. Program. Ser. A **106**, 403–421 (2006)
11. Dai, Y.H., Liao, L.Z.: R-linear convergence of the Barzilai and Borwein gradient method. IMA J. Numer. Anal. **26**, 1–10 (2002)
12. Dai, Y.H., Liao, L.Z.: A new first-order neural network for unconstrained nonconvex optimization. Research Report, Academy of Mathematics and Systems Science, Chinese Academy of Sciences (1999)
13. Dai, Y.H., Yang, X.Q.: A new gradient method with an optimal stepsize property. Comput. Optim. Appl. **33**, 73–88 (2006)
14. Elman, H.C., Golub, G.H.: Inexact and preconditioning Uzawa algorithms for saddle point problems. SIAM J. Numer. Anal. **36**, 1645–1661 (1994)
15. Grippo, L., Lampariello, F., Lucidi, S.: A nonmonotone line search technique for Newton's method. SIAM J. Numer. Anal. **23**, 707–716 (1986)
16. Raydan, M.: On the Barzilai and Borwein choice of steplength for the gradient method. IMA J. Numer. Anal. **13**, 321–326 (1993)
17. Raydan, J.: The Barzilai and Borwein gradient method for the large scale unconstrained minimization problem. SIAM J. Optim. **7**, 26–33 (1997)
18. Serafini, T., Zanghirati, G., Zanni, L.: Gradient projection methods for quadratic programs and applications in training support vector machines. Optim. Methods Softw. **20**, 347–372 (2005)
19. Vrahatis, M.N., Androulakis, G.S., Lambrinos, J.N., Magoulas, G.D.: A class of gradient unconstrained minimization algorithms with adaptive stepsize. J. Comput. Appl. Math. **114**, 367–386 (2000)
20. Wright, S.J., Nowak, R.D., Figueiredo, M.A.T.: Sparse reconstruction by separable approximation. IEEE Trans. Signal Process. **57**, 2479–2493 (2009)

Necessary Optimality Conditions for the Control of Partial Integro-Differential Equations

Leonhard Frerick, Ekkehard W. Sachs, and Lukas A. Zimmer

Abstract In this paper we derive necessary optimality conditions for optimization problems with partial integro-differential equations. We use the concept of mild solutions coming from semi-group theory for evolution equations. The application considered is a model for a cell adhesion process which leads to a two-dimensional system of nonlinear partial integro-differential equations. The objective function is of tracking type with the coefficients in the integral operators as unknown control or design variables. We derive necessary optimality conditions in the form of an adjoint system of partial integro-differential equations.

Keywords Partial integro-differential equations • Optimal control • Applications

1 Introduction

Non-local models in the form of partial integro-differential equations (PIDEs) arise in various fields and become more and more important.

In mechanics, the peridynamics theory was introduced in order to model surfaces with cracks. In finance, in particular for option pricing, existing models were extended with Lévy processes in order to model jumps, like those that occurred during the financial crisis in 2007. Recent work on numerical treatment of such jump-diffusion PIDEs or the corresponding calibration problems can be found, for example, in Andersen and Andreasen [1], Matache et al. [10], Cont and Voltchkova [4], Briani et al. [3], Sachs and Strauss [11], or Schu [12]. Many biological models benefit from non-local terms. Biological applications of PIDEs are discussed, for example, in Armstrong et al. [2] or Gerisch [7].

First, we take a closer look at a motivating example: In biology, cell adhesion describes the binding between two cells or between a cell and the extracellular matrix through certain proteins, called cell-adhesion molecules. Cell adhesion is responsible for tissue formation, tissue stability and—in case of loss of the adhesion—cell breakdown. In 1962, Steinberg showed that two different cell

L. Frerick • E.W. Sachs (✉) • L.A. Zimmer
FB IV – Mathematik, University of Trier, 54286 Trier, Germany
e-mail: frerick@uni-trier.de; sachs@uni-trier.de; lukasa.zimmer@online.de

M. Al-Baali et al. (eds.), *Numerical Analysis and Optimization*, Springer Proceedings in Mathematics & Statistics 134, DOI 10.1007/978-3-319-17689-5_4

populations can aggregate in four different ways: mixing, engulfment, partial engulfment and complete sorting [2]. Steinberg proposed that the way they aggregate depends on differences in the cell's adhesion properties and the cell's surface tension (the differential adhesion hypothesis).

Armstrong et al. published a continuous model that describes cell adhesion in 2006, where they stress that all previous models were discrete ones [2]. Continuous models have an advantage over their discrete counterparts since these models can handle huge cell populations—they are scalable. Furthermore, it is quite difficult to perform analysis for discrete models. Armstrong et al. describe the adhesion driven cell-movement with a non-local term, which results in an integro-differential equation.

Without considering cell birth and cell death, mass conservation implies

$$u_t(t,x) = -J_x(t,x)$$

for the variation of cell concentration u in x over time. Armstrong et al. split up the flux of the cells J in

$$\text{random diffusion } J^{(d)} = -D(u_x) \text{ and adhesive forces } J^{(a)} = \frac{\phi}{R}uF,$$

where D is the diffusion coefficient, ϕ is a viscosity related constant, R the sensing radius of the cells and F the force that is acting on the cells within that radius. The force acting on the cell at x, that is created by a cell at position $x+y$, is given by

$$f(x) = \alpha g(u(x+y))\omega(y),$$

where g describes the nature of the forces and their dependence on the local cell density at $x+y$. The authors provide two possible examples for g: a simple linear one ($g(u) = u$) and one of logistic form ($g(u) = u(1-u/M)$ for $u < M$ and 0 otherwise). The function $\omega(y)$ describes how the direction and magnitude of the force alters according to y (thus, ω is an odd function), a simple example would be $\omega(y) = sign(y)$. In that case, ω only provides the direction, not the magnitude of the force. α is a positive parameter reflecting the strength of the adhesive force between the cells. The total force F is derived as the sum of the local forces

$$F(x) = \int_{-R}^{R} \alpha g(u(x+y)\omega(y)\,dy.$$

Together with random diffusion, we obtain the model of Armstrong et al. in one dimension and for one population:

$$u_t = Du_{xx} - (uK(u))_x = Du_{xx} - u_xK(u) - uK(u)_x \tag{1}$$

with

$$K(u)(x) = \frac{\phi}{R}\int_{-R}^{R} \alpha g(u(x+y)\omega(y)\,dy.$$

With the two transformations $\tau := \frac{D}{R^2}t$ and $\xi := \frac{x}{R}$, a non-dimensional version can be formulated (see [2, Section 2]): If u solves (1), then

$$v(\tau,\xi) := \frac{R\phi}{D}u(\frac{R^2}{D}\tau, R\xi)$$

is the solution of

$$v_\tau = v_{\xi\xi} - (v\kappa(v))_\xi, \tag{2}$$

with

$$\kappa(v)(\xi) = \alpha\int_{-1}^{1} v(\xi+\zeta)\omega(\zeta)\,d\zeta.$$

The remaining non-dimensional parameter α is a measure for the adhesion strength. Armstrong et al. showed that, if α is below a certain threshold, no cell aggregation will occur.

Finally, we present the model of Armstrong et al. for two populations in two space dimensions:

$$\begin{aligned} u_t &= \Delta u - \nabla\cdot(uK_u(u,v)),\\ v_t &= \Delta v - \nabla\cdot(vK_v(u,v)), \end{aligned} \tag{3}$$

with

$$\begin{aligned} K_u(u,v) = \int_0^1\int_0^{2\pi} & r\eta(\theta)[S_u g_{uu}(u(x+r\eta(\theta)),v(x+r\eta(\theta)))\Omega_{uu}(r)\\ &+ Cg_{uv}(u(x+r\eta),v(x+r\eta(\theta)))\Omega_{uv}(r)]\,d\theta\,dr,\\ K_v(u,v) = \int_0^1\int_0^{2\pi} & r\eta(\theta)[S_v g_{vv}(u(x+r\eta(\theta)),v(x+r\eta(\theta)))\Omega_{vv}(r)\\ &+ Cg_{vu}(u(x+r\eta),v(x+r\eta(\theta)))\Omega_{vu}(r)]\,d\theta\,dr \end{aligned}$$

and the outer unit normal η. When observing two cell populations, a distinction is made between homogeneous and heterogeneous cell adhesion. The parameters S_u, S_v and C represent the self-adhesive strength of u, the self-adhesive strength of v and the cross-adhesive strength between u and v, respectively. Armstrong et al. find suitable parameter combinations for the system (3) to model all four different cell aggregations that Steinberg proposed.

Optimal Control Problem with Non-local Operators

We consider a control problem for the one-dimensional two population model. The objective of the control problem is to determine the optimal adhesion parameters to model an observed cell aggregation. Instead of constant adhesion parameters, we consider them to be time-dependent.

Armstrong et al. simulate the model on an interval with periodic boundary conditions. Hence, we choose $\Omega = (a,b)$ with $a,b \in \mathbb{R}$, $a < b$. We choose a least square function, which results in the following control problem:

$$\frac{1}{2}\int_{\Omega} \left(u(T,x) - u_{obs}(x)\right)^2 + \left(v(T,x) - v_{obs}(x)\right)^2 dx. \tag{4a}$$

The functions $u(\cdot,\cdot)$ and $v(\cdot,\cdot)$ solve the initial value problem

$$\begin{aligned} u_t &= u_{xx} - (uK_u(u,v))_x, \quad u(0,x) = u_0(x), \quad u(t,a) = u(t,b),\\ v_t &= v_{xx} - (vK_v(u,v))_x, \quad v(0,x) = v_0(x), \quad v(t,a) = v(t,b), \end{aligned} \tag{4b}$$

with periodic boundary conditions and integral operators

$$K_u(u,v)(t,x) = \int_{-1}^{1} S_u(t)u(t,x+y)\omega(y) + C(t)v(t,x+y)\omega(y)\,dy,$$

$$K_v(u,v)(t,x) = \int_{-1}^{1} S_v(t)v(t,x+y)\omega(y) + C(t)u(t,x+y)\omega(y)\,dy,$$

where $\omega \in L^1([-1,1])$ is a given desired odd function and u_0, $v_0 \in H^1(\Omega)$ are initial values.

The functions u_{obs} and v_{obs} are cell aggregations that have been observed at time T. The goal is to choose the parameter functions S_u, S_v and C in such a way that the solutions u and v of the integro-differential equation system (4b) are closest to the observed cell aggregations at time T.

In order to be able to compute the optimal parameters, we derive and formulate necessary optimality condition.

The outline of the paper is as follows. In the next section, we will formulate a control problem in a Banach space subject to a semilinear evolution equation and will introduce the concept of a mild solution of such an equation. We derive an approach to formulate necessary first order optimality conditions using the adjoint equation in Banach spaces. Section 3 provides our main result, the formulation of the necessary optimality conditions for the two population adhesion model in one dimension. We conclude with a summary of the results.

2 Necessary Optimality Conditions in Banach Spaces

In the following, let X and L be real Banach spaces and let a time interval $[0,T]$ be given. X is the range space of the abstract function u, L is the space of the control function λ and $\Lambda \subset L$ is a nonempty, closed and convex set of admissible control functions. We require the control functions in $L = (C(0,T))^d$ to be continuous in order to apply the existence and uniqueness theorems from literature.

In this section, we consider a control problem for a semilinear evolution equation

$$u_t + Au = F(u,\lambda), \quad t \in (0,T), \quad u(0) = u_0 \tag{5}$$

with an abstract function u with $u(t) \in X$ and control function $\lambda \in \Lambda$.

Let the operator $A\colon D(A) \subset X \to X$ be the generator of an analytic semigroup. For $\alpha \in [0,1)$ we define $X^\alpha = D(A^\alpha)$. The space $(X^\alpha, \|\cdot\|_\alpha)$, with $\|u\|_\alpha := \|A^\alpha u\|$, is a Banach space. Let

$$F\colon X^\alpha \times \mathbb{R}^d \to X \tag{6}$$

be a semilinear mapping depending on the values of the state $u(t)$ and the control $\lambda(t)$. We have $D(A) = X^1 \subset X^\alpha = D(A^\alpha)$, hence, F is of lower order than A, and Equation (5) is indeed semilinear.

Furthermore, let there be a $t_0 > T > 0$ such that $t \mapsto \|F(u(t),\lambda(t))\|$ is integrable on $(0,t_0)$ for a continuous function $u\colon [0,T] \to X^\alpha$ and $\lambda \in L$. In conclusion, the inequality $\int_0^{t_0} \|F(u(s),\lambda(s))\|\,ds < \infty$ holds.

Finally, let an initial value $u_0 \in X^\alpha$ be given.

Let $Z := C([0,T],X^\alpha)$ be the space of continuous functions $u\colon [0,T] \to X^\alpha$. The space Z equipped with the corresponding uniform norm, $\|u\|_Z = \sup_{t\in[0,T]} \|u(t)\|_{X^\alpha}$, is a Banach space.

A local solution $u \in Z \cap C^1((0,T),X)$ of (5) with $u(t) \in D(A)$ on $(0,T)$ is called a strong solution. We introduce a weaker concept, the mild solution.

Definition 1. A function $u \in Z = C((0,T),X^\alpha)$ is called a mild solution of (5) on $(0,T)$ if it solves the integral equation

$$u(t) = e^{-tA}u_0 + \int_0^t e^{-(t-s)A} F(u(s),\lambda(s))\,ds \tag{7}$$

for all $t \in (0,T)$.

Remark 1. If A does not meet the condition $\operatorname{Re}\sigma(A) > 0$, let $X^\alpha = D(A_1^\alpha)$, with $A_1 := A + aI$ and $a > 0$ being the smallest value fulfilling $\operatorname{Re}\sigma(A_1) > 0$. The norms $\|\cdot\|_\alpha$ are equivalent for different values of a.

For a broader introduction to semigroup theory, we refer to Engel and Nagel [5, 6], and Henry [8]. We refer to the latter for existence results for semilinear evolution equations in the subsequent discussion.

We obtain the control problem

$$\begin{aligned} \min_{\lambda \in \Lambda} J(u,\lambda) &= \int_0^T g(t,u(t),\lambda(t))\,dt + h(u(T)) \\ \text{s.t. } G(u,\lambda) &= 0, \quad (u,\lambda) \in Z \times \Lambda, \end{aligned} \tag{8}$$

with a continuous and convex objective functional $J\colon Z \times L \to \mathbb{R}$ with $h\colon X^\alpha \to \mathbb{R}$ and $g\colon [0,T] \times X^\alpha \times \mathbb{R}^d \to \mathbb{R}$. The constraint $G\colon Z \times L \to Z$,

$$G(u,\lambda)(t) = u(t) - e^{-tA}u_0 - \int_0^t e^{-(t-s)A} F(u(s),\lambda(s))\,ds = 0,$$

represents the integral equation (7). In brevity, the spaces were chosen as follows:

$$Z = C([0,T],X^\alpha), \quad \Lambda \subset L = (C(0,T))^d.$$

To formulate the necessary optimality conditions, we assume the following:

Assumption 1. (i) J is Fréchet-differentiable on $Z \times L$.
(ii) F is continuously Fréchet-differentiable on $X^\alpha \times \mathbb{R}^d$.
(iii) F is locally Lipschitz-continuous with respect to u.
(iv) There exists a $K \in C[0,\infty)$ such that for every $t > 0$

$$\|F(u,\lambda)\|_X \leq K(t)(1 + \|u\|_{X^\alpha})$$

Remark 2. The local existence and uniqueness of a mild solution of (5) follows from the continuity of $\lambda \in L$ and Assumption 1 (ii) and (iii): for fixed $\lambda \in L$ define $\tilde{F}_\lambda\colon (0,T) \times X^\alpha \to X$, $\tilde{F}_\lambda(t,u) = F(u,\lambda(t))$ and consider the evolution equation $u_t + Au = \tilde{F}_\lambda(t,u)$ (cf. Henry [8, Section 3.3]). Assumption 1 (iv) yields the global existence (cf. Henry [8, Corollary 3.3.5])

Assumption 1 (ii) yields the continuous Fréchet-differentiability of G on $Z \times L$ with derivative $dG(u,\lambda)\colon Z \times L \to Z$, for a detailed proof see Zimmer [13]. Furthermore, we obtain for the partial derivative with respect to u the following result.

Theorem 1. *Let Assumption 1* (ii) *be true. Then the partial derivative*

$$d_uG(u,\lambda)\colon Z \to Z, \quad d_uG(u,\lambda)(\Delta u) = \Delta u - \int_0^{\cdot} e^{-(\cdot - s)A} d_uF(u(s),\lambda(s))\Delta u(s)\,ds,$$

is an isomorphism.

Proof. Consider for given right-hand side $v \in Z$ the inhomogeneous equation

$$\Delta u = \int_0^{\cdot} e^{-(\cdot - s)A} d_uF(u(s),\lambda(s))\Delta u(s)\,ds + v. \tag{9}$$

If $v = 0$, then the uniqueness of solutions yields that $\Delta u = 0$ is the only solution. For $0 \neq v \in Z$ we look at

$$y = \int_0^{\cdot} e^{-(\cdot - s)A} d_u F(u(s), \lambda(s))(y(s) + v(s))\, ds. \tag{10}$$

Define the mapping $\tilde{F}\colon (0,T) \times X^{\alpha} \to X,\ \tilde{F}(t,y) = d_u F(u,\lambda)(y + v(t))$, which is continuous and linear, hence, Lipschitz-continuous in y. Thus, a unique mild solution $y \in Z$ exists and replacing $\Delta u := y + v$ in (10) shows that Δu satisfies Equation (9). Therefore, $d_u G(u,\lambda)$ is an isomorphism. □

With a similar argument as in the proof of Theorem 1 we can show that $H : Z \to Z$,

$$H(\phi)(t) = \int_0^t e^{-(t-s)A} \phi(s)\, ds,$$

is an isomorphism. Furthermore,

$$H^*(\phi)(s) = \int_s^T e^{-(t-s)A^*} \phi(t)\, dt$$

which can be proofed using Fubini's theorem.

We derive the necessary optimality conditions based on a result by Zowe and Kurcyusz [14]:

Theorem 2. *Let Assumptions 1 be true for an optimal state and optimal control* $(\bar{u},\bar{\lambda}) \in Z \times \Lambda$*, then*

$$\big(d_\lambda J(\bar{u},\bar{\lambda}) + d_\lambda G(\bar{u},\bar{\lambda})^*(l)\big)(\lambda - \bar{\lambda}) \geq 0, \quad \lambda \in \Lambda, \tag{11}$$

where $l \in Z^*$ *is the unique solution of the adjoint equation*

$$d_u G(\bar{u},\bar{\lambda})^* l = -d_u J(\bar{u},\bar{\lambda}). \tag{12}$$

Alternatively, $l \in Z^*$ *is uniquely defined by*

$$l(v) = -d_u J(\bar{u},\bar{\lambda})(d_u G(\bar{u},\bar{\lambda})^{-1} v) \quad \forall v \in Z. \tag{13}$$

Proof. Zowe and Kurcyusz consider in [14] the following optimization problem

$$\min f(x), \quad x \in C, \quad g(x) \in K$$

where $f : X \to \mathbb{R}$, $g : X \to Y$ with a closed convex set $C \subset X$ and a cone $K \subset Y$. In our case this reads as

$$\min J(u,\lambda), \quad \lambda \in \Lambda, \quad G(u,\lambda) = 0,$$

with $J: Z \times L \to R,\ \ G: Z \times L \to Z$, $\Lambda \subset L$ is a closed convex set and

$$X = Z \times L,\ \ Y = Z,\ \ C = Z \times \Lambda,\ \ K = \{0_Z\}.$$

A point $\bar{x}$ is a regular point in the sense of Zowe and Kurcyusz, if

$$g'(\bar{x})C(\bar{x}) - K(g(\bar{x})) = Y$$

where

$$C(\bar{x}) = \{\lambda(c - \bar{x})\colon c \in C, \lambda \geq 0\}, \quad K(y) = \{k - \lambda y\colon k \in K, \lambda \geq 0\}.$$

In our case we have

$$C(\bar{u}, \bar{\lambda}) = Z \times \Lambda(\bar{\lambda}), \quad K(G(\bar{u}, \bar{\lambda})) = 0_Z.$$

This means that we have a regular point $(\bar{u}, \bar{\lambda})$, if

$$d_u G(\bar{u}, \bar{\lambda})Z + d_\lambda G(\bar{u}, \bar{\lambda})\Lambda(\bar{\lambda}) = Z.$$

In other words, for each $v \in Z$ we have to find a $u \in Z$ and $\mu \in \Lambda(\bar{\lambda})$ such that

$$d_u G(\bar{u}, \bar{\lambda})u = v - d_\lambda G(\bar{u}, \bar{\lambda})\mu. \tag{14}$$

Choosing, for example, $\mu = 0$, this holds due to Theorem 1 in our paper.

By [14, Theorem 4.1] there exists a Lagrange multiplier $y^* \in Y^*$, i.e. by definition [14, (1.1)] we have

(i) $y^* \in K^+ = \{y^* \in Y^*\colon \langle y^*, k\rangle \geq 0\ \ \forall k \in K\}$,
(ii) $\langle y^*, g(\bar{x})\rangle = 0$,
(iii) $f'(\bar{x}) - g'(\bar{x})^* y^* \in C(\bar{x})^+$.

In our case this leads to an $\tilde{l} \in K^+ = Z^*$ where (ii) holds trivially since $g(\bar{x}) = 0$. For the third condition note that

$$C(\bar{u}, \bar{\lambda})^+ = Z^+ \times \Lambda(\bar{\lambda})^+ = \{0_Z\} \times \{\lambda^* \in L^*\colon \langle \lambda^*, \lambda - \bar{\lambda}\rangle \geq 0\ \ \forall \lambda \in \Lambda\}.$$

Hence,

$$d_u J(\bar{u}, \bar{\lambda}) - d_u G(\bar{u}, \bar{\lambda})^* \tilde{l} = 0$$

and

$$\langle d_\lambda J(\bar{u}, \bar{\lambda}) - d_\lambda G(\bar{u}, \bar{\lambda})^* \tilde{l}, \lambda - \bar{\lambda}\rangle \geq 0\ \ \forall \lambda \in \Lambda.$$

Setting $l = -\tilde{l} \in K^- = Z^*$ concludes the proof of Equations (11) and (12).

With u and v from Equation (14) (with $\mu = 0$) and the invertibility of $d_uG(\bar{u},\bar{\lambda})$ we obtain

$$-d_uJ(\bar{u},\bar{\lambda})(d_uG(\bar{u},\bar{\lambda})^{-1}v) = -d_uJ(\bar{u},\bar{\lambda})u = (d_uG(\bar{u},\bar{\lambda})^*l)u$$
$$= l(d_uG(\bar{u},\bar{\lambda})u) = l(v).$$

□

Remark 3. Zowe and Kurzyucz provide in [14, Theorem 4.1] the existence of a nonempty bounded set of Lagrange multipliers for a regular point $\bar{x}$. To apply their result to our control problem (8), the surjectivity of $d_uG(\bar{u},\bar{\lambda})$ would be sufficient. However, the injectivity implies the uniqueness of the Lagrange multiplier $l \in Z^*$.

The next result is based on a similar result of Hu and Peng [9, Theorem 3.3], for semilinear stochastic evolution equations.

Theorem 3. *The linear functional $l \in Z^*$ in the setting outlined above is given by*

$$l(v) = \int_0^T \langle \rho(t), v(t) \rangle \, dt \tag{15}$$

where $H^{-*}(p)(t) = \rho(t)$ (i.e. $p(t) = H^*(\rho)(t) = \int_t^T e^{-(s-t)A^*}\rho(s)\,ds$) and $p \in Z$,

$$p(s) = e^{-(T-s)A^*}(-d_uh(\bar{u}(T))) + \int_s^T e^{-(t-s)A^*} d_uF(\bar{u}(t),\bar{\lambda}(t))^* p(t)\,dt, \tag{16}$$

is the mild solution of the final value problem

$$-p_t + A^*p = d_uF(\bar{u},\bar{\lambda})^*p, \quad t \in (0,T), \quad p(T) = -d_uh(\bar{u}(T)).$$

Proof. From (13) follows

$$l(v) = -d_uJ(\bar{u},\bar{\lambda})(\Delta u) = -\langle d_uh(\bar{u}(T)), \Delta u(T) \rangle, \tag{17}$$

where $\Delta u \in Z$ solves the equation

$$\Delta u(t) = \int_0^t e^{-(t-s)A}\big(d_uF(\bar{u}(s),\bar{\lambda}(s))\Delta u(s)\big)\,ds + v(t).$$

For brevity we set $\bar{F}(s) = d_uF(\bar{u}(s),\bar{\lambda}(s))$. Setting

$$v(t) = -H(\bar{F}(\cdot)\Delta u(\cdot))(t) + \hat{v}(t)$$

with H discussed after Theorem 2 we obtain

$$\Delta u(t) = \int_0^t e^{-(t-s)A}\hat{v}(s)\,ds. \tag{18}$$

Equations (17) and (18) lead to

$$\begin{aligned}
l(\hat{v}) &= l(v) + l(H(\bar{F}(\cdot)\Delta u(\cdot))) \\
&= -\langle d_u h(u(T)), \Delta u(T)\rangle + \int_0^T \langle \rho(t), H(\bar{F}(\cdot)\Delta u(\cdot))(t)\rangle\,dt \\
&= -\langle d_u h(u(T)), \Delta u(T)\rangle + \int_0^T \langle H^{-*}(p)(t), H(\bar{F}(\cdot)\Delta u(\cdot))(t)\rangle\,dt \\
&= -\langle d_u h(u(T)), \Delta u(T)\rangle + \int_0^T \langle p(t), \bar{F}(t)\Delta u(t)\rangle\,dt \\
&= -\int_0^T \langle d_u h(\bar{u}(T)), e^{-(T-s)A}\hat{v}(s)\rangle\,ds + \int_0^T \langle \bar{F}(t)^* p(t), \Delta u(t)\rangle\,dt \\
&= -\int_0^T \langle d_u h(\bar{u}(T)), e^{-(T-s)A}\hat{v}(s)\rangle\,ds + \int_0^T \langle \bar{F}(t)^* p(t), \int_0^t e^{-(t-s)A}\hat{v}(s)\,ds\rangle\,dt
\end{aligned}$$

and applying Fubini's theorem to the second term yields

$$\begin{aligned}
l(\hat{v}) &= -\int_0^T \langle e^{-(T-s)A^*} d_u h(\bar{u}(T)), \hat{v}(s)\rangle\,ds \\
&\quad + \int_0^T \langle \int_s^T e^{-(t-s)A^*}\bar{F}(t)^* p(t)\,dt, \hat{v}(s)\rangle\,ds \\
&= \int_0^T \langle e^{-(T-s)A^*}(-d_u h(\bar{u}(T))) + \int_s^T e^{-(t-s)A^*}\bar{F}(t)^* p(t)\,dt, \hat{v}(s)\rangle\,ds.
\end{aligned}$$

By comparing the above to (15) we obtain (16) which concludes the proof. □

3 Necessary Optimality Conditions for the Non-local Adhesion Model

Given the results of the previous section, our aim in this section is to calculate the necessary optimality conditions for the non-local adhesion model for two populations in one space dimension (4).

Let Ω be a real interval (a,b). We set

$$X := (L^2(\Omega))^2, \quad L := (C(0,T))^3, \quad \text{and convex and closed } \Lambda \subset L.$$

Set $\alpha = 1/2$ and let the operator $A \colon D(A) \subset (L^2(\Omega))^2 \to (L^2(\Omega))^2$ be defined as the self-adjoint extension of $-D_{xx}$ with domain of definition $D(A) = (H^2(\Omega))^2 \subset (L^2(\Omega))^2$. Then, A is the generator of an analytic semigroup and $X^{1/2} = D(A^{1/2}) = (H^1(\Omega))^2$. Therefore, we choose

$$Z = C([0,T],(H^1(\Omega))^2).$$

Let $\nu := (u,v) \in Z$ be a vector of the functions u and v and let $\lambda := (S_u, S_v, C) \in L$ be a vector of time-dependent adhesion parameters.

The operator $\mathscr{K} \colon (L^2(\Omega))^2 \times \mathbb{R}^3 \to (L^2(\Omega))^{2\times 2}$ is defined as

$$\mathscr{K}(\nu,\lambda) = \begin{pmatrix} K(u,v,S_u,C) & 0 \\ 0 & K(v,u,S_v,C) \end{pmatrix},$$

with $K \colon (L^2(\Omega))^2 \times \mathbb{R}^2 \to L^2(\Omega)$,

$$K(u,v,S,C)(x) = \int_{-1}^{1} Su(x+y)\omega(y) + Cv(x+y)\omega(y)\,dy.$$

Finally, we define the semilinear mapping $F \colon (H^1(\Omega))^2 \times \mathbb{R}^3 \to (L^2(\Omega))^2$ as set in (6) as

$$F(\nu,\lambda) := -D_x\big(\mathscr{K}(\nu,\lambda)\nu\big).$$

Lemma 1. *The mappings*

$$K \colon (L^2(\Omega))^2 \times \mathbb{R}^3 \to L^2(\Omega) \text{ and } K \colon (H^1(\Omega))^2 \times \mathbb{R}^3 \to H^1(\Omega)$$

are well defined and linear. The mappings

$$\mathscr{K} \colon (L^2(\Omega))^2 \times \mathbb{R}^3 \to (L^2(\Omega))^{2\times 2},\ \mathscr{K} \colon (H^1(\Omega))^2 \times \mathbb{R}^3 \to (H^1(\Omega))^{2\times 2}$$

and

$$F \colon (H^1(\Omega))^2 \times \mathbb{R}^3 \to (L^2(\Omega))^2$$

are well defined. F is locally Lipschitz-continuous with respect to ν and continuously Fréchet-differentiable on $(H^1(\Omega))^2 \times \mathbb{R}^3$ with

$$dF(\nu,\lambda)(\Delta\nu,\Delta\lambda) = -D_x\big(\mathscr{K}(\nu,\lambda)\Delta\nu\big) - D_x\big(\mathscr{K}(\Delta\nu,\lambda)\nu\big) - D_x\big(\mathscr{K}(\nu,\Delta\lambda)\nu\big).$$

We omit a detailed proof but add several comments.

Remark 4. (i) The continuity of the control functions comes here into play, because it ensures $\int_0^{t_0} \|F(u(s),\lambda(s))\| \, ds < \infty$, and with the results of Henry [8], we derive the local existence of a unique solution of (4b).
(ii) The periodic boundary conditions of (4) have also to be included in the definition of the spaces. Consider for $k \in \mathbb{N}$ $(H^k_\Omega)^2$ the space of $|\Omega|$-periodic functions whose restrictions to Ω are in $(H^k(\Omega))^2$. Without loss of generality, let $\Omega = [-\pi, \pi]$. For $s \geq 0$ the Sobolev-space of 2π-periodic functions is defined as

$$(\hat{H}^s_{2\pi})^2 := \{v \in (L^2_{2\pi})^2 : \|v\|_{(\hat{H}^s_{2\pi})^2} := \sum_{k=-\infty}^{\infty} (1+|k|^2)^s |\hat{v}(k)|^2 < \infty\},$$

where $\hat{v}$ is the Fourier transform of v. Since the Fourier transform is an isometry, $(\hat{H}^s_{2\pi})^2$ is a Hilbert-space for $s \geq 0$ and $(\hat{H}^k_{2\pi})^2 = (H^k_{2\pi})^2$ for $k \in \mathbb{N}$. We obtain that $(H^1(\Omega))^2$ together with the boundary conditions is equal to $(\hat{H}^1_{2\pi})^2$ and $(H^1_\Omega)^2$.

We can now formulate (4) in the framework of (8)

$$\begin{aligned}
&\min_{\lambda \in \Lambda} J(v,\lambda) = \frac{1}{2} \|v(T;\lambda) - v_{obs}\|^2_{(L^2(\Omega))^2} \\
&\text{s.t. } G(v,\lambda) = v - e^{-\cdot A} v_0 - \int_0^{\cdot} e^{-(\cdot - s)A} F(v(s),\lambda(s)) \, ds = 0 \\
&\qquad (v,\lambda) \in C([0,T],(H^1(\Omega))^2) \times \Lambda.
\end{aligned} \tag{19}$$

It is easy to see that Assumption 1 (i) is fulfilled for a tracking type objective function with

$$dJ(v,\lambda)(\Delta v,\Delta\lambda) = \int_\Omega \big(v(T,x;\lambda) - v_{obs}(x)\big)^{\mathsf{T}} \Delta v(T,x) \, dx. \tag{20}$$

Assumptions 1 (ii) and (iii) are met by Lemma 1 and we obtain the partial derivatives of G,

$$d_v G(v,\lambda)(\Delta v) = \Delta v - \int_0^{\cdot} e^{-(\cdot - s)A} [D_x\big(\mathscr{K}(v,\lambda)\Delta v\big) + D_x\big(\mathscr{K}(\Delta v,\lambda)v\big)] \, ds$$

and

$$d_\lambda G(v,\lambda)\Delta\lambda = -\int_0^{\cdot} e^{-(\cdot - s)A} D_x\big(\mathscr{K}(v,\Delta\lambda)v\big) \, ds.$$

We need to verify Assumption 1 (iv). For the norm of F follows componentwise

$$\begin{aligned}
\|\big(F(v,\lambda)\big)_1\|_{L^2(\Omega)} &\leq \|K(u,v,\lambda)D_x u\|_{L^2(\Omega)} + \|K(D_x u, D_x v,\lambda)u\|_{L^2(\Omega)} \\
&\leq c_1 \|D_x(u)\|_{L^2(\Omega)} + c_2 \|u\|_{L^2(\Omega)}
\end{aligned}$$

$$\leq c_3\big(\|D_x(u)\|_{L^2(\Omega)} + \|u\|_{L^2(\Omega)}\big)$$
$$= c_3\|u\|_{H^1(\Omega)}.$$

Analogously for the second component $\|\big(F(v,\lambda)\big)_2\|_{L^2(\Omega)} \leq c_4\|v\|_{H^1(\Omega)}$. Finally, $\|F(v,\lambda)\|_{(L^2(\Omega))^2}$ satisfies

$$\|F(v,\lambda)\|_{(L^2(\Omega))^2} \leq c_5\|v\|_{(H^1(\Omega))^2} \leq c_5(1+\|v\|_{(H^1(\Omega))^2}).$$

Theorem 2 provides a Lagrange multiplier l and we can formulate the necessary optimality conditions of the control problem (19). First, we find a representation of the adjoint $l \in Z^* = \big(C([0,T],(H^1(\Omega))^2)\big)^*$.

Lemma 2. *The linear functional $l \in \big(C([0,T],(H^1(\Omega))^2)\big)^*$ in the setting outlined above is given by*

$$l(\phi) = \int_0^T \int_\Omega \pi(t,x)^\top \phi(t,x)\,dx\,dt \tag{21}$$

where $\pi = (p,q) \in L^2([0,T],(H^1(\Omega))^2)$ is the solution of the final value problem

$$\begin{aligned} -\pi_t &= D_{xx}\pi + \mathscr{K}(v,\lambda)D_x\pi - \mathscr{K}(D_x\pi\cdot v,\lambda)\mathrm{id} \\ \pi(T,x) &= v_{obs}(x) - v(T,x), \quad \pi(t,a) = \pi(t,b) \end{aligned} \tag{22}$$

for all $(t,x) \in (0,T)\times\Omega$. Here id *is the identity and $a\cdot b$ is the Hadamard product or entrywise product of vectors a and b. The problem* (22) *is equivalent to*

$$\begin{aligned} -p_t &= p_{xx} + p_xK_u(u,v) - K_u(p_xu,q_xv), \\ -q_t &= q_{xx} + q_xK_v(u,v) - K_v(p_xu,q_xv), \\ p(T,x) &= u_{obs}(x) - u(T,x), \quad p(t,a) = p(t,b), \\ q(T,x) &= v_{obs}(x) - v(T,x), \quad q(t,a) = q(t,b). \end{aligned}$$

Proof. Since $A = -D_{xx}$ is a self-adjoint operator, Theorem 3 yields

$$\begin{aligned} -\pi_t &= D_{xx}\pi + d_uF(\bar{\boldsymbol{u}},\bar{\lambda})^*\pi, \quad t\in(0,T) \\ \pi(T) &= v_{obs} - v(T,x), \quad \pi(t,a) = \pi(t,b). \end{aligned}$$

In the following we will derive a representation of $d_uF(\bar{\boldsymbol{u}},\bar{\lambda})^*\pi$. We consider

$$\langle \pi, d_uF(\bar{\boldsymbol{u}},\bar{\lambda})\Delta v\rangle = \int_\Omega -\pi(x)^\top\Big(D_x\big(\mathscr{K}(v,\lambda)\Delta v\big) + D_x\big(\mathscr{K}(\Delta v,\lambda)v\big)\Big)\,dx.$$

In the following, we isolate Δv in every summand. We present the procedure for the first component only, the same approach can be applied to the second component analogously.

$$\langle p,(d_uF(\bar{\boldsymbol{u}},\bar{\lambda})\Delta v)_1\rangle$$
$$= \int_\Omega -p(x)\Big(D_x\big(\Delta u(x)K_u(u,v)(x)\big)+D_x\big(u(x)K_u(\Delta u,\Delta v)(x)\big)\Big)\,dx.$$

Integration by parts and the periodic boundary condition of p and Δu yield

$$\int_\Omega -p(x)D_x\big(\Delta u(x)K_u(u,v)(x)\big)\,dx = \int_\Omega D_xp(x)K_u(u,v)(x)\Delta u(x)\,dx.$$

With

$$\int_\Omega f(x)g(x+y)\,dx = \int_\Omega f(z-y)g(z)\,dz = \int_\Omega f(x-y)g(x),$$

for $f,\, g\in L^2(\Omega)$ and with the anti-symmetry of ω we derive

$$\int_{-1}^{1}\int_\Omega D_xp(x)S_uu(x)\Delta u(x+y)\omega(y)\,dx\,dy$$
$$=\int_{-1}^{1}\int_\Omega D_xp(x-y)S_uu(x-y)\Delta u(x)\omega(y)\,dx\,dy$$
$$=\int_\Omega -\Delta u(x)\int_{-1}^{1} S_uD_xp(x+y)u(x+y)\omega(y)\,dy\,dx.$$

Therefore,

$$\int_\Omega D_xp(x)u(x)K_u(\Delta u,\Delta v)(x)\,dx$$
$$=\int_\Omega -\Delta u(x)\int_{-1}^{1} S_uD_xp(x+y)u(x+y)\omega(y)\,dy$$
$$-\Delta v(x)\int_{-1}^{1} CD_xp(x+y)u(x+y)\omega(y)\,dy\,dx.$$

Eventually, with Fubini's lemma, integration by parts, the equation above and the periodicity of u, v and Δu we obtain

$$\int_\Omega -p(x)D_x\big(u(x)K_u(\Delta u,\Delta v)(x)\big)\,dx\,dt = \int_\Omega D_xp(x)u(x)K_u(\Delta u,\Delta v)(x)\,dx$$

$$= \int_\Omega \Delta u(x) \int_{-1}^{1} -S_u D_x p(x+y) u(x+y) \omega(y)\, dy$$

$$+ \Delta v(x) \int_{-1}^{1} -C D_x p(x+y) u(x+y) \omega(y)\, dy\, dx.$$

Overall, we have

$$\begin{aligned} \langle p, (d_u F(\bar{\boldsymbol{u}}, \bar{\lambda}) \Delta \nu)_1 \rangle = \int_\Omega \Big(& \Delta u(x) \big(D_x p(x) K_u(u,v)(x) \\ & - \int_{-1}^{1} S_u D_x p(x+y) u(x+y) \omega(y)\, dy \big) \\ & - \Delta v(x) \int_{-1}^{1} C D_x p(x+y) u(x+y) \omega(y)\, dy \Big)\, dx, \end{aligned}$$

$$\begin{aligned} \langle q, (d_u F(\bar{\boldsymbol{u}}, \bar{\lambda}) \Delta \nu)_2 \rangle = \int_\Omega \Big(& \Delta v(x) \big(D_x q(x) K_v(u,v)(x) \\ & - \int_{-1}^{1} S_v D_x q(x+y) v(x+y) \omega(y)\, dy \Big) \\ & - \Delta u(x) \int_{-1}^{1} C D_x q(x+y) v(x+y) \omega(y)\, dy \Big)\, dx \end{aligned}$$

and altogether

$$\begin{aligned} \langle \pi, d_u F(\bar{\boldsymbol{u}}, \bar{\lambda}) \Delta \nu \rangle = \int_\Omega & \Delta \nu(x)^\mathsf{T} \big(\mathscr{K}(\nu, \lambda)(x) D_x \pi(x) \\ & - \mathscr{K}(D_x \pi \cdot \nu, \lambda)(x) \mathrm{id}(x) \big)\, dx \ = \langle \Delta \nu, d_u F(\bar{\boldsymbol{u}}, \bar{\lambda})^* \pi \rangle. \end{aligned}$$

Hence, $d_u F(\bar{\boldsymbol{u}}, \bar{\lambda})^* \pi = \mathscr{K}(\nu, \lambda) D_x \pi - \mathscr{K}(D_x \pi \cdot \nu, \lambda)$. □

Since in our case $d_\lambda J(\nu, \lambda)(\Delta \lambda) = 0$, the inequality (11) reduces to

$$0 \le d_\lambda G(\nu, \lambda)^*(l)(\Delta \lambda) = l\big(d_\lambda G(\nu, \lambda)(\Delta \lambda)\big),$$

with $\Delta \lambda = \lambda - \bar{\lambda}$. Substituting $\phi = d_\lambda G(\nu, \lambda)(\Delta \lambda)$ with its actual representation leads to

$$l\big(D_x(\mathscr{K}(\nu, \Delta\lambda)\nu)\big) = \int_0^T \int_\Omega \pi(t,x)^\mathsf{T} \Big(D_x\big(\mathscr{K}(\nu, \Delta\lambda)(t,x) \nu(t,x)\big) \Big)\, dx\, dt.$$

The Riesz representation theorem yields componentwise

$$
\begin{aligned}
(l(\phi))_1 &= \int_0^T \int_\Omega p(t,x)\Big(D_x\big(u(t,x)K(u,v,\Delta S_u,\Delta S_v,\Delta C)(t,x)\big)\Big)\,dx\,dt \\
&= \int_0^T \Big(\int_\Omega \int_{-1}^1 p(t,x)D_x\big(u(t,x)u(t,x+y)\big)\,\omega(y)\,dy\,dx\Delta S_u(t) \\
&\quad + \int_\Omega \int_{-1}^1 p(t,x)D_x\big(u(t,x)v(t,x+y)\big)\,\omega(y)\,dy\,dx\Delta C(t)\Big)\,dt \\
&= \int_0^T \Big(\int_\Omega p(t,x)\hat{\mathscr{K}}_{\boldsymbol{u}}(\nu)(t,x)\,dx\Big)^{\mathsf{T}}\Delta\lambda(t)\,dt \\
&= \langle \int_\Omega p(x)\hat{\mathscr{K}}_{\boldsymbol{u}}(\nu)(x)\,dx, \Delta\lambda\rangle_{(C(0,T))^3},
\end{aligned}
$$

with $\hat{\mathscr{K}}_{\boldsymbol{u}}(\nu)(x) \in (C(0,T))^3$ defined as

$$
\hat{\mathscr{K}}_{\boldsymbol{u}}(\nu)(t,x) = \begin{pmatrix} D_x\big(u(t,x)K(u,v,\boldsymbol{e}_1)(t,x)\big) \\ 0 \\ D_x\big(u(t,x)K(u,v,\boldsymbol{e}_3)(t,x)\big) \end{pmatrix}
$$

where e_i is the ith unit vector in $(C(0,T))^3$. Analogously, for the second component holds

$$
(l(\phi))_2 = \langle \int_\Omega q(x)\hat{\mathscr{K}}_v(\nu)(x)\,dx, \Delta\lambda\rangle_{(C(0,T))^3}
$$

with

$$
\hat{\mathscr{K}}_v(\nu)(t,x) = \begin{pmatrix} 0 \\ D_x\big(v(t,x)K(v,u,\boldsymbol{e}_2)(t,x)\big) \\ D_x\big(v(t,x)K(v,u,\boldsymbol{e}_3)(t,x)\big) \end{pmatrix}.
$$

The adjoint $l(\phi)$ can be represented with $\hat{\mathscr{K}}(\nu)(x) \in (L^2(0,T)^{3\times 2})$,

$$
\hat{\mathscr{K}}(\nu)(t,x) = \big(\,\hat{\mathscr{K}}_{\boldsymbol{u}}(\nu)(t,x)\;\;\hat{\mathscr{K}}_v(\nu)(t,x)\,\big)\,,
$$

as

$$
\begin{aligned}
l(d_\lambda G(\nu,\lambda)(\Delta\lambda)) &= \int_0^T \Big(\int_\Omega \hat{\mathscr{K}}(\nu)(t,x)\pi(t,x)\,dx\Big)^{\mathsf{T}}\Delta\lambda(t)\,dt \\
&= \langle \int_\Omega \hat{\mathscr{K}}(\nu)(x)\pi(x)\,dx, \Delta\lambda\rangle_{(C(0,T))^3}.
\end{aligned}
$$

As a result, we obtain the necessary optimality conditions for the control problem (19):

Corollary 1. *Given optimal controls $\bar{\lambda} \in \Lambda$ and the corresponding optimal state $\bar{v} \in L^2([0,T],(H^1(\Omega))^2)$, the equation*

$$\langle \int_\Omega \hat{\mathscr{K}}(\bar{v})(x)\pi(x)\,dx, \lambda - \bar{\lambda}\rangle \geq 0 \tag{23}$$

holds for all $\lambda \in \Lambda$, where $\pi \in L^2([0,T],(H^1(\Omega))^2)$ is a solution of the final value problem

$$-\pi_t = D_{xx}\pi + \mathscr{K}(v,\lambda)D_x\pi - \mathscr{K}(D_x\pi \cdot v,\lambda)\mathrm{id}$$
$$\pi(T,x) = v_{\boldsymbol{obs}}(x) - v(T,x), \quad \pi(t,a) = \pi(t,b).$$

Armstrong et al. suggest for the numerical treatment of their model (4b) to use an explicit finite volume method over time to obtain a system of ODEs. For the diffusion term, a central differencing scheme is used, while Armstrong et al. use a high order upwinding scheme with flux limiting for the advection term. The integral is calculated directly by summing over the enclosed points and the time integration uses an explicit trapezoidal scheme [2]. The same scheme could be used to treat the adjoint final value problem numerically and is work in progress.

4 Summary

We introduced a system of nonlinear parabolic partial integro-differential equations as it appears in an application in biology. We considered an optimization problem, where the objective function is of tracking type functional and the controls are time-dependent adhesion parameter functions. In the sequel we derived necessary optimality conditions for an optimal point which include adjoint differential equations. It turned out that these are also partial integro-differential equations type which evolve backwards in time with final conditions coming from the tracking type of objective function.

References

1. Andersen, L., Andreasen, J.: Jump-diffusion processes: volatility smile fitting and numerical methods for option pricing. Rev. Deriv. Res. **4**(3), 231–262 (2000)
2. Armstrong, N.J., Painter, K.J., Sherratt, J.A.: A continuum approach to modelling cell-cell adhesion. J. Theor. Biol. **243**(1), 98–113 (2006)
3. Briani, M., Natalini, R., Russo, G.: Implicit–explicit numerical schemes for jump–diffusion processes. Calcolo **44**(1), 33–57 (2007)

4. Cont, R., Voltchkova, E.: A finite difference scheme for option pricing in jump diffusion and exponential Lévy models. SIAM J. Numer. Anal. **43**(4), 1596–1626 (2005)
5. Engel, K., Nagel, R.: One-Parameter Semigroups for Linear Evolution Equations. Springer, New York (2000)
6. Engel, K., Nagel, R.: A Short Course on Operator Semigroups. Springer, New York (2006)
7. Gerisch, A.: On the approximation and efficient evaluation of integral terms in PDE models of cell adhesion. J. Numer. Anal. **30**, 173–194 (2010)
8. Henry, D.: Geometric Theory of Semilinear Parabolic Equations. Lecture Notes in Mathematics, vol. 61. Springer, Heidelberg (1981)
9. Hu, Y., Peng, S.: Maximum principle for semilinear stochastic evolution control systems. Stoch. Stoch. Rep. **33**(3–4), 159–180 (1990)
10. Matache, A.-M., von Petersdorff, T., Schwab, C.: Fast deterministic pricing of options on Lévy driven assets. Math. Model. Numer. Anal. **38**(1), 37–72 (2004)
11. Sachs, E., Strauss, A.: Efficient solution of a partial integro-differential equation in finance. Appl. Numer. Math. **58**, 1687–1703 (2008)
12. Sachs, E., Schneider M., Schu, M.: Adaptive trust-region POD methods in PIDE-constrained optimization. In: Leugering, G., et al. (eds.) Trends in PDE Constrained Optimization. International Series of Numerical Mathematics, vol. 165, pp. 303–326. Birkhäuser, Heidelberg (2014)
13. Zimmer, L.: Notwendige Optimalitätsbedingungen von partiellen Integro-Differentialgleichungen. Diploma thesis, Universität Trier (2012)
14. Zowe, J., Kurcyusz, S.: Regularity and stability for the mathematical programming problem in Banach spaces. Appl. Math. Optim. **5**, 49–62 (1979)

The AMPL Modeling Language: An Aid to Formulating and Solving Optimization Problems

David M. Gay

Abstract Optimization problems arise in many contexts. Sometimes finding a good formulation takes considerable effort. A modeling language, such as AMPL, facilitates experimenting with formulations and simplifies using suitable solvers to solve the resulting optimization problems. AMPL lets one use notation close to familiar mathematical notation to state variables, objectives, and constraints and the sets and parameters that may be involved. AMPL does some problem transformations and makes relevant problem information available to solvers. The AMPL command language permits computing and displaying information about problem details and solutions returned by solvers. It also lets one modify problem formulations and solve sequences of problems. AMPL addresses both continuous and discrete optimization problems and offers some constraint-programming facilities for the latter. More generally, AMPL permits stating and solving problems with complementarity constraints. For continuous problems, AMPL makes first and second derivatives available via automatic differentiation. The freely available AMPL/solver interface library (ASL) facilitates interfacing with solvers. This paper gives an overview of AMPL and its interaction with solvers and discusses some problem transformations and implementation techniques. It also looks forward to possible enhancements to AMPL.

Keywords Mathematical Programming • Linear Programming • Nonlinear Programming • Automatic Differentiation

1 Introduction

Science is all about models and data—theories (models) that explain observed data and make predictions about data that may be observed later. Science makes engineering possible and has led to many developments that heavily influence modern human life. Many kinds of models are useful. Some involve mathematical

D.M. Gay (✉)
AMPL Optimization Inc., 900 Sierra Place SE, Albuquerque, NM 87108-3379, USA
e-mail: dmg@ampl.com
http://www.ampl.com

M. Al-Baali et al. (eds.), *Numerical Analysis and Optimization*, Springer Proceedings in Mathematics & Statistics 134, DOI 10.1007/978-3-319-17689-5_5

structures, such as distributions or differential equations, to which one can devote lifetimes of study. Simpler models, involving only finite numbers of variables, equations, inequalities, and objectives and described by finitely many parameters and sets, have a surprisingly wide range of uses. When one studies a new area, choices for suitable models may be far from obvious, and it may be necessary to try many models. Statistics is largely about comparing candidate models and, particularly with exploratory data analysis, finding suitable ones.

Algebraic modeling languages, such as the AMPL language considered in this paper, facilitate formulating, comparing, changing, and deriving results from a subset of the class of "simpler" models outlined above in which equations, inequalities, objectives, and derived sets and parameters are expressed algebraically. In short, AMPL is focused on *mathematical programming* problems, such as constrained optimization problems of the form

$$\text{minimize}\, f(x) \tag{1a}$$

$$\text{s.t. } \ell \le c(x) \le u, \tag{1b}$$

with $x \in \mathbb{R}^n$ and $c : \mathbb{R}^n \to \mathbb{R}^m$, possibly with some components of x restricted to integer values.

2 AMPL Design Principles

AMPL is meant to make it easy to transcribe models from mathematical notation, such as one might write by hand on paper or white board, to the AMPL language. We sought to make the language both close to elementary algebraic notation and easy to enter on an ordinary computer keyboard. As explained below, AMPL was created at Bell Labs, in the then Computing Sciences Research Center, where such languages as C [26, 27], aC++ [29], and awk [1, 2] had been created, so AMPL uses some of the same notational conventions as these languages, such as square brackets for subscripts. Models often have sets of variables and constraints, and AMPL allows one to have various kinds of subscripted entities. In model entities, such as constraints and objectives, all subscripting is explicit, in part to make meaning of these entities clear. AMPL is a declare-before-use language, so one can read a model from top to bottom without worrying about the meaning of something whose properties are given later.

An AMPL model can represent a whole class of problems. For example, a linear objective might be specified by the declarations

```
set S;
var x{S} >= 0;
param p{S};
minimize Cost: sum{i in S} p[i]*x[i];
```

in which the objective is named "`Cost`" and is a transcription of

$$\sum_{i \in S} p_i x_i.$$

Thus a model can involve sets (such as `S`) over which entities, such as parameters and variables, e.g., `p` and `x`, are indexed, and can be stated without regard to the values that its sets and parameters will have in a particular problem to be solved, i.e., an *instance*. The AMPL language consists of three sub-languages: one for declarations, such as the `set`, `var`, `param`, and objective (`minimize`) declarations above, a simplified language in "data sections" for giving values to sets and parameters, and a command language for modifying values, solving problems, and writing results in various ways. While AMPL permits commingling declarations and instance data, AMPL also makes it easy to separate pure models from instance data.

AMPL does not solve problems by itself (except when AMPL's problem simplifications—its *presolve*—result in a solved problem), but instead writes files with full details of the problem instances to be solved and invokes separate solvers. The AMPL/solver interface library [19], whose source is freely available, provides problem details to *solver interfaces*, which interact with particular solvers to find solutions and return them to AMPL.

Often one needs to solve sequences of problems, with the solution of one problem providing data used in the next problem. Sometimes this involves updating set and parameter values. AMPL only instantiates or recomputes problem entities as needed, effectively using lazy evaluation to help speed up processing.

While parts of the AMPL language are general purpose, other parts, such as the *presolve* phase and computation of reduced costs, are tailored to mathematical programming. AMPL is meant for use with both linear and nonlinear problems; its internal use of sparse data structures allows AMPL to be useful with some very large problem instances.

3 AMPL History

AMPL arose in part because of Karmarkar's linear-programming algorithm [24]. At the time, there was much interest at the Computing Sciences Research Center in "little languages," e.g., for graphing data, solving least-squares problems, drawing figures, etc. While Karmarkar's algorithm seemed to promise faster solutions of some linear programming problems, we thought a "little language" to express such problems would help make the algorithm useful in practice. I had known Bob Fourer since the mid-1970s, when we both worked at the NBER Computer Research Center in Cambridge, Massachusetts—Bob had done his undergraduate work at the nearby Massachusetts Institute of Technology. He had subsequently obtained a Ph.D. at Stanford University under George Dantzig and had published a nice paper

[10] arguing for modeling languages. Bob was now a professor at Northwestern University and, as I learned when I saw him at a meeting, was coming up for a sabbatical. My management arranged for Bob to spend his sabbatical at Bell Labs in the 1985–1986 academic year, during which he, Brian Kernighan, and I worked on the first version of AMPL. (We were aware of GAMS [5], but GAMS was not yet generally available and, anyway, we thought we could do a better language design. Such other modeling languages as AIMMS [4] and MPL [28] came along later.) Brian wrote the first implementation of AMPL; I wrote a preprocessor to transform data sections to a simpler, now defunct, format for the original AMPL processor.

Our first technical report on AMPL [13] appeared in 1987. In revised form, it eventually appeared in *Management Science* [14]. By then, I had written a new implementation to facilitate various extensions we had in mind, such as handling nonlinearities.

Since the late 1970s I had been aware of Kedem's work [25] on forward automatic differentiation (AD), which provides a mechanical way to compute analytically correct derivatives, and I was thinking of adding such facilities to AMPL. I mentioned this to Andreas Griewank when I saw him in 1988 at the International Symposium on Mathematical Programming (ISMP) in Tokyo, and he told me about the more efficient "reverse" automatic differentiation. (He has subsequently written much more about AD; see [22] for pointers to AD history and [23] for more on AD in general.) Reverse AD computes a function and its gradient with work proportional to that of computing the function alone, whereas forward AD, like straightforward symbolic differentiation, can turn a function evaluation involving n arithmetic operations into a computation involving $O(n^2)$ operations. Both avoid the truncation errors inherent in finite differences. Ever since the Tokyo ISMP, I have been a fan of reverse AD. AMPL itself uses reverse AD to compute nonlinear reduced costs, but most AD happens in the solver interface library. See [17] for more on first derivative computations in this regard and [18] for some details of finding and exploiting partially separable structure when doing Hessian (second derivative) computations.

By the early 1990s we had enough material to write a book on AMPL [15]. We continued adding facilities to AMPL and added much new material to the second edition [16] of the book.

The "dot-com bubble burst" of 2001 threw a monkey wrench into AMPL development, but did cause creation of the AMPL Optimization company. Eventually I went to work at Sandia National Labs in Albuquerque, New Mexico, where I worked on AMPL support after hours (and without pay). Brian became a professor at Princeton. The three co-authors continued (and continue) to interact via E-mail. When we got an NSF SBIR grant for some new work on AMPL, I left Sandia to work for the AMPL company (and get some pay). Bob Fourer retired somewhat later from Northwestern University and now also works full time for the AMPL company.

4 Some Simple Declarations and Commands

Here is a simple example of some declarations, commands, and a little data section:

```
param p;
param q = p + 10;
data; param p := 2.5;
display p, q;
```

The third line is the data section, which gives a value to p that is used in the "display" command, which produces output

```
p = 2.5
q = 12.5
```

Data sections are good for conveying single values as well as tables of data, but data sections have relaxed quoting rules and other simplifications that preclude the appearance of expressions. The "let" command, by contrast, can involve general expressions. For example,

```
let p := 17; display p, q;
```

gives

```
p = 17
q = 27
```

Notice that q was automatically recomputed.

AMPL can be used in batch mode (reading from a file) or interactive mode (reading from the standard input). Prompts are given in interactive mode. Doing the above exercise in interactive mode, one would see

```
ampl: param p;
ampl: param q = p + 10;
ampl: data; param p := 2.5;
ampl: display p, q;
p = 2.5
q = 12.5
ampl: let p := 17; display p, q;
p = 17
q = 27
```

5 Simple Sets

To illustrate some simple sets and an error, here is a continuation of the above interactive-mode session.

```
ampl: set A; set B;
ampl: set C = p .. q;
```

```
ampl: display A;
Error executing "display" command:
        no data for set A
ampl: data; set A := a b c; set B := c d;
ampl data: display A, B, C;
set A := a b c;

set B := c d;

set C := 17 18 19 20 21 22 23 24 25 26 27;
```

The prompt "ampl data:" indicates data-section mode; the "display" command causes AMPL to revert to model/command reading mode. Here are examples of some set operations:

```
ampl: display A intersect B, A union B;
set A inter B := c;

set A union B := a b c d;

display A diff B, A symdiff B;
set A diff B := a b;

set A symdiff B := a b d;
```

6 Iterated and Recursive Expressions

Often it is useful to use iterated expressions, such as iterated sums. Here are some iterated expressions and a recursive definition, illustrated with the help of "print" commands.

```
ampl: print sum {i in 1..4} i;
10
ampl: print prod {i in 1..4} i;
24
ampl: param fac{ i in 1..9 }
ampl? = if i == 1 then 1 else i*fac[i-1];
ampl: print max{i in 1..9}
ampl? abs(fac[i] - prod{j in 2..i} j);
0
ampl: display fac, {i in 1..9} prod{j in 2..i} j;
:    fac   prod{j in 2 .. i} j    :=
1      1                1
2      2                2
3      6                6
4     24               24
```

```
5      120          120
6      720          720
7     5040         5040
8    40320        40320
9   362880       362880
;
```

7 Example Model: diet.mod

The diet model in the AMPL book [16] provides a short but complete example of a model for choosing what foods to buy. The model involves sets NUTR and FOOD of nutrients and foods, subscripted parameters f_min, f_max, and cost that specify minimum and maximum amounts of each food to buy and how much one unit of each food costs, a doubly subscripted parameter amt that tells how many units of each nutrient are provided by one unit of each food, and subscripted parameters n_min and n_max that give lower and upper bounds on the amounts of each nutrient that the foods we buy are to provide. The objective is to satisfy the nutritional requirements at minimal cost by choosing suitable values for the decision variables Buy.

```
set NUTR;
set FOOD;

param cost {FOOD} > 0;
param f_min {FOOD} >= 0;
param f_max {j in FOOD} >= f_min[j];

param n_min {NUTR} >= 0;
param n_max {i in NUTR} >= n_min[i];

param amt {NUTR,FOOD} >= 0;

var Buy {j in FOOD} >= f_min[j], <= f_max[j];

minimize Total_Cost:
            sum {j in FOOD} cost[j] * Buy[j];

subject to Diet {i in NUTR}:
   n_min[i] <= sum{j in FOOD} amt[i,j]*Buy[j]
            <= n_max[i];

data; set NUTR := A B1 B2 C ;
set FOOD := BEEF CHK FISH HAM MCH MTL SPG TUR ;

param:   cost  f_min  f_max :=
  BEEF   3.19    0     100
```

```
   CHK     2.59     0      100
   FISH    2.29     0      100
   HAM     2.89     0      100
   MCH     1.89     0      100
   MTL     1.99     0      100
   SPG     1.99     0      100
   TUR     2.49     0      100 ;

param:    n_min  n_max :=
    A      700   10000
    C      700   10000
    B1     700   10000
    B2     700   10000 ;

param amt (tr):
              A    C    B1   B2 :=
    BEEF     60   20    10   15
    CHK       8    0    20   20
    FISH      8   10    15   10
    HAM      40   40    35   10
    MCH      15   35    15   15
    MTL      70   30    15   15
    SPG      25   50    25   15
    TUR      60   20    15   10 ;
```

The data section above illustrates some tabular input formats. AMPL also has "table" declarations and "read table" and "write table" commands for reading data from, and writing data to, external repositories, such as databases and spreadsheets.

8 Sample Session

Here is an example of solving the above problem instance.

```
ampl: model diet.mod; data diet.dat;
ampl: solve;
MINOS 5.51: optimal solution found.
6 iterations, objective 88.2
ampl: display Buy;
Buy [*] :=
BEEF   0
 CHK   0
FISH   0
 HAM   0
 MCH  46.6667
 MTL   1.57618e-15
```

```
 SPG   8.42982e-15
 TUR   0
;
```

The resulting menu is not very satisfactory: $46\frac{2}{3}$ packages of macaroni and cheese ("MCH"). We probably want to buy only whole packages, which we can do by using integer variables:

```
ampl: redeclare var Buy{j in FOOD}
ampl? integer >= f_min[j] <= f_max[j];
ampl: solve;
MINOS 5.51: ignoring integrality of 8 variables
MINOS 5.51: optimal solution found.
4 iterations, objective 88.2
```

Since MINOS (the default solver) does not deal with integer variables, we need to use a solver that only allows integer variables to have integer values. Many solvers can do this; here, we use CPLEX:

```
ampl: option solver cplex; solve;
CPLEX 12.6.1.0: optimal integer solution;
                objective 88.44
4 MIP simplex iterations
0 branch-and-bound nodes
ampl: display Buy;
Buy [*] :=
BEEF   0
 CHK   2
FISH   0
 HAM   0
 MCH  43
 MTL   1
 SPG   0
 TUR   0
;
```

9 Analyzing Infeasibility

Formulating a good model is often an iterative process: we repeatedly try a formulation, examine its consequences, then modify it. As a simple example, the diet above is still not very satisfactory, so we could change the data to provide positive lower bounds on the amounts of each food bought. Here is file "diet2.dat" from the AMPL book:

```
set NUTR := A B1 B2 C NA CAL ;
set FOOD := BEEF CHK FISH HAM MCH MTL SPG TUR ;
```

```
param:    cost  f_min  f_max :=
  BEEF    3.19    2     10
  CHK     2.59    2     10
  FISH    2.29    2     10
  HAM     2.89    2     10
  MCH     1.89    2     10
  MTL     1.99    2     10
  SPG     1.99    2     10
  TUR     2.49    2     10   ;

param:    n_min  n_max :=
   A       700   20000
   C       700   20000
   B1      700   20000
   B2      700   20000
   NA        0   40000
   CAL   16000   24000 ;

param amt (tr):
            A    C    B1   B2    NA   CAL :=
   BEEF    60   20   10   15   938   295
   CHK      8    0   20   20  2180   770
   FISH     8   10   15   10   945   440
   HAM     40   40   35   10   278   430
   MCH     15   35   15   15  1182   315
   MTL     70   30   15   15   896   400
   SPG     25   50   25   15  1329   370
   TUR     60   20   15   10  1397   450 ;
```

By using a “reset data” command, we can keep the current model but associate a fresh set of data with it.

```
ampl: reset data; data diet2.dat;
ampl: solve;
CPLEX 12.6.1.0: integer infeasible.
1 MIP simplex iterations
0 branch-and-bound nodes
No basis.
```

There are various approaches to diagnosing infeasibility. Sometimes it is helpful just to see which constraints are infeasible and what variables are at lower or upper bound at the variable values where the solver detected infeasibility. For example,

```
ampl: option solver minos; solve;
MINOS 5.51: ignoring integrality of 8 variables
MINOS 5.51: infeasible problem.
9 iterations
ampl: display Diet.lb, Diet.body, Diet.ub, Diet.slack;
```

```
:    Diet.lb    Diet.body Diet.ub      Diet.slack    :=
A        700     1993.09    20000    1293.09
B1       700      841.091   20000     141.091
B2       700      601.091   20000     -98.9086
C        700     1272.55    20000     572.547
CAL    16000    17222.9     24000    1222.92
NA         0    40000       40000       7.27596e-12
;
```

Here, `Diet.lb`, `Diet.body`, and `Diet.ub` correspond to ℓ, $c(x)$ and u in (1b), and the constraint slack `Diet.slack` corresponds to $\min(u-c(x), c(x)-\ell)$. Most of the constraints are satisfied as inequalities (i.e., they have positive slacks), but the B2 constraint has a decidedly negative slack, while the NA (sodium) constraint is essentially satisfied as an equality (with a slack of about 7.3×10^{-12}) and `Diet.body` is approximately at its upper bound. Increasing the upper bound on the sodium constraint might help:

```
ampl: let n_max['NA'] := 50000; solve;
MINOS 5.51: ignoring integrality of 8 variables
MINOS 5.51: optimal solution found.
5 iterations, objective 118.0594032
```

so allowing more sodium is one way to remove the infeasibility.

Another way to diagnose infeasibility is by finding an *irreducible infeasible set* (IIS) of constraints and variable bounds that are mutually inconsistent; see [6, 30] and the references therein for more details. Some solvers nowadays have facilities for finding an IIS. With CPLEX, for example,

```
option cplex_options 'iisfind=1'; solve;
```

would also implicate the B2 and sodium constraints.

10 A Nonlinear Example

AMPL allows general nonlinear expressions in constraints and objectives. The "largest small hexagon" problem [21] provides a small example of an interesting nonlinear optimization problem. Here is a lightly edited variant of a little AMPL model, "pgon.mod," that describes the problem and has long been available as http://www.netlib.org/ampl/models/pgon.mod:

```
# Maximum area for unit-diameter polygon of N sides.
# The following model started as a GAMS model by
# Francisco J. Prieto.

param N integer > 0 default 6;
set I = 1..N;
```

```
param pi = 4*atan(1.);

var rho{i in I} <= 1, >= 0  # polar radius (distance
                            # to fixed vertex)
              := 4*i*(N + 1 - i)/(N+1)**2;

var theta{i in I} >= 0  # polar angle (measured from
                        # fixed direction)
              := pi*i/N;

subject to cd{i in I, j in i+1 .. N}:
      rho[i]**2 + rho[j]**2
            - 2*rho[i]*rho[j]*cos(theta[j]-theta[i])
      <= 1;

subject to ac{i in 2..N}:
        theta[i] >= theta[i-1];

subject to fix_theta: theta[N] = pi;
subject to fix_rho:   rho[N] = 0;

maximize area:
      .5*sum{i in 2..N}
            rho[i]*rho[i-1]*sin(theta[i]-theta[i-1]);
```

The # character introduces a comment that extends to the end of the line. The "`:=` *expression*" phrases specify initial guesses for the variables. Perhaps surprisingly, the solution is not the regular N-gon. Figure 1 depicts a solution for N = 6.

11 Slices

AMPL's basic indexing notation introduces one new dummy variable for each component of the tuples that comprise a set. For example,

```
set S dimen 2;
```

declares a set of pairs, and

```
{(i,j) in S}
```

is an "indexing" in which dummy variables `i` and `j` assume the values of the first and second components of each pair in the set. Sometimes one wants a "slice" of a set of tuples, i.e., an indexing in which some components are given by expressions valid in the context of the indexing. For example,

```
s.t. c{a in A}:
    sum{(i,j) in S: i == a} x[i,j] == 1;
```

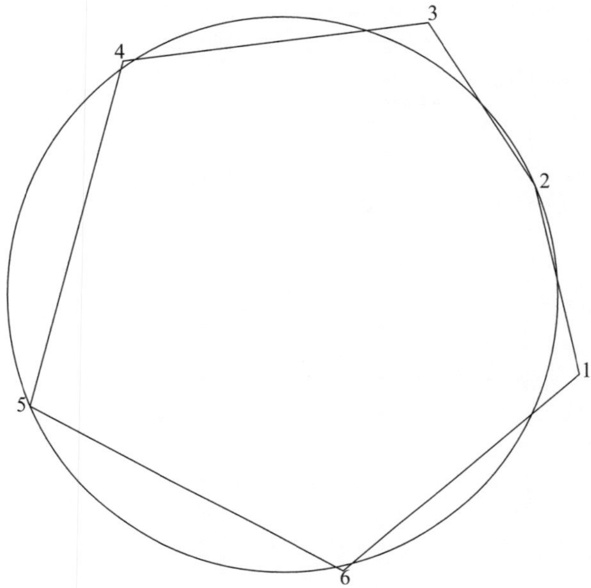

Fig. 1 Solution of `pgon.mod` for $N = 6$

is a constraint declaration with a sum that effectively involves a slice. AMPL's slice notation allows one to put desired values directly into the indexing notation. The above example has the same effect as

```
s.t. c{a in A}:
     sum{(a,j) in S} x[a,j] == 1;
```

but the latter is easier to read and can be much faster, since internally S is split into a set of one-dimensional sets. For a set S of n members, this can turn an $O(n^2)$ computation into an $O(n)$ computation. A few years ago I saw an example in which changing the former to the latter reduced problem instantiation time from four hours to a minute.

12 Iterated Unions

In various contexts, it is useful to construct sets by iterating over computed set expressions and forming their union. For example, given the declaration

```
set A dimen 2;
```

of a set of pairs, the declaration

```
set J = union{(i,j) in A} {j};
```

forms the set J of second components of the pairs in A. For forming iterated unions of singleton sets, such as `{j}` above, the `setof` operator provides simpler syntax that achieves the same effect:

```
set J = setof{(i,j) in A} j;
```

As an example where `setof` is useful, here is a little model for choosing a convex combination of classifiers that is "best" in a least-squares sense.

```
set A dimen 2;  # (observation, classifier) pairs
param E{A};     # signed, weighted predictions
set I = setof {(i,j) in A} i;   # observations
set J = setof {(i,j) in A} j;   # classifiers
param y{I} in {1,-1};      # y[i] = 1 ==> "yes",
                           # -1 ==> "no"
var x{J} >= 0;             # weights on classifiers
set B = {(i,j) in A: y[i]*E[i,j] < 0};
# B = mis-classified pairs
minimize errsq: sum{i in I}
            (sum{(i,j) in B} y[i]*E[i,j]*x[j])^2;
s.t. convex: sum{i in J} x[i] = 1;
```

More elaborate iterated unions are sometimes useful. For example, the following fragment from a mesh-untangling model declares the set of directed edges of some boxes.

```
set P;    # points
set Boxes within {P,P,P,P,P,P,P,P};
set Edges = union {(a,b,c,d,e,f,g,h) in Boxes} {
                       (a,b), (a,d), (a,e),
                       (b,c), (b,a), (b,f),
                       (c,d), (c,b), (c,g),
                       (d,a), (d,c), (d,h),
                       (e,h), (e,f), (e,a),
                       (f,e), (f,g), (f,b),
                       (g,f), (g,h), (g,c),
                       (h,g), (h,e), (h,d)};
```

Products of matrices appear surprisingly rarely in the mathematical programming problems one sees in practice, but the sparse product of sparse matrices is easily expressed with the help of an iterated union (via *setof*) and slice notation:

```
set IJ dimen 2; param A{IJ};
set JK dimen 2; param B{JK};
set IK = setof{(i,j) in IJ, (j,k) in JK} (i,k);
param C{(i,k) in IK} =
     sum{(i,j) in IJ: (j,k) in JK} A[i,j]*B[j,k];
```

13 AMPL Flexibility Goals

We have sought to make AMPL useful in various contexts. For developing models, it can be helpful to use AMPL interactively, typing commands at it. For longer computations, "batch" mode, in which AMPL reads everything from specified files, can be convenient. We have long had some experimental graphical user interfaces (GUIs) and have recently put considerable effort into developing a new "integrated development environment" (IDE); see http://ampl.com/products/ide/.

The AMPL language itself is primitive recursive, but AMPL has facilities for importing libraries of functions implemented in other languages. A README file about these facilities and some examples appear in http://www.netlib.org/ampl/solvers/funclink. (At times, http://www.ampl.com/netlib/ampl/solvers/funclink may be more up to date.) A library that includes more than 300 functions from the GNU Scientific Library (http://www.gnu.org/software/gsl/) is available in source and binary form at http://ampl.com/resources/extended-function-library/. AMPL's imported function facilities also allow AMPL to import "table handlers" for reading data from and writing data to external repositories, such as spreadsheets and databases. Details on using the table facilities appear in http://ampl.com/resources/database-and-spreadsheet-table-handlers/. Details on writing your own table handlers are in http://ampl.com/NEW/TABLES/. The *tableproxy* table handler permits accessing data on remote machines and facilitates mixing 32- and 64-bit versions of AMPL and data providers on the same machine. See http://ampl.com/NEW/TABLEPROXY/.

In various ways, we have sought to make it convenient for AMPL to interact with its host environment (operating system). A general "shell" command allows one to run arbitrary programs. AMPL's printing commands (`print` for unformatted printing, `printf` for formatted printing, and `display` for labeled printing) can have their output directed to files, which may either be created afresh or appended to. The `remove` command is for deleting files. "Pipe" functions provide a simple way for AMPL to interact with external programs: AMPL writes function arguments to the standard input of an external program, and the program returns the function value by writing to its standard output. A program implementing a "pipe" function must flush its output buffers before reading new function arguments, which can be awkward.

The currently popular operating systems all provide an "environment" of name-value pairs that programs can see and manipulate. The names are "environment variables." AMPL's "option" command operates on these environment variables and exports them to solvers (which are invoked as separate processes) and "shell" commands (which also are invoked as separate processes). AMPL's behavior is affected by some options. When starting execution, AMPL acquires values for these options from the incoming environment if present there and provides default values for them if not. Most solvers also are affected by environment variable values. Conventionally, the AMPL interface to a solver named `mysolver` would look at the environment variable named `mysolver_options`, which could be specified in an AMPL session by "`option mysolver_options`" commands, such as

```
option cplex_options 'advance=2 lpdisplay=1 \
                      prestats = 1 \
                      primalopt'
                     " aggregate=1 aggfill=20";


option solver knitro,
            knitro_options "maxit=30";
```

Strings may be quoted by single or double quotes. For option values, adjacent strings are concatenated.

Currently under development is "AMPL API," another way for AMPL to interact with external programs. See http://ampl.com/products/api/.

14 Interaction with Solvers

AMPL's "`solve`" command proceeds by writing a ".nl file" (a file whose name ends with ".nl") containing

- problem statistics
- coefficients for linear expressions
- expression graphs for nonlinear expressions
- initial guesses (primal and dual)
- suffixes (builtin or user declared).

Solvers return solution results to AMPL by writing a ".sol" file for AMPL to read. This file contains a "solve_message" and status code and may contain updated primal and dual variable values. It may also contain suffix values, which are auxiliary values associated with individual variables, constraints, objectives and problems, such as basis status for variables and constraints.

15 Problem Transformations

AMPL's presolve phase [11] derives and propagates bounds with directed roundings and may fix variables, remove constraints (e.g., inequalities that are never tight), resolve complementarities, turn nonlinear expressions into linear expressions (after fixing relevant variables), simplify convex piecewise-linear expressions, and convert nonconvex piecewise-linear expressions into equivalent systems of integer variables and SOS-2 [3] constraints. It also processes "defined variables," which in effect are named common expressions. For example, the declarations

```
param N integer > 0;
set I = 1 .. N;
var x{I}; var y{I};
var dot = sum{i in I} x[i]*y[i];
```

declares independent variables x and y and defined-variable dot, which is the inner product of x and y. Constraints and objectives could involve dot, but the solver would only see x and y as independent variables.

16 Spline Example

A referee asked about splines. I do not recall anyone wanting to use splines with AMPL, but the following illustration of constructing a spline approximation provides an example of using some of the facilities sketched above. We will use an imported function called `bspline` that, given a spline degree, a set of breakpoints and weights on B-spline basis functions (see chapter X of [7]) and a point x sufficiently within the breakpoints that all relevant basis functions are defined, computes the value of the spline at x and the first derivatives of this value with respect to x, the weights, and the breakpoints. The derivatives facilitate choosing the weights to fit specified data. The derivatives are handled by the ASL and do not explicitly appear in the following model.

```
param N default 3;   # degree of splines
param ND;            # ND+1 = number of data points
set SD = 0 .. ND;    # indices of data points
param xd{SD};        # ordinates of data points
param fd{SD};        # function values at data points

check{i in 1 .. ND}: xd[i-1] < xd[i];

param NI >= 1;          # number of intervals for
                        # x in bspline(n,x,...)
set SK = -N .. NI + 3;  # indices of knots
set SW = 1 .. NI + N;   # indices of B-spline weights
param wrange = xd[ND] - xd[0];
param b0{i in SK} = xd[0] + i*wrange/NI;
var b{i in SK} := b0[i];    # spline knots
var w{i in SW};             # spline weights

function bspline;
var s{i in SD} =
   bspline(N, xd[i], {j in SK} b[j], {j in SW} w[j]);

minimize ssq: sum{i in SD} 0.5*(fd[i] - s[i])^2;
```

```
s.t. resid{i in SD}: s[i] == fd[i];

problem SSQ: b, w, ssq;
problem NLS: b, w, resid; option presolve 0;
```

It might be good to add constraints that would keep the breakpoints ordered, but for the solvers used in the sample session shown below, this turns out not to be needed. To find values for `b` and `w` so `bspline(n, xd[i],...)` approximates `fd[i]` in a least-squares sense, we can either use an unconstrained solver with problem SSQ or a least-squares solver with problem NLS; least-squares solvers, such as `nl2` (discussed in [19] and based on NL2SOL [8]) solve equations in a least-squares sense. For such solving, it is often necessary to turn AMPL's presolve off to prevent it from satisfying some equations exactly.

For an example session, let us fit a cubic spline to the sine function. Suppose the above model appears in file `bspline.mod` and that file `sine.fit` contains

```
model splined.mod;
param pi = 4*atan(1);
data;
param ND := 21;  param NI := 5;

let{i in SD} xd[i] := 2*pi*(i/ND);
let{i in SD} fd[i] := sin(xd[i]);
fix{i in -N .. -1} b[i];
fix{i in NI+1 .. NI+N} b[i];
```

Here is a session fitting the data both ways with the above model and setup files:

```
ampl: include spline.fit
ampl: load bspline.dll;
ampl: option solver nl2; solve;
nl2:    Relative Function Convergence;
        function = 5.40485704e-06
        RELDX = 8.12e-05; PRELDF = 1.94e-11;
        NPRELDF = 1.94e-11
        19 func. evals; 16 grad. evals
ampl: printf "%.3g\en", max{i in SD} abs(s[i]-fd[i]);
0.00109
ampl: problem SSQ;
ampl: option reset_initial_guesses 1, solver snopt;
ampl: solve;
SNOPT 7.2-8 : Optimal solution found.
80 iterations, objective 5.404937356e-06
Nonlin evals: obj = 67, grad = 66.
ampl: printf "%.3g\en", max{i in SD} abs(s[i]-fd[i]);
0.00109
```

Both solvers achieved about the same residual sum of squares and maximum fit error on the set of sample points.

The problem of choosing `b` and `w` to fit best in a least-squares sense is a separable nonlinear least-squares problem [20], as the `w` variables appear linearly, and a separable solver probably would be faster and somewhat more robust. At any rate, after determining `b` and `w`, we could fix them (causing them to retain their current values and be treated as parameters) and deal with some application where the spline just found would be useful.

Source for `bspline.dll` is too long to include with this paper, but is available as

http://www.ampl.com/netlib/ampl/solvers/examples/bspline.c

17 Implementation Techniques

AMPL's implementation is an exercise in practical computer science. Parsing proceeds via the venerable Unix tools *lex* and *yacc*, which build up expression graphs that are subsequently manipulated. Declared names are associated with unique "symbols" found by hashing. Hashing is also used in a "compile" phase to find common expressions. The compile phase lifts invariant subexpressions out of inner loops. With the help of dependency graphs, entities are only instantiated or updated when needed—lazy evaluation. When appropriate, cleanup routines are registered, so they can be invoked either when an operation completes normally or when it is interrupted by an error, such as an invalid subscript or missing data. Error handling proceeds via *longjmp*. Some things are reference-counted, and sparse-matrix techniques make processing large, sparse models feasible. AMPL is written (and debugged) in C++, but for porting to various platforms, the AMPL source code is converted to portable C with the help of *cfront* (the original C++ "compiler").

18 Wish List

There are many improvements we hope to make to AMPL and its associated ASL (solver-interface library). Just when and whether these improvements will be available remains to be seen. Functions expressed directly in AMPL would turn AMPL from a primitive-recursive language to a Turing-complete language. When conveyed to solvers via the ASL, they would allow providing callbacks to solvers, e.g., for influencing branching decisions in integer programming. They would also find some use in AMPL models. Ordered sets of tuples would sometimes be useful. While AMPL already facilitates solving sequences of related problems, updating entities could sometimes be done more efficiently. AMPL has long had some facilities for constraint programming, but allowing variables in subscripts remains to be done. When there is just one objective (for multi-objective optimization,

AMPL allows one to declare several objectives, including indexed collections of objectives), AMPL's presolve could exploit duality. (It already does reductions for complementarity.) AMPL has long permitted some declarations related to stochastic programming, but corresponding extensions to the ASL need to be completed and examples of their use need to be created. Facilities supporting semi-definite programming and multi-level optimization would be useful. We have long wanted AMPL to be able to carry on two-way conversations with solvers, so after a problem has been solved, a slightly modified problem could be conveyed just by telling the solver of changes to the existing problem. Units (of distance, time, charge, etc.) might help catch or avoid some mistakes. For some mathematical research, such data types as rational, complex, and complex rational could be helpful. Facilities for parallel evaluations in the ASL would be useful. Constructs for parallelism might also be useful in AMPL itself.

19 Other AMPL Facilities

This paper provides an overview of AMPL, but gives little or no detail about various useful AMPL facilities:

- drop, restore (affecting what constraints and objectives a solver sees)
- fix, unfix (affecting the variables a solver sees)
- named problems and environments
- suffixes
- tables and table handlers
- column-generation syntax (e.g., node and arc)
- complementarity constraints [9]
- subscripted sets versus tuples
- constraint programming [12]

The AMPL web site (http://www.ampl.com) provides pointers to more detail on the above topics, including

- the AMPL book (and free PDF files for it)
- examples (models and data)
- descriptions of new facilities
- a new IDE
- a new API
- *Try AMPL!* and NEOS for free web-based use
- course licenses
- trial licenses
- downloads
  - student binaries
  - ASL (solver-interface library) source

- example solver interfaces
- "standard" table handler (binaries, source)
- papers, reports, talk slides

20 Concluding Remarks

Mathematical programming models, such as (1), are useful in many contexts. Formulating good models is often an iterative process: you test a formulation, assess how well it works, modify it, and test again. The AMPL modeling language can assist in this endeavor. Its associated interface library (ASL) provides automatically derived details to solvers, such as sparsity information and derivatives.

References

1. Aho, A.V., Weinberger, P.J., Kernighan, B.W.: Awk — a pattern scanning and processing language. Softw. Pract. Exp. **9**, 267–279 (1979)
2. Aho, A.V., Weinberger, P.J., Kernighan, B.W.: The AWK Programming Language. Addison-Wesley, Reading (1988)
3. Beale, E.M.L., Tomlin, J.A.: Special facilities in a general mathematical system for non-convex problems using ordered sets of variables. In: Lawrence, J. (ed.) Proceedings of the Fifth International Conference on Operational Research, pp. 447–454. Tavistock, London (1970)
4. Bisschop, J., Entriken, R.: AIMMS, The Modeling System. Paragon Decision Technology, Haarlem (1993)
5. Bisschop, J., Meeraus, A.: Selected aspects of a general algebraic modeling language. In: Iracki, K., Malanowski, K., Walukiewicz, S. (eds.) Optimization Techniques, Part 2. Lecture Notes in Control and Information Sciences, vol. 23, pp. 223–233. Springer, Berlin (1980)
6. Chinneck, J.W., Dravnieks, E.W.: Locating minimal infeasible constraint sets in linear programs. ORSA J. Comput. **3**(2), 157–168 (1991)
7. de Boor, C.: A Practical Guide to Splines. Applied Mathematical Sciences, vol. 27. Springer, New York (1978)
8. Dennis, J.E., Jr., Gay, D.M., Welsch, R.E.: An adaptive nonlinear least-squares algorithm. ACM Trans. Math. Softw. **7**, 348–368 (1981)
9. Ferris, M.C., Fourer, R., Gay, D.M.: Expressing complementarity problems in an algebraic modeling language and communicating them to solvers. SIAM J. Optim. **9**(4), 991–1009 (1999)
10. Fourer, R.: Modeling languages versus matrix generators for linear programming. ACM Trans. Math. Softw. **9**(2), 143–183 (1983)
11. Fourer, R., Gay, D.M.: Experience with a primal presolve algorithm. In: Hager, W.W., Hearn, D.W., Pardalos, P.M. (eds.) Large Scale Optimization: State of the Art, pp. 135–154. Kluwer Academic, Dordrecht (1994)
12. Fourer, R., Gay, D.M.: Extending an algebraic modeling language to support constraint programming. INFORMS J. Comput. **14**(4), 322–344 (2002)
13. Fourer, R., Gay, D.M., Kernighan, B.W.: AMPL: A mathematical programming language. Technical Report Computing Science Technical Report No. 133, AT&T Bell Laboratories, Murray Hill, NJ, Jan 1987 (revised June 1989)

14. Fourer, R., Gay, D.M., Kernighan, B.W.: A modeling language for mathematical programming. Manag. Sci. **36**(5), 519–554 (1990)
15. Fourer, R., Gay, D.M., Kernighan, B.W.: AMPL: A Modeling Language for Mathematical Programming. The Scientific Press, South San Francisco, California (1993)
16. Fourer, R., Gay, D.M., Kernighan, B.W.: AMPL: A Modeling Language for Mathematical Programming. Brooks/Cole–Thomson Learning, Pacific Grove, CA (2003)
17. Gay, D.M.: Automatic differentiation of nonlinear ampl models. In: Griewank, A., Corliss, G.F. (eds.) Automatic Differentiation of Algorithms: Theory, Implementation, and Application, pp. 61–73. SIAM, Philadelphia (1991)
18. Gay, D.M.: More AD of nonlinear ampl models: computing hessian information and exploiting partial separability. In: Corliss, G.F. (ed.) Computational Differentiation: Applications, Techniques, and Tools. SIAM, Philadelphia (1996)
19. Gay, D.M.: Hooking your solver to AMPL. Technical Report Technical Report 97-4-06, Computing Sciences Research Center, Bell Laboratories, Murray Hill (1997)
20. Golub, G.H., Pereyra, V.: The differentiation of pseudo-inverses and nonlinear least-squares problems whose variables separate. SIAM J. Numer. Anal. **10**, 413–432 (1973)
21. Graham, R.L.: The largest small hexagon. J. Comb. Theory A **18**, 165–170 (1975)
22. Griewank, A.: On automatic differentiation. In: Iri, M., Tanabe, K. (eds.) Mathematical Programming, pp. 83–107. Kluwer Academic, Boston (1989)
23. Griewank, A., Walther, A.: Evaluating Derivatives. SIAM, Philadelphia (2008)
24. Karmarkar, N.: A new polynomial-time algorithm for linear programming. Combinatorica **4**, 373–395 (1984)
25. Kedem, G.: Automatic differentiation of computer programs. ACM Trans. Math. Softw. **6**(2), 150–165 (1980)
26. Kernighan, B.W., Ritchie, D.M.: The C Programming Language. Prentice-Hall, Upper Saddle River (1978)
27. Kernighan, B.W., Ritchie, D.M.: The C Programming Language. Prentice-Hall, Upper Saddle River (1988)
28. Kristjansson, B.: MPL — Modelling System Quick Guide. Maximal Software, Reykjavik (1991)
29. Stroustrup, B.: The C++ Programming Language. Addison-Wesley, Reading (1986)
30. van Loon, J.: Irreducibly inconsistent systems of linear inequalities. Eur. J. Oper. Res. **8**, 283–288 (1981)

An Interior-Point ℓ_1-Penalty Method for Nonlinear Optimization

Nick I.M. Gould, Dominique Orban, and Philippe L. Toint

Abstract We describe a mixed interior/exterior-point method for nonlinear programming that handles constraints by way of an ℓ_1-penalty function. The penalty problem is reformulated as a smooth inequality-constrained problem that always possesses bounded multipliers, and that may be solved using interior-point techniques as finding a strictly feasible point is trivial. If finite multipliers exist for the original problem, exactness of the penalty function eliminates the need to drive the penalty parameter to infinity. If the penalty parameter needs to increase without bound and if feasibility is ultimately attained, a certificate of degeneracy is delivered. Global and fast local convergence of the proposed scheme are established and practical aspects of the method are discussed.

Keywords ℓ_1-Penalty • Interior point • Elastic variables • Nonconvex optimization

1 Introduction

A typical nonlinear programming problem is to

$$\underset{x \in \mathbb{R}^n}{\text{minimize}}\ f(x) \quad \text{subject to } c_{\mathcal{E}}(x) = 0,\ c_{\mathcal{I}}(x) \geq 0, \tag{1}$$

N.I.M. Gould
Computational Science and Engineering Department, Rutherford Appleton Laboratory, Chilton, Oxfordshire OX11 0QX, UK
e-mail: nick.gould@stfc.ac.uk

D. Orban (✉)
Département de Mathématiques et Génie Industriel, GERAD and École Polytechnique de Montréal, Montréal, QC, Canada
e-mail: Dominique.Orban@gerad.ca

P.L. Toint
Department of Mathematics, University of Namur, 61, rue de Bruxelles, 5000 Namur, Belgium
e-mail: Philippe.Toint@fundp.ac.be

M. Al-Baali et al. (eds.), *Numerical Analysis and Optimization*, Springer Proceedings in Mathematics & Statistics 134, DOI 10.1007/978-3-319-17689-5_6

involving smooth, possibly nonlinear and nonconvex, equality and inequality constraints. Here $f : \mathbb{R}^n \to \mathbb{R}$, $c_{\mathscr{E}} : \mathbb{R}^n \to \mathbb{R}^{n_{\mathscr{E}}}$ and $c_{\mathscr{I}} : \mathbb{R}^n \to \mathbb{R}^{n_{\mathscr{I}}}$, where $\mathscr{E} = \{1,\ldots,n_{\mathscr{E}}\}$ and $\mathscr{I} = \{n_{\mathscr{E}}+1,\ldots,n_{\mathscr{E}}+n_{\mathscr{I}}\}$. We propose an infeasible interior-point approach for (1) that embeds the set of variables into a higher dimensional space for which the constraints have a nonempty and easily locatable interior.

A common way to solve (1) is to build the corresponding ℓ_1-penalty function and to

$$\underset{x\in\mathbb{R}^n}{\text{minimize}}\ \phi^{\mathrm{P}}(x;\nu) := f(x) + \nu\vartheta^{\mathrm{P}}(x), \qquad \vartheta^{\mathrm{P}}(x) := \sum_{i\in\mathscr{E}} |c_i(x)| + \sum_{i\in\mathscr{I}} c_{\mathscr{I}}^{-}(x), \tag{2}$$

where $c_{\mathscr{I}}^{-}(x) := \max[0, -c_{\mathscr{I}}(x)]$ componentwise, for some sufficiently large penalty parameter $\nu > 0$. As we will see in Section 2, (2) is equivalent to the *smooth* problem

$$\begin{aligned} &\underset{x\in\mathbb{R}^n, s\in\mathbb{R}^{n_{\mathscr{C}}}}{\text{minimize}}\ \phi^{\mathrm{S}}(x,s;\nu) := f(x) + \nu\vartheta(x,s) \\ &\text{subject to } c_i(x) + s_i \geq 0,\ s_i \geq 0,\ i \in \mathscr{C} \end{aligned} \qquad \vartheta(x,s) = \sum_{i\in\mathscr{E}} (c_i(x) + 2s_i) + \sum_{i\in\mathscr{I}} s_i \tag{3}$$

involving $n_{\mathscr{C}}$ additional *elastic* variables $s \in \mathbb{R}^{n_{\mathscr{C}}}$—i.e., penalized slack variables—where $\mathscr{C} := \mathscr{E} \cup \mathscr{I}$ and $n_{\mathscr{C}} := n_{\mathscr{E}} + n_{\mathscr{I}}$. This problem only involves inequality constraints, and it is trivial to pick s sufficiently large so that (x,s) is strictly feasible for (3). Note also that had (1) been a convex optimization problem, (3) inherits this property. In other words, adding elastic variables preserves convexity. This is at variance with other types of infeasible methods, such as those based on the addition of slack variables, e.g., [5, 36].

An immediate possibility is to apply an interior-point method to (3), i.e.,

$$\underset{x\in\mathbb{R}^n,\ s\in\mathbb{R}^{n_{\mathscr{C}}}}{\text{minimize}}\ \phi^{\mathrm{B}}(x,s;\mu,\nu) := \phi^{\mathrm{S}}(x,s;\nu) - \mu\sum_{i\in\mathscr{C}} \log(c_i(x) + s_i) - \mu\sum_{i\in\mathscr{C}} \log s_i, \tag{4}$$

for a sequence of barrier parameters, $\{\mu^k\}$, converging to zero from above. A theoretical investigation of the properties of ϕ^{B} and the problem (4) forms the basis of Section 2. The global and local convergence properties of two standard trust-region methods for solving (4), for fixed (μ,ν) are considered in Section 3. Section 4 provides global and local convergence properties of the method. Algorithmic variations, improvements, and extensions are described in Sections 5 and 6. Numerical experience is reported in Section 7 and conclusions drawn in Section 8.

The use of the transformation to the ℓ_1-penalty function to solve (1) is, of course, well known. The equivalence between the optimality conditions for nonconvex nonlinear programming problems and related penalty functions was first reported by Pietrzykowski [33], and the results subsequently strengthened by Charalambous [7],

Han and Mangasarian [24], Coleman and Conn [9], Bazaraa and Goode [3] and Huang and Ng [28]. See also [14, Chapters 12 and 14]. In Section 2, we shall see how this equivalence is inherited by (3).

Although a constraint qualification condition is not required to conduct the convergence analysis, degeneracy is indicated by a diverging sequence of penalty parameters in the following sense. If the penalty parameter diverges yet feasibility is attained, our method delivers a certificate of degeneracy by explicitly providing Fritz-John multipliers. The relation satisfied by those multipliers characterizes failure of the Mangasarian and Fromovitz constraint qualification condition.

The approach taken in this paper has its genesis in the work of Mayne and Polak [31], more recently extended by Herskovits [26], Lawrence and Tits [29] and Tits et al. [34], all of whom also reformulate (1) so as only to involve inequality constraints. Indeed, our basic approach coincides with theirs on setting s to zero. However, we prefer not to do this, as the resulting problem then has no obvious initial feasible point. Armand et al. [2] investigated the reformulation

$$\underset{x\in\mathbb{R}^n,\, s\in\mathbb{R}^{n_{\mathscr{I}}}}{\text{minimize}}\ f(x) \quad \text{subject to } c_{\mathscr{I}}(x)+s\geq 0,\ s=0, \tag{5}$$

for convex, inequality-constrained problems in which the resulting equality constraints $s=0$ are handled by penalization. This idea was refined by Armand [1] to give (3) in the convex, inequality-constrained case, which was then solved by minimizing a sequence of (convex) barrier functions like (4).

The present approach is also related to the so-called *elastic mode* used by Boman [4] and in the SNOPT package of Gill et al. [17], where it is used in a sequential quadratic programming framework as a fallback strategy to relax the constraints in case the current quadratic subproblem appears to be infeasible, unbounded or to have unbounded multipliers. In such a case, once the elastic mode has been triggered, it persists until convergence.

A related approach is investigated by Chen and Goldfarb [8] with an ℓ_2 exact penalty function in a linesearch context. In that approach, a sequence of equality-constrained problems must be solved.

Other methods with an interior-point flavour include the primal–dual filter method of Wächter and Biegler [36], implemented in the IPOPT package, the primal–dual trust-region and linesearch methods of Byrd et al. [5] and Waltz et al. [37], implemented in the commercial package KNITRO, and the primal–dual linesearch method of Vanderbei and Shanno [35] implemented in the commercial package LOQO. Those methods typically add slack variables to convert general inequality constraints into bound constraints.

The notation used in the sequel is as follows. If $q\geq 0$ and $v\in\mathbb{R}^q$, we shall denote its i-th component by a subscript v_i. If $\mathscr{S}\subseteq\{1,\ldots,q\}$, we write $v_{\mathscr{S}}$ for the subvector of v whose components are the v_i, $i\in\mathscr{S}$. Likewise, if $M\in\mathbb{R}^{q\times p}$, $M_{\mathscr{S}}$ is the submatrix of M whose rows are indexed by $\mathscr{S}$. In an algorithmic context, the value taken by the vector v at iteration k will be denoted by a superscript v^k and its i-th component is v_i^k. A sequence indexed by the set $\mathbb{N}$ of nonnegative integers whose general term is v^k is denoted $\{v^k\}$ and a subsequence indexed by the infinite index set $\mathscr{K}\subseteq\mathbb{N}$ is denoted $\{v^k\}_{\mathscr{K}}$.

As exceptions to the above, if $e_{\mathscr{E}}$ and $e_{\mathscr{I}}$ are vectors of ones of dimension $n_{\mathscr{E}}$ and $n_{\mathscr{I}}$, respectively, we define two vectors $e^0_{\mathscr{E}} = (e_{\mathscr{E}}, 0)$ and $e^0_{\mathscr{I}} = (0, e_{\mathscr{I}})$ in $\mathbb{R}^{n_{\mathscr{C}}}$, and let $e = e^0_{\mathscr{E}} + e^0_{\mathscr{I}}$. Wherever appropriate, the notation e_p denotes the vector of all ones in $\mathbb{R}^p$ and similarly, 0_p denotes the zero vector of $\mathbb{R}^p$. In addition, $I_{\mathscr{E}}$ and $I_{\mathscr{I}}$ are identity matrices of dimensions $n_{\mathscr{E}}$ and $n_{\mathscr{I}}$, respectively.

2 Equivalent Smooth Formulation of the Exact Penalty

It is well known that (2) may be reformulated as a smooth problem [16, §4.2.3]. To see this, consider first an equality constraint $c_i(x) = 0$. The penalty contribution from this constraint, $\nu|c_i(x)|$, may be expressed as $\nu[r_i + s_i]$, where $c_i(x) = r_i - s_i$ and $(r_i, s_i) \geq 0$, or alternatively as $\nu[c_i(x) + 2s_i]$, where $c_i(x) + s_i \geq 0$ and $s_i \geq 0$. Now turning to an inequality constraint $c_i(x) \geq 0$, its penalty contribution, $\nu \max(-c_i(x), 0)$, may be expressed as νs_i, where $c_i(x) = r_i - s_i$ and $(r_i, s_i) \geq 0$, or alternatively as νs_i, where $c_i(x) + s_i \geq 0$ and $s_i \geq 0$. Thus the minimization of ϕ^{P} may be expressed as (3). Notice that for given x, any set of values $s_i \geq \max(-c_i(x), 0)$ provides an initial feasible point for the enlarged feasible region involving (x, s), and that this point lies in the strict interior if $s_i > \max(-c_i(x), 0)$ for all $i \in \mathscr{C}$. The central idea of this paper will then be to apply a primal–dual interior-point method to solve (3). A number of equivalent smooth reformulations of the penalty problem appear in [23].

A nonlinear problem of the form (1) is said to satisfy the Mangasarian and Fromovitz [30] constraint qualification (MFCQ) at a feasible point x^* if the vectors $\{\nabla c_i(x^*)\}_{i \in \mathscr{E}}$ are linearly independent and if there exists a direction $d \neq 0$ such that

$$\nabla c_i(x^*)^T d = 0 \quad \text{for } i \in \mathscr{E} \quad \text{and} \quad \nabla c_i(x^*)^T d < 0 \quad \text{for } i \in \mathscr{A},$$

where $\mathscr{A} = \{i \in \mathscr{I} \mid c_i(x^*) = 0\}$ is the set of active indices at x^*.

We denote the full vector of constraints by $c : \mathbb{R}^n \to \mathbb{R}^{n_{\mathscr{C}}}$. The Lagrangian associated with problem (1) is

$$L(x, \lambda) = f(x) - c_{\mathscr{E}}(x)^T \lambda_{\mathscr{E}} - c_{\mathscr{I}}(x)^T \lambda_{\mathscr{I}}, \tag{6}$$

where $\lambda_{\mathscr{E}} \in \mathbb{R}^{n_{\mathscr{E}}}$, $\lambda_{\mathscr{I}} \in \mathbb{R}^{n_{\mathscr{I}}}_+$ and $\lambda = (\lambda_{\mathscr{E}}, \lambda_{\mathscr{I}})$. A vector $z = (x, \lambda)$ is a *first-order critical point* for (1) if it satisfies the Karush-Kuhn-Tucker (KKT) conditions

$$\nabla f(x) - J^T_{\mathscr{E}}(x)\lambda_{\mathscr{E}} - J^T_{\mathscr{I}}(x)\lambda_{\mathscr{I}} = 0 \tag{7a}$$

$$C_{\mathscr{I}}(x)\lambda_{\mathscr{I}} = 0 \tag{7b}$$

$$c_{\mathscr{E}}(x) = 0 \tag{7c}$$

$$c_{\mathscr{I}}(x), \lambda_{\mathscr{I}} \geq 0. \tag{7d}$$

Here and elsewhere $J_{\mathcal{E}}(x)$ and $J_{\mathcal{I}}(x)$ are the Jacobian matrices of $c_{\mathcal{E}}(x)$ and $c_{\mathcal{I}}(x)$, respectively, while a capitalized (e.g.) $C_{\mathcal{I}}(x)$ denotes the diagonal matrix whose entries are the components of the vector (e.g.) $c_{\mathcal{I}}(x)$. Under a constraint qualification condition, and in particular under the MFCQ, conditions (7) are necessary for optimality of z.

If x^* is a first-order critical point for (1), let Λ^* be the set of all associated Lagrange multipliers, i.e, the (possibly empty) set of all vectors $(\lambda_{\mathcal{E}}, \lambda_{\mathcal{I}})$ satisfying (7). The MFCQ being satisfied at x^* is equivalent to Λ^* being nonempty and bounded [15].

If x is feasible for (1), we say that it is a *Fritz-John* point if there exist $(\gamma, \lambda_{\mathcal{E}}, \lambda_{\mathcal{I}}) \neq (0,0,0)$ with $\gamma \geq 0$ such that $(x, \gamma, \lambda_{\mathcal{E}}, \lambda_{\mathcal{I}})$ satisfies (7) with (7a) replaced by

$$\gamma \nabla f(x) - J_{\mathcal{E}}^T(x)\lambda_{\mathcal{E}} - J_{\mathcal{I}}^T(x)\lambda_{\mathcal{I}} = 0. \tag{8}$$

It is easy to see that if $\gamma > 0$, $(x, \lambda_{\mathcal{E}}/\gamma, \lambda_{\mathcal{I}}/\gamma)$ is in fact first-order critical. This would occur, e.g., if the MFCQ held at x. If, on the other hand, $\gamma = 0$, x is a feasible point where the MFCQ fails to hold [30].

The Lagrangian for problem (3) is

$$\mathcal{L}(x, s, y, u; \nu) = \phi^{\mathrm{s}}(x, s; \nu) - (c(x) + s)^T y - s^T u, \tag{9}$$

where the Lagrange multipliers $y = (y_{\mathcal{E}}, y_{\mathcal{I}}) \in \mathbb{R}_+^{n_{\mathcal{C}}}$ and $u = (u_{\mathcal{E}}, u_{\mathcal{I}}) \in \mathbb{R}_+^{n_{\mathcal{C}}}$ are associated with the constraints $c(x) + s \geq 0$ and $s \geq 0$ of (3), respectively. The vectors $v_{\mathrm{P}} = (x, s)$ and $v_{\mathrm{D}} = (y, u)$ contain primal and dual variables/Lagrange multipliers for (3), respectively.

The gradient of $\phi^{\mathrm{s}}(x, s; \nu)$ may be expressed as

$$\nabla \phi^{\mathrm{s}}(x, s; \nu) = \begin{bmatrix} \nabla f(x) \\ 0 \end{bmatrix} + \nu \nabla \vartheta(x, s) = \begin{bmatrix} \nabla f(x) \\ 0 \end{bmatrix} + \nu \begin{bmatrix} J_{\mathcal{E}}^T(x) e_{\mathcal{E}} \\ e + e_{\mathcal{E}}^0 \end{bmatrix}, \tag{10}$$

while the $2n_{\mathcal{C}} \times (n + n_{\mathcal{C}})$ Jacobian of the constraints of (3) with respect to v_{P} is

$$J^{\mathrm{S}}(v_{\mathrm{P}}) = \begin{bmatrix} J_{\mathcal{E}}(x) & I_{\mathcal{E}} & 0 \\ J_{\mathcal{I}}(x) & 0 & I_{\mathcal{I}} \\ 0 & I_{\mathcal{E}} & 0 \\ 0 & 0 & I_{\mathcal{I}} \end{bmatrix} = \begin{bmatrix} J(x) & I_{\mathcal{C}} \\ 0 & I_{\mathcal{C}} \end{bmatrix}, \tag{11}$$

where the $n_{\mathcal{C}} \times n$ Jacobian matrix of the full vector of constraint functions $c(x)$ is such that $J(x)^T = \left[J_{\mathcal{E}}(x)^T \; J_{\mathcal{I}}(x)^T \right]$. There is an intimate connection between the Lagrange multipliers λ for (1) and the multipliers y for (3). To keep later results concise, we formalize this as follows.

Definition 1. For a given, fixed, value $\nu \geq 0$ of the penalty parameter, and given vectors x, y and λ, we define the *shifted multipliers*

$$y(\lambda,\nu) := (\lambda_{\mathscr{E}} + \nu e_{\mathscr{E}}, \lambda_{\mathscr{I}}) = \lambda + \nu e^0_{\mathscr{E}} \tag{12a}$$

$$\text{and} \quad \lambda(y,\nu) := (y_{\mathscr{E}} - \nu e_{\mathscr{E}}, y_{\mathscr{I}}) = y - \nu e^0_{\mathscr{E}}, \tag{12b}$$

i.e., the vectors where the multipliers corresponding to the nonlinear equality constraints of (1) and (3) have been shifted by $\pm\nu e_{\mathscr{E}}$.

We may now express the KKT conditions for (3) as

$$\nabla f(x) - J^T(x)\lambda(y,\nu) = 0 \tag{13a}$$

$$\nu e_{\mathscr{E}} - (y_{\mathscr{E}} - \nu e_{\mathscr{E}}) - u_{\mathscr{E}} = 0 \tag{13b}$$

$$\nu e_{\mathscr{I}} - y_{\mathscr{I}} - u_{\mathscr{I}} = 0 \tag{13c}$$

$$(C(x)+S)y = 0 \tag{13d}$$

$$Su = 0 \tag{13e}$$

$$c(x)+s, s, y, u \geq 0. \tag{13f}$$

A first-order solution of

$$\underset{x\in\mathbb{R}^n,\ s\in\mathbb{R}^{n_{\mathscr{C}}}}{\text{minimize}} \ \vartheta(x,s) \quad \text{subject to } c(x)+s \geq 0 \quad \text{and} \quad s \geq 0 \tag{14}$$

is attained at a point (x,s) for which

$$\begin{array}{rr} J^T(x)(\bar{y} - e^0_{\mathscr{E}}) = 0 & e - (\bar{y} - e^0_{\mathscr{E}}) - \bar{u} = 0 \\ (C(x)+S)\bar{y} = 0 & S\bar{u} = 0 \\ (c(x)+s, s) \geq 0 & (\bar{y},\bar{u}) \geq 0 \end{array} \tag{15}$$

where $\bar{y}$ and $\bar{u}$ are Lagrange multipliers associated with the inequality constraints $c(x)+s \geq 0$ and $s \geq 0$, respectively. It is important to recognize that such an x is also a critical point for the infeasibility measure (2), $\vartheta^{\mathrm{P}}(x)$, for the true constraints.

Theorem 1. *If (x,s) satisfies (15), then x is a first-order critical point of $\vartheta^{P}(x)$.*

Proof. A first-order critical point for $\vartheta^{\mathrm{P}}(x)$ satisfies $J^T(x)\lambda = 0$, where the generalized gradient λ satisfies

$$\lambda_i = \begin{cases} -1 & \text{if } c_i > 0 \ (i \in \mathscr{E}) \\ 0 & \text{if } c_i > 0 \ (i \in \mathscr{I}) \\ 1 & \text{if } c_i < 0 \end{cases} \tag{16}$$

$$\lambda_i \in \begin{cases} [-1,1] & \text{if } c_i = 0 \ (i \in \mathscr{E}) \\ [0,1] & \text{if } c_i = 0 \ (i \in \mathscr{I}) \end{cases} \tag{17}$$

(see, for example, [11, Example 11.4.1].) Let $(\bar{y}, \bar{u})$ satisfy (15), and define $(\lambda, u) = (\bar{y} - e^0_{\mathscr{E}}, \bar{u})$ so that (15) becomes

$$J^T(x)\lambda = 0, \quad (18a) \qquad \lambda + u = e, \quad (18d)$$

$$(C(x)+S)(\lambda + e^0_{\mathscr{E}}) = 0, \quad (18b) \qquad Su = 0 \quad (18e)$$

$$c(x) + s, s \geq 0, \quad (18c) \qquad \lambda + e^0_{\mathscr{E}}, u \geq 0, \quad (18f)$$

We have established (18a), so it remains to show that the given λ satisfies (17).

Firstly, then, consider an index i for which $c_i(x) + s_i > 0$. In this case (18b) shows that $\lambda_i = -1$ if $i \in \mathscr{E}$ or $\lambda_i = 0$ if $i \in \mathscr{I}$. In either case, (18d) then ensures that $u_i > 0$, and hence $c_i(x) > 0$ since necessarily (18e) shows that $s_i = 0$. These are the first two possibilities in (17). Since $c_i(x) + s_i \geq 0$, it remains to consider indices for which $c_i(x) + s_i = 0$. In this case $c_i(x) = -s_i$ and thus (18e) implies that $c_i(x)u_i = 0$. If $s_i \neq 0$, $c_i(x) < 0$ so that $u_i = 0$, and hence $\lambda_i = 1$ from (18d). This is the third possibility in (17). By contrast, if $s_i = 0$, then immediately $c_i = 0$. But (18d) and (18f) ensure that $\lambda_i \in [-1, 1]$ if $i \in \mathscr{E}$ and $\lambda_i \in [0, 1]$ if $i \in \mathscr{E}$ for any i, giving the final two possibilities in (17). □

Note that for any $x \in \mathbb{R}^n$, x is feasible for (1) if and only if $(x, 0)$ is feasible for (3) and $c_{\mathscr{E}}(x) = 0$.

The reformulated problem (3) is surprisingly regular, for we have the following result.

Theorem 2. *Suppose that (x, s) is a feasible point for (3) and that c is continuously differentiable in an open neighbourhood of x. Then* MFCQ *is satisfied at (x, s).*

Proof. Let $d = (0_n, -e)$. There are no equality constraints, and checking the remaining requirement that $J_{\mathscr{A}}(x, s)d < 0$ for active constraints is trivial given the form (11) of $J^{\mathrm{S}}(v_{\mathrm{P}})$. □

As a consequence, all sets of Lagrange multipliers associated with first-order critical points are bounded.

Note that the MFCQ condition is satisfied at *every* feasible (x, s) and not only at local solutions of (3), regardless of any constraint qualification being satisfied for (1). Of course Theorem 2 may have been anticipated, since the same is true for (2)—for this problem, the set of corresponding sub-gradients of the non-differentiable constraint norms is automatically bounded [14, §14.3].

Since any constraint qualification is a property of the algebraic description of a feasible set, Theorem 2 holds true for (14) as well. There thus always exist multipliers satisfying (15).

Some of the results we will establish later require that the far stronger linear independence constraint qualification (LICQ)—that the rows of (11) corresponding to active indices are independent—be satisfied for (3). To obtain LICQ on (3), one may unfortunately need to have as strong an assumption as the active constraint gradients being linearly independent over the whole feasible set.

In the following, we are taking advantage of MFCQ being satisfied for (3) but for generality, are not tacitly assuming that MFCQ is satisfied for (1).

We start with the following fundamental assumption.

Assumption 2.1. *The functions f, $c_{\mathscr{E}}$ and $c_{\mathscr{I}}$ are continuously differentiable over an open set containing the feasible set of (1).*

We now examine the relationships between stationary points of (1) and (3). The following results are adaptations or variations of results of [31]. Our first result gives an important property of solutions to (3).

Theorem 3. *If Assumption 2.1 is satisfied, if (v_P, v_D) is a first-order critical point for (3) with fixed penalty parameter $\nu > 0$ and if $c_{\mathscr{E}}(x) = 0$ and $c_{\mathscr{I}}(x) \geq 0$, then $s = 0$.*

Proof. If $i \in \mathscr{E}$, $c_i(x) = 0$ and from (13d), we have $s_i y_i = 0$. It cannot be that $s_i > 0$ since then $y_i = 0$ and (13b) would imply $u_i = 2\nu$ and consequently (13e) gives that $s_i = 0$, which is a contradiction. Therefore $s_{\mathscr{E}} = 0$. For $i \in \mathscr{I}$, if $c_i(x) = 0$, as before (13c) and (13e) guarantee that $s_i = 0$. Otherwise, $c_i(x) > 0$ and (13f), (13d), (13c) and (13e) successively imply that $c_i(x) + s_i > 0$, that $y_i = 0$, that $u_i = \nu$ and finally that $s_i = 0$. Hence we also have $s_{\mathscr{I}} = 0$, which completes the proof. □

This first result confirms intuition about the reformulation that led to (3), namely that all the elastic variables should eventually vanish if a critical point which is also feasible for (1) has been found.

The following result connects systems (7) and (13) and parallels results from [31] and [34, Proposition 3].

Theorem 4. *If Assumption 2.1 is satisfied, if (v_P, v_D) is a first-order critical point for (3) with fixed penalty parameter $\nu > 0$ and if $c_{\mathscr{E}}(x) = 0$ and $c_{\mathscr{I}}(x) \geq 0$, then the shifted vector $(x, \lambda(y, \nu))$ from (12b) is a first-order critical point for (1).*

Proof. Primal feasibility with respect to the linear constraints and non-negativity of x follows directly from the assumption. The dual feasibility condition (13a) readily implies that (7a) is satisfied with the given multipliers. The feasibility conditions (7c), (7d) are satisfied by (13f) and our assumptions. Moreover, Theorem 3 gives that $s = 0$, and hence (13d) implies (7c) as $\lambda_{\mathscr{I}}(y, \nu) = y_{\mathscr{I}}$ by definition. □

Conversely, we now show that provided there exist finite Lagrange multipliers for (1) and for sufficiently large values of the penalty parameter, every stationary point of (1) is a stationary point of (3).

Theorem 5. *If Assumption 2.1 is satisfied, suppose x^* is a first-order critical point for (1) for which the Lagrange multipliers λ^* are finite. Then for all $\nu \geq \|\lambda^*\|_\infty$, the shifted primal–dual vector (v_P, v_D), where $v_P = (x^*, 0)$ and $v_D = (y(\lambda^*, \nu), \nu e - \lambda^*)$ from (12a), is a first-order critical point for (3).*

Proof. Because $\lambda^* \geq 0$, the smallest value of ν for which $\lambda^*_{\mathscr{E}} + \nu e_{\mathscr{E}} \geq 0$, $\nu e_{\mathscr{E}} - \lambda^*_{\mathscr{E}} \geq 0$ and $\nu e_{\mathscr{I}} - \lambda^*_{\mathscr{I}} \geq 0$ is given by $\|\lambda^*\|_\infty$. For any $\nu \geq \|\lambda^*\|_\infty$, the proof is completed by a straightforward verification that the given primal–dual vector satisfies (13) using the assumed conditions (7). □

Note that Theorem 5 deals with one particular critical point and one particular, possibly out of many, vector of Lagrange multipliers associated with it. A standard, but stronger, assumption to ensure boundedness of the multipliers in Theorem 5 is to impose MFCQ on (1) [15].

3 The Full Algorithm

As we have already suggested, an appealing way to solve the reformulated problem (3) is to (approximately) minimize a sequence of *logarithmic barrier* functions (4) for a sequence $\{\mu^k\}$ of positive barrier parameters whose limit is zero and, in this case, a possibly increasing sequence $\{\nu^k\}$ of positive penalty parameters.

For convenience, we define (primal) first-order Lagrange multiplier estimates

$$y(x,s) := \mu(C(x)+S)^{-1}e, \qquad u(s) := \mu S^{-1}e, \tag{19}$$

where, as before, a capital letter denotes the diagonal matrix whose diagonal is the vector denoted by the corresponding lowercase letter. Using these multiplier estimates, the gradient of the barrier function with respect to $v_{\text{P}} = (x,s)$ is

$$\nabla\phi^{\text{B}}(v_{\text{P}};\mu,\nu) = \begin{bmatrix} \nabla f(x) - J^T(x)(y(x,s) - \nu e^0_{\mathscr{E}}) \\ \nu e - (y(x,s) - \nu e^0_{\mathscr{E}}) - u(s) \end{bmatrix}. \tag{20}$$

Given fixed values of the barrier and penalty parameters $\mu, \nu \geq 0$, primal and dual vectors $v_{\text{P}} = (x,s)$ and $v_{\text{D}} = (y,u)$ and primal–dual vector $v = (v_{\text{P}}, v_{\text{D}})$, we also define the primal–dual function $\Phi : \mathbb{R}^{n+3n_{\mathscr{C}}} \to \mathbb{R}^{n+3n_{\mathscr{C}}}$ as

$$\Phi(v;\mu,\nu) := \begin{bmatrix} \nabla f(x) - J^T(x)(y - \nu e^0_{\mathscr{E}}) \\ \nu e - (y - \nu e^0_{\mathscr{E}}) - u \\ (C(x)+S)y - \mu e \\ Su - \mu e \end{bmatrix}. \tag{21}$$

As is well known, the first-order criticality conditions for (4) are equivalently described by the *primal–dual* system

$$\Phi(v;\mu,\nu) = 0, \qquad (c(x)+s, s, y, u) \geq 0. \tag{22}$$

In addition, observe that the KKT conditions (13) for (3) are simply (22) with $\mu = 0$.

Algorithm 3.1 Prototype algorithm—outer iteration

Step 0. Let the forcing functions $\varepsilon^{\mathrm{D}}(\cdot)$, $\varepsilon^{\mathrm{C}}(\cdot)$ and $\varepsilon^{\mathrm{U}}(\cdot)$ be given, and let $\kappa_\nu > 0$. Choose $x^0 \in \mathbb{R}^n$, $s^0 \in \mathbb{R}_+^{n_{\mathscr{C}}}$ such that $c(x^0) + s^0 > 0$, initial dual estimates $y^0, u^0 \in \mathbb{R}_+^{n_{\mathscr{C}}}$, and penalty and barrier parameters ν^0 and $\mu^0 > 0$, and set $k = 0$.

Step 1. Inner Iteration: choose a suitable scaling norm $\|.\|_{P^{k+1}}$ and find a new primal–dual iterate $v^{k+1} = (x^{k+1}, s^{k+1}, y^{k+1}, u^{k+1})$ satisfying

$$\left\| \begin{bmatrix} \nabla f(x^{k+1}) - J^T(x^{k+1})(y^{k+1} - \nu^k e^0_{\mathscr{E}}) \\ \nu^k e - (y^{k+1} - \nu^k e^0_{\mathscr{E}}) - u^{k+1} \end{bmatrix} \right\|_{[P^{k+1}]} \leq \varepsilon^{\mathrm{D}}(\mu^k) \tag{23a}$$

$$\|(C(x^{k+1}) + S^{k+1})y^{k+1} - \mu^k e\| \leq \varepsilon^{\mathrm{C}}(\mu^k) \tag{23b}$$

$$\|S^{k+1}u^{k+1} - \mu^k e\| \leq \varepsilon^{\mathrm{U}}(\mu^k) \tag{23c}$$

$$\left(c(x^{k+1}) + s^{k+1}, s^{k+1}\right) > 0 \tag{23d}$$

$$\text{and} \quad \left(\nu^k[e + e^0_{\mathscr{E}}] + \kappa_\nu e, \nu^k[e + e^0_{\mathscr{E}}] + \kappa_\nu e\right) \geq \left(y^{k+1}, u^{k+1}\right) > 0 \tag{23e}$$

by (for example) approximately minimizing (4).

Step 2. Select a new barrier parameter, $\mu^{k+1} \in (0, \mu^k]$ such that $\lim_{k\to\infty} \mu^k = 0$. If necessary, adjust the penalty parameter, ν^k. Increment k by one, and return to Step 1.

We call $\varepsilon(\cdot)$ a *forcing function* if $\varepsilon(\mu) > 0$ for all $\mu > 0$ and $\varepsilon(\mu) \downarrow 0$ as $\mu \downarrow 0$ [32]. Since the Hessian of the logarithmic barrier function (4) can be highly ill-conditioned, it is vital that we dynamically (and implicitly) scale the variables to mitigate this effect. At iteration k, we shall measure variables using a norm, say $\|\cdot\|_{P^k}$, designed to achieve this, and gradients in the dual norm, denoted $\|\cdot\|_{[P^k]}$. We shall return to this shortly. We summarize our algorithm as Algorithm 3.1.

The required upper bounds on the dual variables (y^{k+1}, u^{k+1}) in (23e) are simply those ultimately implied by (13b), (13c) and (13f), with a little elbow room provided by $\kappa_\nu > 0$ to allow for finite termination of the inner iteration. Crucially, although the primal multiplier estimates $y^{k+1} = y(x^{k+1}, s^{k+1})$ and $u^{k+1} = u(s^{k+1})$ might be used in (23), there is no necessity that this be so.

The update of the barrier parameter in Step 2 may follow traditional rules but should ultimately allow for a superlinear decrease if fast asymptotic convergence is sought—this issue is addressed in Section 4.3.

The penalty parameter update appears in Step 2 of the algorithm for clarity, but a practical implementation might make provision for updates of ν^k *inside* the inner iteration and possibly to allow occasional decreases of ν. A suitable update for the penalty parameter is less obvious, but we shall discuss alternatives in Section 3.2.

3.1 *The Trust-Region Inner Iteration*

In order to address concretely the practical aspects of Algorithm 3.1, we use trust-region models that incorporate exact second-order derivative information. In order to be able to do this, we must replace Assumption 2.1 by the following assumption.

Assumption 3.1. *The functions f, $c_{\mathcal{E}}$ and $c_{\mathcal{I}}$ are twice continuously differentiable over an open set covering all iterates encountered by Algorithm 3.1.*

Given a strictly feasible point v_{P}, a typical primal–dual interior-point trust-region method for solving (4) attempts to find an improved point $v_{\mathrm{P}} + d = (x + d_x, s + d_s)$, where $d = (d_x, d_s)$ approximately solves the primal–dual subproblem

$$\underset{d \in \mathscr{B}(\Delta)}{\text{minimize}}\ \nabla_{v_{\mathrm{P}}} \phi^{\mathrm{B}}(v_{\mathrm{P}}; \mu, \nu)^T d + \tfrac{1}{2} d^T H^{\mathrm{PD}}(v) d \quad \text{subject to } \|d\|_P \le \Delta, \tag{24}$$

where $\|\cdot\|_P$ is an appropriate scaling norm, the primal–dual Hessian is defined by

$$H^{\mathrm{PD}}(v) = \begin{bmatrix} H(x, \lambda(y, \nu)) + J^T(x)\Theta(v)J(x) & J^T(x)\Theta(v) \\ \Theta(v)J(x) & \Theta(v) + US^{-1} \end{bmatrix}, \tag{25}$$

with

$$\Theta(v) = Y(C(x) + S)^{-1}, \tag{26}$$

for some suitable strictly positive primal–dual multiplier estimates u and y, where

$$H(x, \lambda) = \nabla_{xx} f(x) - \sum_{i \in \mathscr{C}} \lambda_i \nabla_{xx} c_i(x) = \nabla_{xx} L(x, \lambda) \tag{27}$$

is the Hessian of the Lagrangian (6), and $\lambda(y, \nu)$ is defined by (12b). Under standard assumptions on these estimates and as convergence occurs, the difference between $\nabla^2 \phi^{\mathrm{B}}(v_{\mathrm{P}}; \mu, \nu)$ and $H^{\mathrm{PD}}(v)$ is insignificant [11, Theorem 13.9.1].

Besides the step-computing procedure, our trust-region algorithm is quite standard. The step d is accepted or rejected based on how much of the reduction in (4) predicted by (24) is actually achieved—a poor prediction results in a reduction in the trust-region radius, Δ, while an accurate one may be rewarded by an increase in Δ. Since the logarithmic barrier function is undefined outside (or on the boundary) of the shifted feasible region $\{(x, s) \mid c(x) + s \ge 0 \quad \text{and} \quad s \ge 0\}$, any step $v_{\mathrm{P}} + d_{\mathrm{P}}$ outside this region is automatically rejected, and the trust-region radius reduced. See [11, Chapter 13] for more details. Unlike other trust-region interior-point methods such as KNITRO [5], no direct attempt is made to enforce feasibility by imposing extra constraints on the trust-region subproblem.

We may find an approximation to the solution to (24) using the Generalized Lanczos Trust-Region GLTR method of [19]. This method requires that, at each iteration, we solve "preconditioning" systems of the form (now dropping suffices ^{PD})

$$K(v)d \equiv \begin{bmatrix} P+J^T(x)\Theta(v)J(x) & J^T(x)\Theta(v) \\ \Theta(v)J(x) & \Theta(v)+US^{-1} \end{bmatrix} \begin{bmatrix} d_x \\ d_s \end{bmatrix} = \begin{bmatrix} r_x \\ r_s \end{bmatrix} \equiv r \tag{28}$$

for appropriate right-hand sides r and where $\Theta(v)$ is defined in (26). Here P is a suitable "preconditioning" approximation to H, and can range from the naive ($P = I$) to the sophisticated ($P = H$), but must be chosen so that $K(v)$ is positive definite. As [10] explains, the preconditioner used defines the scaling norm appropriate for the trust-region in (24) and the dual norm appropriate to measure progress towards dual feasibility. In particular the dual norm satisfies $\|r\|^2_{[P]} = d^T r$, where d is the solution to (28).

Of particular concern, however, is that the matrix $J^T(x)\Theta(v)J(x)$ in (28) might be dense, making a direct factorization of $K(v)$ unviable. Fortunately, upon defining $\xi := \Theta(v)(J(x)d_x + d_s)$, (28) may be rewritten as the sparser

$$\begin{bmatrix} P & J^T(x) \\ J(x) & -\Theta^{-1}(v) - U^{-1}S \end{bmatrix} \begin{bmatrix} d_x \\ \xi \end{bmatrix} = \begin{bmatrix} r_x \\ -U^{-1}Sr_s \end{bmatrix} \tag{29}$$

where we recover $d_s = -U^{-1}S\xi + U^{-1}Sr_s$.

Significantly $K(v)$ is positive definite if and only if (29) has precisely $n_{\mathscr{C}}$ negative eigenvalues [18], so we can ensure that P is appropriate whenever an inertia-calculating factorization (such as those given by the codes `MA27` and `MA57` of the [25]) is used.

3.2 Updating Dual Variables and the Penalty Parameter

Given newly computed primal values v_{P}^+, we follow [10] and project candidate dual variables v_{D}^+ componentwise into the box $[(y^{\mathrm{L}}, u^{\mathrm{L}}), (y^{\mathrm{U}}, u^{\mathrm{U}})]$, where

$$\begin{aligned} y^{\mathrm{L}} &= \kappa_l \min\left[e, y, \mu^k (C(x^+) + S^+)^{-1} e\right], \\ y^{\mathrm{U}} &= \max\left[\kappa_u e, y, \kappa_u (\mu^k)^{-1} e, \kappa_u \mu^k (C(x^+) + S^+)^{-1} e\right], \end{aligned} \tag{30}$$

$$u^{\mathrm{L}} = \kappa_l \min\left[e, u, \mu^k (S^+)^{-1} e\right], \quad u^{\mathrm{U}} = \max\left[\kappa_u e, u, \kappa_u (\mu^k)^{-1} e, \kappa_u \mu^k (S^+)^{-1} e\right],$$

in order to ensure that the multipliers remain sufficiently positive and suitably bounded. Here $0 < \kappa_l < 1 < \kappa_u$, and values $\kappa_l = \frac{1}{2}$ and $\kappa_u = 10^{20}$ have proved to be satisfactory. Note that the primal estimates (19), $v_{\mathrm{D}}^+ = v_{\mathrm{D}}(v_{\mathrm{P}}^+)$, naturally lie in the interval. However, as we have just mentioned, we usually prefer to use primal–dual estimates $v_{\mathrm{D}}^+ = v_{\mathrm{D}} + d_{\mathrm{D}}$ of the dual variables, where d_{D} is the correction to the dual variable estimates obtained from the trust-region subproblem (24). Our convergence analysis is actually independent of how this is done, so long as the resulting estimates lie in the box above.

The purpose of the penalty parameter is to force satisfaction of the equality and inequality constraints for (1). Introducing decreasing sequences $\{\eta^k_{\mathscr{E}}\}$ and $\{\eta^k_{\mathscr{I}}\}$ converging to zero, we increase ν^k whenever

$$\|c_{\mathscr{E}}(x^k)\| > \eta^k_{\mathscr{E}} \qquad \text{or} \qquad \|c^-_{\mathscr{I}}(x^k)\| > \eta^k_{\mathscr{I}}. \tag{31}$$

Refining further, we also increase ν^k whenever

$$\|y^{k+1} - \nu^k e^0_{\mathscr{E}}\| \leq \gamma \nu^k \tag{32}$$

is violated, for some preset $\gamma \in (0,1)$. Then one possibility is to update ν^k using

$$\nu^{k+1} = \begin{cases} \max[\tau_1 \nu^k, \nu^k + \tau_2] & \text{if (31) is satisfied or (32) is violated,} \\ \nu^k & \text{otherwise,} \end{cases} \tag{33}$$

for some preset constants $\tau_1 > 1$ and $\tau_2 > 0$, following rules suggested by Mayne and Polak [31] and Conn et al. [11].

Remarkably, the convergence results of Section 4.2 are independent of the particular form of the sequences $\{\eta^k_{\mathscr{E}}\}$ and $\{\eta^k_{\mathscr{I}}\}$ besides the fact that they are sequences of positive numbers converging to zero. In practice, all such sequences might not be equally efficient and those converging to zero at a reasonable rate should be chosen.

4 Convergence Analysis

In this section, we discuss the convergence properties of Algorithm 3.1 for the solution of (1). We consider, in turn, the global convergence of the inner iteration, of the outer iteration, and fast local convergence issues. In order to derive suitable convergence results for the convergence of our interior-point method, we make the following additional assumptions.

Assumption 4.1. *The logarithmic barrier function $\phi^B(x,s;\mu,\nu)$ for problem (3), defined in (4), is bounded below over the set $\{(x,s) \mid c(x)+s \geq 0,\ s \geq 0\}$ for all $\mu > 0$*

Assumption 4.2. *The iterates remain in a region Ω over which the first and second derivatives $\nabla f(x)$, $\nabla_{xx} f(x)$, $\nabla c_i(x)$ and $\nabla_{xx} c_i(x)$ ($i \in \mathscr{C}$) remain uniformly bounded.*

4.1 Convergence of the Inner Iteration

Each inner iteration—Step 1 of Algorithm 3.1—proceeds by computing a vector of primal $v^k_{\mathrm{P}} = (x^k, s^k)$ and dual variables $v^k_{\mathrm{D}} = (y^k, u^k)$ satisfying (23) by means of the method described in [10]. We devote this section to verifying that the assumptions required by this method are satisfied in the present case, and to recalling the main convergence properties of the resulting inner iteration. We shall only be concerned with exact derivatives of the quantities involved, but the aforementioned inner iteration makes provision for inexact Hessian matrices provided they satisfy appropriate regularity and asymptotic properties.

As we already mentioned, we must require the following condition on the preconditioning matrices P^k chosen during Step 1 of Algorithm 3.1.

Assumption 4.3. *Each preconditioning matrix P^k is both bounded from above in norm, and such that the smallest eigenvalue of the matrix K from the system (28) is uniformly positive for all iterates encountered.*

For simplicity, we consider the matrix P^k fixed during an inner iteration, although this need not be the case [10]. Let an outer iteration index be denoted by k and the successive values taken by a generic vector v during the inner iterations corresponding to this outer iteration be denoted by $v^{k,j}$, $j = 1,2,\ldots$ The following assumption introduces upper bounds on the sequences of multipliers.

Assumption 4.4. *For all $k \geq 0$, there exists a constant $\kappa^D(k)$ depending only on k such that*

$$y^{k,j} \leq \kappa^D(k)\max((C(x^{k,j})+S^{k,j})^{-1}e,e) \qquad u^{k,j} \leq \kappa^D(k)\max((S^{k,j})^{-1}e,e). \tag{34}$$

In view of MFCQ, requiring that the Lagrange multipliers remain bounded is very reasonable for fixed (μ^k, v^k). Indeed, if (23e) were to be imposed for every inner iteration, Assumption 4.4 would automatically be satisfied.

Armed with the above assumptions, the next result corresponds to [10, Theorem 2].

Theorem 6. *Under Assumptions 2.1–4.4, the inner iteration procedure corresponding to outer iteration k of Algorithm 3.1 generates a sequence $\{(x^{k,j},s^{k,j})\}$ satisfying*

$$\lim_{j\to\infty} \|\nabla\phi^B(v_P^{k,j};\mu^k,v^k)\|_{[P^k]} = \lim_{j\to\infty} \|\nabla\phi^B(v_P^{k,j};\mu^k,v^k)\| = 0.$$

Proof. It is readily verified that Assumptions 3.1–4.4 imply Assumptions A1–A8 of [10] and thus global convergence of the inner iteration. Theorem 2 of [10] concludes the proof. □

Theorem 6 shows that the inner-iteration termination test will be satisfied after a finite number of iterations if primal multiplier estimates $y^{k+1} = y(x^{k+1},s^{k+1})$ and $u^{k+1} = u(s^{k+1})$ are used. If we plan to use other dual variables, we require an extra assumption, namely that the primal–dual estimates converge to their ideal, primal, values when convergence takes place.

Assumption 4.5. *The inner iteration produces dual sequences* $\{u^{k,j}\}$ *and* $\{y^{k,j}\}$ *satisfying* $\lim_{j\to\infty}\|u^{k,j}-\mu^k(S^{k,j})^{-1}e\|=0$ *and* $\lim_{j\to\infty}\|y^{k,j}-\mu^k(C(x^{k,j})+S^{k,j})^{-1}e\|=0$ *whenever* $\lim_{j\to\infty}\|\nabla\phi^B(v_P^{k,j};\mu^k,v^k)\|_{[P^k]}=0$.

With this additional assumption, we obtain the following result.

Theorem 7. *Under Assumptions 3.1–4.5, the inner iteration procedure corresponding to outer iteration k of Algorithm 3.1 generates a sequence* $\{(v_P^k,v_D^k)\}$ *satisfying the stopping conditions (23) after finitely many steps.*

Proof. The stated assumptions allow us to use Theorem 4 of [10] to deduce that the sequence $\{(v_P^{k,j},v_D^{k,j})\}$ generated by Algorithm 3.1 ultimately satisfies $\lim_{j\to\infty}\Phi(v^{k,j};\mu^k,v^k)=0$ and $\lim_{j\to\infty}(c(x^{k,j})+s^{k,j},s^{k,j},y^{k,j},u^{k,j})\geq 0$ and thus indirectly that $\lim_{j\to\infty}\left(y^{k,j}+v^k e^0_{\mathscr{E}},u^{k+1}\right)\leq v^k\left(e+e^0_{\mathscr{E}},e+e^0_{\mathscr{E}}\right)$. Thus (23) is satisfied after finitely many steps, since Lemma 2 of [10] shows that the $\|.\|_{[P^{k+1}]}$ and Euclidean norms are equivalent for fixed k. □

The numerical method suggested in Sections 3.1–3.2 to tackle the inner iteration satisfies the assumptions stated here, and thus guarantees global convergence of each inner iteration.

4.2 Convergence of the Outer Iteration

We now study the convergence of the outer iteration algorithm. We concentrate on the case where the penalty parameter is updated as in Section 3.2. Our first task is to show that although we are measuring the violation of dual feasibility in (23a) in the $\|\cdot\|_{[P^{k+1}]}$ norm, this actually allows us to make deductions in the Euclidean norm. To do this, we need to be slightly more restrictive in the choice of our forcing functions ε^{D}, ε^{C} and ε^{U}, and we make the following assumption.

Assumption 4.6. *The forcing functions* ε^D, ε^C *and* ε^U *satisfy the bounds*

$$\varepsilon^C(\mu)\leq\kappa_c\mu,\quad \varepsilon^U(\mu)\leq\kappa_c\mu,\quad \varepsilon^D(\mu)\leq\kappa_d\mu^{\frac{1}{2}+\gamma^k}, \tag{35}$$

for some constants $\kappa_c\in(0,1)$ *and* $\kappa_d>0$ *and sequence* $\{\gamma^k\}>0$.

We then have the following result.

Lemma 1. *Suppose that the iterates* $v^{k+1}=(x^{k+1},s^{k+1},y^{k+1},u^{k+1})$ *are generated by Algorithm 3.1, and that Assumptions 4.2, 4.3 and 4.6 hold. Then there exist constants* $\mu_{\max}$ *and* $\kappa>0$ *for which* $\|v\|\leq\kappa(v^k+\kappa_v)/\sqrt{\mu^k}\|v\|_{[P^{k+1}]}$ *for all* $\mu^k\leq\mu_{\max}$ *and all vectors v, and, additionally,* $\|v\|\leq\kappa(v^k+\kappa_v)(\mu^k)^{\gamma^k}$ *whenever* $\|v\|_{[P^{k+1}]}\leq\varepsilon^D(\mu^k)$.

Proof. The requirements (23b) and (35) imply that $(c_i(x^{k+1})+s_i^{k+1})y_i^{k+1}\geq(1-\kappa_c)\mu^k$. Combining this bound with the required upper bound from (23e) reveals

$$c_i(x^{k+1}) + s_i^{k+1} \geq \frac{(1-\kappa_c)\mu^k}{y_i^{k+1}} \geq \frac{(1-\kappa_c)\mu^k}{2\nu^k + \kappa_\nu} > \frac{(1-\kappa_c)\mu^k}{2(\nu^k + \kappa_\nu)}. \tag{36}$$

Similarly, (23c) and (23e) and (35) give that $s_i^{k+1} \geq (1-\kappa_c)\mu^k/(2\nu^k + \kappa_\nu) > (1-\kappa_c)\mu^k/(2(\nu^k + \kappa_\nu))$. But the form of the Jacobian in (11) together with Assumptions 4.2, 4.3 and 4.6 are sufficient to allow us to invoke [10, Lemma 4.1] to deduce that

$$\|v\|_{[P^{k+1}]} \geq \kappa_2 \min\left(\min_{i\in\mathscr{C}} \frac{c_i(x^{k+1}) + s_i^{k+1}}{\sqrt{\mu^k}}, \min_{i\in\mathscr{C}} \frac{s_i^{k+1}}{\sqrt{\mu^k}}, 1 \right) \|v\| \tag{37}$$

for some $\kappa_2 > 0$ and all v. Combining (36)–(37), we see that

$$\|v\|_{[P^{k+1}]} \geq \kappa_2 \min\left(\frac{(1-\kappa_c)\sqrt{\mu^k}}{2(\nu^k + \kappa_\nu)}, 1 \right) \|v\| \geq \frac{\kappa_2(1-\kappa_c)\sqrt{\mu^k}}{2(\nu^k + \kappa_\nu)} \|v\|$$

for all $\mu^k \leq \mu_{\max} := (2\kappa_\nu/(1-\kappa_c))^2$, which is the first required bound when $\kappa := 2/(\kappa_2(1-\kappa_c))$. The remaining bound follows directly from the first and (35). □

In the following results, we shall be concerned with limit points $v_{\rm P}^* = (x^*, s^*)$ and $v_{\rm D}^* = (y^*, u^*)$, of the primal and dual sequences respectively, generated by Algorithm 3.1. In order to easily make connections with Theorem 4, we shall be using the *shifted* limit point $(x, \lambda(y^*, \nu^*))$ as defined in (12b).

We first consider the case where the penalty parameter remains bounded.

Lemma 2. *Suppose Assumptions 3.1 and 4.2–4.6 hold, Algorithm 3.1 generates infinite sequences $\{v_{\rm P}^k\}$ and $\{v_{\rm D}^k\}$, and the penalty parameter ν^k is updated finitely many times to reach its final value ν^*. Then the sequence $\{(s^k, y^k, u^k)\}$ is bounded. Moreover, if $\{x^k\}$ has a limit point and if $(v_{\rm P}^*, v_{\rm D}^*)$ is any limit point of $\{v^k\}$, then $s^* = 0$ and the shifted limit point $(x^*, \lambda(y^*, \nu^*))$ is a first-order critical point for (1).*

Proof. By assumption, there exists a positive integer k^* such that $\nu^k = \nu^*$ for all $k \geq k^*$. The updating rule (33) then implies that $\|c_{\mathscr{E}}(x^k)\| \leq \eta_{\mathscr{E}}^k$ and $\|c_{\mathscr{I}}^-(x^k)\| \leq \eta_{\mathscr{I}}^k$. Consequently, $\lim_{k\to\infty} c_{\mathscr{E}}(x^k) = 0$ and $\lim_{k\to\infty} c_{\mathscr{I}}(x^k) \geq 0$.

We first show that $\{s^k\}$ is bounded. Assume by contradiction that $s_i^k \to \infty$ for some $i \in \mathscr{C}$ along some subsequence. By using the forcing property of the functions $\varepsilon^{\rm D}(\cdot)$, $\varepsilon^{\rm U}(\cdot)$ and $\varepsilon^{\rm C}(\cdot)$, Lemma 1 and the fact that $\mu^k \downarrow 0$, from (23c), we must have $u_i^k \to 0$ and from (23a), $\{y_i^k\}$ must be bounded. Hence, (23b) imposes $c_i(x^k) \to -\infty$, which is a contradiction. Thus $\{s^k\}$ must be bounded. Moreover, for all $k \geq k^*$, (23e) implies that $\{(y^k, u^k)\}$ satisfies the bounds $(y_i^k, u_i^k) \in [0, \kappa_\nu + 2\nu^*]$ for $i \in \mathscr{E}$ and $(y_i^k, u_i^k) \in [0, \kappa_\nu + \nu^*]$ for $i \in \mathscr{I}$.

Suppose that $\lim_{k\in\mathscr{K}} v^k = (v_{\rm P}^*, v_{\rm D}^*)$. Along the subsequence defined by $\mathscr{K}$, (23b)–(23d), the forcing property of the function $\varepsilon^{\rm D}(\cdot)$, Lemma 1 and the fact that $\mu^k \downarrow 0$ together guarantee that

$$\lim_{k\in\mathcal{K}} \begin{bmatrix} \nabla f(x^{k+1}) - J^T(x^{k+1})(y^{k+1} - \nu^k e^0_{\mathcal{E}}) \\ \nu^k e - (y^{k+1} - \nu^k e^0_{\mathcal{E}}) - u^{k+1} \end{bmatrix}$$

$$= \begin{bmatrix} \nabla f(x^*) - J^T(x^{k+1})(y^* - \nu^* e^0_{\mathcal{E}}) \\ \nu^k e - (y^* - \nu^* e^0_{\mathcal{E}}) - u^* \end{bmatrix} = 0$$

as well as $(C(x^*) + S^*)y^* = 0$ and $S^*u^* = 0$. Thus $(v^*_{\mathrm{P}}, v^*_{\mathrm{D}})$ satisfies (22) with $\mu = 0$ and the assumptions of Theorems 3 and 4. □

Next, we consider the consequences of an unbounded penalty parameter.

Lemma 3. *Suppose Assumptions 3.1 and Assumptions 4.2–4.6 hold. Let $\{v^k_{\mathrm{P}}\}$ and $\{v^k_{\mathrm{D}}\}$ be sequences generated by Algorithm 3.1. Assume the penalty parameter ν^k is updated infinitely many times at iterations $k \in \mathcal{K}$. Then the subsequence $\{(y^k, u^k)\}_{\mathcal{K}}$ is unbounded. In addition any limit point $v^*_{\mathrm{P}} = (x^*, s^*)$ of $\{v^k_{\mathrm{P}}\}$ solves (14) and x^* is a first-order critical point of $\vartheta^{\mathrm{P}}(x)$.*

Proof. Along $\mathcal{K}$, (33) implies $\nu^{k+1} \geq \nu^k + \tau_2$ with $\tau_2 > 0$ and thus $\{\nu^k\}_{\mathcal{K}} \to \infty$. Since ν^k is nondecreasing, the whole sequence $\{\nu^k\} \to \infty$.

Now suppose that $\{v^k_{\mathrm{D}}\}_{\mathcal{K}}$ is bounded and thus has a limit point v^*_{D}. In particular, there are vectors y^* and u^* such that $\{y^k_{\mathcal{E}}\}_{\mathcal{K}'} \to y^*$ and $\{u^k\}_{\mathcal{K}'} \to u^*$ for some $\mathcal{K}' \subseteq \mathcal{K}$, and thus both $\|y^k\| \leq 2\|y^*\|$ and $\|u^k\| \leq 2\|u^*\|$ for all sufficiently large $k \in \mathcal{K}'$. But then the triangle inequality, the stopping condition (23a) and Lemma 1 give that

$$\sqrt{n_{\mathcal{C}}}\nu^{k-1} - (\|y^k\| + \|u^k\|) \leq \|\nu^{k-1}e - y^k_{\mathcal{E}} - u^k_{\mathcal{E}}\| \leq \kappa(\nu^{k-1} + \kappa_\nu)(\mu^{k-1})^{\gamma^k}$$

and this combines with the bounds on $\|y^k\|$ and $\|u^k\|$ to give

$$\begin{aligned} (\sqrt{n_{\mathcal{C}}} - \kappa(\mu^{k-1})^{\gamma^k})\nu^{k-1} &\leq (\|y^k\| + \|u^k\|) + \kappa\kappa_\nu(\mu^{k-1})^{\gamma^k} \\ &\leq 2(\|y^*\| + \|u^*\|) + \kappa\kappa_\nu(\mu^{k-1})^{\gamma^k} \end{aligned} \tag{38}$$

for all sufficiently large $k \in \mathcal{K}'$. Taking the limit of (38) as $k \to \infty$ then contradicts the unboundedness of $\{\nu^{k-1}\}$. Thus $\{v^k_{\mathrm{D}}\}_{\mathcal{K}}$ is unbounded.

To prove the second part of the lemma, we now suppose that $\{v^k_{\mathrm{P}}\}$ has a limit point v^*_{P}. Define $\bar{y}^{k+1} = y^{k+1}/\nu^k$ and $\bar{u}^{k+1} = u^{k+1}/\nu^k$. Then the stopping rules (23) and Lemma 1 give

$$\left\| \begin{bmatrix} \frac{1}{\nu^k}\nabla f(x^{k+1}) - J^T(x^{k+1})(\bar{y}^{k+1} - e^0_{\mathcal{E}}) \\ e - (\bar{y}^{k+1} - e^0_{\mathcal{E}}) - \bar{u}^{k+1} \end{bmatrix} \right\| \leq \kappa_p(\mu^k)^{\gamma^k} \tag{39a}$$

$$\left\| (C(x^{k+1}) + S^{k+1})\bar{y}^{k+1} - \frac{\mu^k}{\nu^k}e \right\| \leq \frac{\varepsilon^{\mathrm{C}}(\mu^k)}{\nu^k}. \tag{39b}$$

$$\left\| S^{k+1}\bar{u}^{k+1} - \frac{\mu^k}{\nu^k} e \right\| \le \frac{\varepsilon^{\mathrm{U}}(\mu^k)}{\nu^k} \tag{39c}$$

$$\left(c(x^{k+1}) + s^{k+1}, s^{k+1}\right) > 0 \tag{39d}$$

$$\left(\left[1 + \frac{\kappa_\nu}{\nu^0}\right] e + e^0_{\mathscr{E}}, \left[1 + \frac{\kappa_\nu}{\nu^0}\right] e + e^0_{\mathscr{E}}\right) \ge (\bar{y}^{k+1}, \bar{u}^{k+1}) > 0 \tag{39e}$$

where $\kappa_p := \kappa(1 + \kappa_\nu/\nu^0)$. Since (39e) implies that $(\bar{y}^{k+1}, \bar{u}^{k+1})$ is bounded, there is a subsequence $\mathscr{K}' \subseteq \mathscr{K}$ for which $\lim_{k \in \mathscr{K}' \to \infty} (\bar{y}^{k+1}, \bar{u}^{k+1}) = (y^*, u^*)$. Taking limits of (39) as $k \in \mathscr{K}' \to \infty$ (and thus $\mu^k \to 0$ and $\nu^k \to \infty$) shows that (x^*, s^*, y^*, u^*) satisfies (15), and hence (x^*, s^*) is a first-order critical point of $\vartheta(x,s)$ subject to $c(s) + s \ge 0$ and $s \ge 0$. The remaining result follows directly from Theorem 1. □

Finally, (32) yields a certificate of failure of the MFCQ whenever the penalty parameter diverges and yet the iterates approach a feasible point.

Lemma 4. *Suppose that Assumptions 3.1 and 4.2–4.6 hold. Let $\{v^k_{\scriptscriptstyle P}\}$ and $\{v^k_{\scriptscriptstyle D}\}$ be sequences generated by Algorithm 3.1. Assume (31) holds for only a finite number of iterations but the penalty parameter ν^k is updated infinitely many times at iterations $k \in \mathscr{K}$. If, in addition, the sequence $\{v^k_{\scriptscriptstyle P}\}$ has a limit point $v^*_{\scriptscriptstyle P}$, x^* is a feasible Fritz-John point of (1) and therefore the MFCQ fails to hold at x^*.*

Proof. As in Lemma 3, the sequences $\{y^k\}_{\mathscr{K}}$ and $\{u^k\}_{\mathscr{K}}$ are unbounded, and from our assumptions, $\|c_{\mathscr{E}}(x^k)\| \le \eta^k_{\mathscr{E}}$ and $\|c_{\mathscr{I}}(x^k)^-\| \le \eta^k_{\mathscr{I}}$ for infinitely many $k \in \mathscr{K}$. By taking limits, we see that x^* is feasible.

Since $\{\nu^k\} \to +\infty$ but increases in ν^k are not due to lack of progress towards feasibility, (32) must be violated infinitely many times. Let $\alpha^{k+1} := \max\{\|y^{k+1}_{\mathscr{E}} - \nu^k e_{\mathscr{E}}\|, \|y^{k+1}_{\mathscr{I}}\|\}$. We have from (32) that $\alpha^{k+1} = \Omega(\nu^k)$ for all $k \in \mathscr{K}$. We now define $\bar{y}^{k+1}_{\mathscr{E}} := (y^{k+1}_{\mathscr{E}} - \nu^k e_{\mathscr{E}})/\alpha^{k+1}$, $\bar{y}^{k+1}_{\mathscr{I}} := y^{k+1}_{\mathscr{I}}/\alpha^{k+1}$, and $\bar{u}^{k+1} := u^{k+1}/\alpha^{k+1}$. By construction, $\|(\bar{y}^{k+1}_{\mathscr{E}}, \bar{y}^{k+1}_{\mathscr{I}})\|_\infty = 1$ for all $k \in \mathscr{K}$. Let $\bar{y}^* = (\bar{y}^*_{\mathscr{E}}, \bar{y}^*_{\mathscr{I}})$ be a limit point of the latter sequence. Upon scaling the stopping conditions (23) by α^{k+1} and taking limits as $k \to \infty$, we see that $\{\bar{u}^{k+1}\}$ must also remain bounded so that, reducing to a further subsequence if necessary, (8) is satisfied with $\gamma = 0$, together with (7c), (7d). Moreover, since $\|\bar{y}^*\|_\infty = 1$ by construction, there is at least one nonzero multiplier, which proves that x^* is a feasible Fritz-John point of (1). Combining those conditions, we obtain

$$\sum_{i \in \mathscr{E}} \bar{y}^*_i \nabla c_i(x^*) + \sum_{i \in \mathscr{A}(x^*)} \bar{y}^*_i \nabla c_i(x^*) = 0,$$

where $\mathscr{A}(x^*)$ is the set of active inequality constraints at x^*. By application of Motzkin's transposition theorem [30], the latter is equivalent to failure of the MFCQ at x^*. □

To summarize, Lemmas 2–4 lead to the following global convergence result.

Theorem 8. *Suppose that Assumptions 3.1 and 4.2–4.6 hold. Let* $\{v_{\text{P}}^k\}$ *and* $\{v_{\text{D}}^k\}$ *be sequences generated by Algorithm 3.1, and that* x^* *is a limit point of* $\{x^k\}$*. Then either* $\{\nu^k\}$ *remains bounded, and* x^* *is a first-order critical point for the nonlinear programming problem (1), or* $\{\nu^k\}$ *diverges, and* x^* *is a first-order critical point of the infeasibility* $\vartheta^{\text{P}}(x)$*.*

4.3 Fast Asymptotic Convergence

We examine in this section the superlinear convergence properties of iterates generated by Algorithm 3.1 in the regular case where LICQ is satisfied for simplicity, although past research suggests that similar convergence properties could be derived under MFCQ [39].

The framework is that of Gould et al. [20, 21]. From Theorem 8, we assume that Algorithm 3.1 generates a sequence $\{v^k\}$ from which a convergent subsequence $\{v^k\}_{\mathscr{K}}$ may be extracted, where $\mathscr{K}$ is an infinite index set, whose limit point $v^* = (v_{\text{P}}^*, v_{\text{D}}^*)$ is feasible, and hence for which the penalty parameter ν^k is only updated finitely many times. We denote its final value by $\nu^* > 0$, and let $\lambda^* = \lambda(y^*, \nu^*)$. We consider indices $k \in \mathscr{K}$ sufficiently large that $\nu^k = \nu^*$ and for related positive quantities α and β, we write $\alpha = O(\beta)$ if there is a constant $\kappa > 0$ such that $\alpha \leq \kappa\beta$ for all β sufficiently small. We write $\alpha = o(\beta)$ if $\alpha/\beta \to 0$ as $\beta \to 0$. We also write $\alpha = \Theta(\beta)$ if $\alpha = O(\beta)$ and $\beta = O(\alpha)$.

From Lemma 2, we have that $s^* = 0$, which enables us to conveniently formulate our assumptions in terms of (1) instead of (3). In particular, all the bound constraints on s in (3) are active and we may thus define the set of active indices in the nonlinear constraints of (3) as $\mathscr{A} \cup \mathscr{E}$ where $\mathscr{A} = \{i \in \mathscr{I} \mid c_i(x^*) = 0\}$. We make the following standard assumptions on (1).

Assumption 4.7. *The gradients* $\{\nabla c_i(x^*) \mid i \in \mathscr{A} \cup \mathscr{E}\}$ *are a linearly independent;*

Assumption 4.8. *The strong second-order sufficiency conditions for (1) are satisfied at* (x^*, λ^*)*, i.e.,* $d^T \nabla_{xx} L(x^*, \lambda^*) d > 0$ *for all* $d \neq 0$ *such that* $\nabla c_i(x^*)^T d = 0$ *for all* $i \in \mathscr{A} \cup \mathscr{E}$*;*

Assumption 4.9. $\|\lambda^*\|_\infty < \nu^*$ *and* $\lambda_i^* > 0$ *for all* $i \in \mathscr{A}$*;*

Assumption 4.10. *The functions* f*,* $c_{\mathscr{E}}(x)$ *and* $c_{\mathscr{I}}(x)$ *are* $\mathscr{C}^3$ *over the intersection of an open neighbourhood of* x^* *with the feasible set of (1).*

Lemma 5. *The penalty problem (3) satisfies* LICQ*, the strong second-order sufficient condition and strict complementarity at* v^* *with a value of the penalty parameter equal to* ν^* *if and only if 4.7–4.9 are satisfied. Moreover, if 4.10 holds, the objective and constraint functions for (3) are* $\mathscr{C}^3$ *in an open neighbourhood of* v_{P}^**.*

Proof. Define the $|\mathscr{A}| \times n$ matrices $J_{\mathscr{A}}(x^*)$ and $E_{\mathscr{A}}$ as the rows of $J_{\mathscr{I}}(x^*)$ and $I_{\mathscr{I}}$ corresponding to indices in $\mathscr{A}$, respectively. The active part of (11) is

$$J^{\mathrm{S}}_{\mathscr{A}}(x^*,0) = \begin{bmatrix} J_{\mathscr{E}}(x^*) & I_{\mathscr{E}} & 0 \\ J_{\mathscr{A}}(x^*) & 0 & E_{\mathscr{A}} \\ 0 & I_{\mathscr{E}} & 0 \\ 0 & 0 & I_{\mathscr{I}} \end{bmatrix}, \tag{40}$$

and has full row rank by Assumption 4.7.

Because the variables s appear linearly in the Lagrangian (9), its Hessian with respect to primal variables $v_{\mathrm{P}} = (x,s)$ is

$$\nabla_{v_{\mathrm{P}}v_{\mathrm{P}}}\mathscr{L}(v;\nu) = \begin{bmatrix} \nabla_{xx}\mathscr{L}(v;\nu) & 0 \\ 0 & 0 \end{bmatrix} = \begin{bmatrix} \nabla_{xx}L(x,\lambda(y,\nu)) & 0 \\ 0 & 0 \end{bmatrix},$$

where $L(x,\lambda)$ is the Lagrangian (6) and $\lambda(y,\nu)$ is defined by (12b), hence imposing the strong second-order sufficient condition on (3) at v^* amounts to Assumption 4.8. The requirement on d follows from (40).

Since $c_{\mathscr{E}}(x^*) = 0$ and $s^* = 0$, strict complementarity on (3) imposes $y_i^* > 0$ for all $i \in \mathscr{A} \cup \mathscr{E}$ and $u_i^* > 0$ for all $i \in \mathscr{C}$. Eliminating u_i^* using (13b), (13c) gives $y^* < \nu^*(e + e_{\mathscr{E}})$, which is in turn equivalent to the bound $\|\lambda^*\|_\infty < \nu^*$ on the multipliers $\lambda^* \equiv \lambda(y^*,\nu^*) = y^* - \nu^* e^0_{\mathscr{E}}$ associated with (1). The final part is immediate. □

Under Assumption 4.9, the central trajectory approaches its end point non-tangentially to active constraints [38]. Differentiating the primal–dual system with respect to μ yields an explicit expression of the *tangent vector* $\dot{v}(\mu)$

$$\nabla_v \Phi(v;\mu,\nu)\dot{v}(\mu) = \begin{bmatrix} 0 \\ 0 \\ -e_{2n_{\mathscr{C}}} \end{bmatrix}. \tag{41}$$

As $\mu \downarrow 0$, this tangent vector converges to a nonzero limit vector $\dot{v}(0)$. As will appear in Theorem 4.3, the individual components of $\dot{v}(0)$ are relevant to fast local convergence issues.

Slightly strengthening (35), we assume in this section that the forcing functions in Algorithm 3.1 have the following asymptotic form

Assumption 4.11. *$\varepsilon^{\mathrm{D}}(\mu^k) = \Theta((\mu^k)^{\gamma^k+1})$ and $\varepsilon^{\mathrm{C,U}}(\mu^k) = \Theta(\mu^k)$, where $0 < \gamma^k < 1$ for all sufficiently large $k \in \mathscr{K}$.*

The only changes in Algorithm 3.1 are that a step is computed according to

$$\nabla_v \Phi(v^k;\mu^k,\nu^*)d^{\mathrm{N}} = -\Phi(v^k;\mu^k,\nu^*), \tag{42}$$

and that the barrier parameter is updated using

$$\mu^{k+1} = \Theta\left((\mu^k)^{\tau^k}\right) \qquad \text{where} \qquad 1+\varepsilon_\tau \le \tau^k \le \frac{2}{1+\gamma^{k+1}} - \varepsilon_\tau. \tag{43}$$

From Assumptions 4.7–4.9, the Jacobian in (42) remains uniformly nonsingular.

Upon defining the set of nonzero components of the tangent vector (41) to the primal–dual central path at v^*,

$$\mathscr{J} = \{i = 1,\ldots,n+2n_{\mathscr{C}} \mid \dot{v}(0)_i \neq 0\}, \tag{44}$$

and under the above assumptions, Algorithm 3.1 fits in the framework of Gould et al. [20, 21] and we obtain the following results, which we state without proof. The first result states that the Newton step d^{N} defined in (42) is strictly feasible and $v^k + d^{\mathrm{N}}$ satisfies the stopping conditions (23) with barrier parameter μ^k.

Theorem 9 ([20], Theorem 6.2). *Under Assumptions 4.7–4.11 for $k \in \mathscr{K}$ sufficiently large, the stopping conditions (23) are satisfied at v^{k+1} with $\mu = \mu^k$, and*

$$\|\Phi(v^{k+1};\mu^k,v^*)\| = o(\mu^k). \tag{45}$$

The next result states the precise rate of convergence, not only in the error in norm, but in some individual components, defined by (44), of the error. It states that the same rate takes place in individual components of the *complementarity residuals*

$$\Phi^{\mathrm{C}}(v;\mu,v) = \begin{bmatrix} (C(x)+S)y - \mu e \\ Su - \mu e \end{bmatrix}. \tag{46}$$

Theorem 10. *Under Assumptions 4.7–4.11, assume that the complete sequence $\{v^k\}$ converges to v^*, then the sequence $\{\Phi(v^{k+1};\mu^k,v^k)\}$ converges to zero and we have the asymptotic expansions*

$$v^{k+1} = v^* + \mu^k \dot{v}(0) + o(\mu^k) \qquad \text{and} \qquad \Phi^{\mathrm{C}}(v^{k+1};\mu^k,v^*) = -\mu^k e + o(\mu^k). \tag{47}$$

As a consequence, the asymptotic convergence rate is described by

$$\frac{|v_i^{k+2} - v_i^*|}{|v_i^{k+1} - v_i^*|^{\tau^k}} = \Theta(1)\ i \in \mathscr{J},\ \text{and}\ \frac{|\Phi_i^{\mathrm{C}}(v^{k+2};\mu^{k+1},v^*)|}{|\Phi_i^{\mathrm{C}}(v^{k+1};\mu^k,v^*)|^{\tau^k}} = \Theta(1)\ i = 1,\ldots,2n_{\mathscr{C}}, \tag{48}$$

for k sufficiently large, where τ^k is as in (43), which implies that the iterates v^{k+1} and the residuals in complementarity converge componentwise Q-superlinearly to their limit, along the given components. The remaining components $i \notin \mathscr{J}$ satisfy $|v_i^{k+1} - v_i^*| = o(\mu^k)$ and $\Phi_i(v^{k+1};\mu^k,v^*) = o(\mu^k)$.

As a consequence of Theorem 4.3, a Q-rate of convergence which is as close to quadratic as desired, and which takes place not only in norm but in all the indicated components, is achievable by constructing the sequence $\{\gamma^k\}$ so it converges to zero, by choosing $\varepsilon_\tau \simeq 0$ and by selecting τ^k equal to its upper bound in (43).

5 Implicit Elastics Variant

Suppose that our inner-iteration trust-region algorithm has produced a new approximation $(x^{k,j}, s^{k,j})$ to the minimizer of the barrier function $\phi^{\mathrm{B}}(x,s;\mu^k,\nu^k)$. Since $\phi^{\mathrm{B}}(x,s;\mu,\nu)$ is a *separable* function of s, we might then aim to improve on $(x^{k,j}, s^{k,j})$ by finding the (global) minimizer $s(x)$ of $\phi^{\mathrm{B}}(x,s;\mu,\nu)$ for the given $x = x^{k,j}$. Replacing $(x^{k,j}, s^{k,j})$ by the improvement $(x^{k,j}, s(x^{k,j}))$ is an example of what is known as a *magical step*, and fortunately the use of such steps does not interfere with global convergence of the underlying algorithm—see, for example, [11, §10.4.1].

To compute the elastics $s(x)$, note that $s(x)$ necessarily satisfies componentwise

$$r(s(x)) \equiv \nabla_s \phi^{\mathrm{B}}(x,s(x);\mu,\nu) = \nu(e + e^0_{\mathscr{E}}) - y(x,s(x);\mu) - u(s(x);\mu) = 0. \tag{49}$$

We summarize the properties of (49) in the next result, whose proof is elementary.

Lemma 6. *Let Assumption 3.1 be satisfied, the function $r(s(x))$ be defined by (49) where x is fixed and the multiplier estimates be given by (19). We then have the following properties:*

1. *$r(s)$ is a separable function of s,*
2. *$r(s(x))$ has a unique root, $s(x)$, for which $(x,s(x))$ lies in the interior of the feasible set of (3),*
3. *$s(x)$ is twice continuously differentiable for* $\max(0,-c_i(x)) < s(x) < \infty$.

In our case, a simple calculation reveals that the magical correction for s is given (componentwise) by

$$s_i^{k,j} = \begin{cases} \dfrac{\mu^k}{2\nu^k} - \dfrac{c_i(x^{k,j})}{2} + \sqrt{\left(\dfrac{c_i(x^{k,j})}{2}\right)^2 + \left(\dfrac{\mu^k}{2\nu^k}\right)^2} & \text{for } i \in \mathscr{E} \\[2ex] \dfrac{\mu^k}{\nu^k} - \dfrac{c_i(x^{k,j})}{2} + \sqrt{\left(\dfrac{c_i(x^{k,j})}{2}\right)^2 + \left(\dfrac{\mu^k}{\nu^k}\right)^2} & \text{for } i \in \mathscr{I}. \end{cases}$$

As we have just suggested, we may improve upon a given (x,s) by replacing it by the "magical" $(x,s(x))$. However, this is somewhat inefficient as x is chosen without regard to what $s(x)$ might result. This suggests a better approach might be to treat the elastic variables as implicitly dependent on x *throughout* the inner iteration.

With this in mind, in this section we present an *implicit elastics* alternative to Algorithm 3.1. Since we know from Lemma 6 that $s(x)$ is (at least) twice continuously differentiable, we might instead minimize

$$\psi(x) \equiv \phi^{\mathrm{B}}(x, s(x); \mu, \nu) \tag{50}$$

solely as a function of the variables x. Here $\phi^{\mathrm{B}}(\cdot)$ is as defined by (4), and we have hidden the dependency of $\psi(\cdot)$ on μ and ν for brevity. In practice, in addition to the reduction in dimension this suggests, the definition of $s(x)$ should help to keep the constraints a comfortable distance from their boundaries, preventing steps from being repeatedly cut back. We now show that a classical trust-region algorithm for the minimization of $\psi(x)$ is well defined.

For future reference, we give the derivatives of (50) in the following result.

Lemma 7. *Under Assumption 4.2, the first and second derivatives of (50) are*

$$\nabla_x \psi(x) = \nabla f(x) - J^T(x)\sigma(x) \quad \textit{and} \tag{51a}$$

$$\begin{aligned}\nabla_{xx}\psi(x) &= H(x,\sigma(x)) + \mu J^T(x)\left[(C(x)+S(x))^2 + S(x)^2\right]^{-1} J(x), \qquad (51b)\\ &= H(x,\sigma(x)) + J^T(x)\left[(C(x)+S(x))Y^{-1}(x) + S(x)U^{-1}(x)\right]^{-1} J(x)\end{aligned}$$

where $H(x,\sigma)$ is given by (27) and we have defined the estimates

$$y(x) := y(x,s(x)) = \mu(C(x)+S(x))^{-1}e, \tag{52a}$$

$$u(x) := u(s(x)) = \mu S^{-1}(x)e \quad \textit{and} \tag{52b}$$

$$\sigma(x) = y(x) - \nu e^0_{\mathscr{E}}. \tag{52c}$$

Proof. Elementary calculations with (49) prove (51a). We note from (49) that $\nabla_x r(s(x)) = 0$, implying $(C(x)+S(x))^{-2}(J(x)+\nabla_x s(x)) = -S^{-2}(x)\nabla_x s(x)$. Extracting $\nabla_x s(x)$ from this identity gives

$$\nabla_x s(x) = -\left[I + (C(x)+S(x))^2 S^{-2}(x)\right]^{-1} J(x),$$

which combines with (52c) to yield

$$\nabla_x \sigma(x) = -\mu(C(x)+S(x))^{-2}(J(x)+\nabla_x s(x)) = \mu S^{-2}(x)\nabla_x s(x)$$

and finally, (51b). The second expression for (51b) follows from (52a)–(52b). □

Note that the second term in the right-hand side of (51b) is positive semi-definite.

A typical primal–dual trust-region method for minimizing $\psi(x)$ computes a correction d to the current solution estimate x so as to (approximately)

$$\underset{d}{\text{minimize}}\ \nabla\psi(x)^T d + \tfrac{1}{2} d^T B(x,\sigma) d \quad \text{subject to } \|d\|_M \le \Delta, \tag{53}$$

where the trust-region radius $\Delta > 0$. The primal–dual approximation $B(x,\sigma)$

$$B^{\text{PD}}(x,\sigma) = H(x,\sigma) + J^T(x)\left[\Theta^{-1}(x) + S(x)U^{-1}\right]^{-1} J(x), \tag{54}$$

where

$$\Theta(x) = Y(C(x)+S(x))^{-1}, \quad u \approx u(x) > 0, \quad y \approx y(x) > 0, \quad \sigma \approx \sigma(x) \tag{55}$$

(c.f. (26)). Note also that $\Theta^{-1}(x) + S(x)U^{-1}$ is a diagonal matrix.

As in Section 3.1, lengths of steps and gradients should be measured in norms that reflect curvature. The trust-region norm $\|w\|_M^2 \equiv \langle w, Mw\rangle$ depends on a suitable symmetric, positive-definite approximation M to $B(x,\sigma)$, and we shall use

$$M = P + J^T(x)\left[\Theta^{-1}(x) + S(x)U^{-1}\right]^{-1} J(x), \tag{56}$$

where as before, P can range from simple ($P = I$) to sophisticated ($P = H(x,\sigma)$). To be specific, we shall assume that, at the termination of the k-th inner-iteration, the following assumption is satisfied.

Assumption 5.1. *Each matrix M_k is defined by (56), where $P = P^k$ satisfies Assumption 4.3.*

The counterpart of the preconditioning system (28) is here that

$$Md_x = r_x \tag{57}$$

for some given r_x. If we define $d_s = -\left[\Theta(x) + US^{-1}(x)\right]^{-1} J(x)\Theta(x)d_x$, we see that (57) is equivalent to (28) in the case that $r_s = 0$. Because $\nabla_{v_P}\phi^{\text{B}}(x,s(x);\mu,\nu) = (\nabla_x\psi(x),0)$ when $s = s(x)$, we may replace (23a) with $\|\nabla\psi(x^{k+1})\|_{M_{k+1}^{-1}} \le \varepsilon^{\text{D}}(\mu^k)$.

The resulting trust-region method is entirely standard, except that any trial value x for which $s(x)$ is undefined or infeasible will be rejected and the trust-region radius retracted.

Identities (49), (55) and (52c) directly imply that the Hessian matrix of the model, $B^{\text{PD}}(x,\sigma)$, is bounded.

Lemma 8. *The Lagrange multiplier estimates satisfy the bounds*

$$0 < y(x) < \nu(e + e_{\mathcal{E}}^0), \quad 0 < u(x) < \nu(e + e_{\mathcal{E}}^0), \quad -\nu e_{\mathcal{E}}^0 < \sigma(x) < \nu e. \tag{58}$$

In view of the required approximations (55) and Lemma 8, we make the further reasonable assumption.

Assumption 5.2. *For given ν, the multiplier estimates y, u and σ are bounded.*

Given this assumption, we now show that our model Hessian remains bounded. To this end, let $\delta_i = 1$ if $i \in \mathscr{E}$ and 0 otherwise.

Lemma 9. *Under Assumptions 4.2 and 5.2, the primal–dual Hessian approximation (54) remains bounded for fixed values of $\mu > 0$ and $\nu > 0$.*

Proof. Since (54) implies that

$$\|B^{\text{PD}}(x,\sigma)\| \le \|H(x,\sigma)\| + \|J(x)\| \|J^T(x)\| \| \left[Y^{-1}(C(x)+S(x)) + U^{-1}S(x)\right]^{-1} \|,$$

and as Assumptions 4.2 and 5.2 ensure that $\|H(x,\sigma)\|$, $\|J(x)\|$ and $\|J^T(x)\|$ are bounded, it remains to show that the (diagonal) entries $s_i(x)/u_i + (c_i(x)+s_i(x))/y_i$ of the diagonal matrix $Y^{-1}(C(x)+S(x)) + U^{-1}S(x)$ are bounded away from zero. But combining (52a) and (52b) with (58) shows that

$$c_i(x) + s_i(x) > \frac{\mu}{\nu(1+\delta_i)} \quad \text{and} \quad s_i(x) > \frac{\mu}{\nu(1+\delta_i)} \quad \text{and}$$

and this together with 5.2 gives the required lower bound on those entries. □

We summarize the results of this section by stating Algorithm 5.1.

The convergence properties of Algorithm 5.1 are summarized in Theorem 11, which we state without proof since this result is a direct parallel of Theorem 8.

Algorithm 5.1 Prototype algorithm—outer iteration (implicit elastics)

Step 0. Let the forcing functions $\varepsilon^{\text{D}}(\cdot)$, $\varepsilon^{\text{C}}(\cdot)$ and $\varepsilon^{\text{U}}(\cdot)$ be given, and let $\kappa_\nu > 0$. Choose $x^0 \in \mathbb{R}^n$, $s^0 \in \mathbb{R}^{n_\mathscr{C}}_+$ such that $c(x^0) + s^0 > 0$, initial dual estimates $y^0, u^0 \in \mathbb{R}^{n_\mathscr{C}}_+$, and penalty and barrier parameters ν^0 and $\mu^0 > 0$, and set $k = 0$.

Step 1. Inner Iteration: find a new primal–dual iterate $(x^{k+1}, s(x^{k+1}), y^{k+1}, u^{k+1})$ satisfying

$$\|\nabla f(x^{k+1}) - J^T(x^{k+1})(y^{k+1} - \nu^k e^0_\mathscr{E})\|_{M^{-1}_{k+1}} \le \varepsilon^{\text{D}}(\mu^k) \quad (59a)$$

$$\|(C(x^{k+1}) + S(x^{k+1}))y^{k+1} - \mu^k e\| \le \varepsilon^{\text{C}}(\mu^k). \quad (59b)$$

$$\|S(x^{k+1})u^{k+1} - \mu^k e\| \le \varepsilon^{\text{U}}(\mu^k) \quad (59c)$$

$$\left(c(x^{k+1}) + s(x^{k+1}), s(x^{k+1})\right) > 0 \quad (59d)$$

$$\text{and} \quad \left(\nu^k[e+e^0_\mathscr{E}] + \kappa_\nu e, \nu^k[e+e^0_\mathscr{E}] + \kappa_\nu e\right) \ge \left(y^{k+1}, u^{k+1}\right) > 0 \quad (59e)$$

for some suitable scaling norm $\|.\|_{M_{k+1}}$ by (for example) approximately minimizing (50).

Step 2. Select a new barrier parameter, $\mu^{k+1} \in (0, \mu^k]$ such that $\lim_{k\to\infty} \mu^k = 0$. Update the penalty parameter ν^k according to the rule (33). Increment k by one, and return to Step 1.

Theorem 11. *Suppose that Assumptions 3.1, 4.2, 4.4–4.6 and 5.1 hold. Suppose that* x^* *is a limit point of the sequence* $\{x^k\}$ *generated by Algorithm 5.1. Then either* $\{\nu^k\}$ *remains bounded, and* x^* *is a first-order critical point for the nonlinear programming problem (1), or* $\{\nu^k\}$ *diverges, and* x^* *is a first-order critical point of the infeasibility* $\vartheta^{\mathrm{P}}(x)$*. In the first case, the multipliers* $\{\sigma(x^k)\}$ *generated converge to* $\lambda(y^*,\nu^*)$ *defined in (12). If additionally* ν^k *is updated whenever (32) is violated, if (31) holds only for a finite number of iterations and if* $\{\nu^k\}$ *diverges,* x^* *is a feasible Fritz-John point for (1) and the* MFCQ *fails to hold at* x^*.

In addition to the reasons mentioned earlier in this section, this alternative is attractive in that it empirically stabilizes the algorithm. In contrast with Algorithm 3.1, it also helps prevent infeasible steps from being generated and repeatedly cut. Indeed, it is easy to see from (55) and (58) that

$$c_i(x)+s_i(x) > \frac{\mu}{\nu(1+\delta_i)} \geq \frac{\mu}{2\nu}, \quad \text{and} \quad s_i(x) > \frac{\mu}{\nu(1+\delta_i)} \geq \frac{\mu}{2\nu}.$$

so long as $s(x)$ exists.

For completeness, in view of Lemma 9 and [10, Theorem 4], it is straightforward to show the following result, which again we state without proof.

Theorem 12. *Under Assumptions 3.1, 4.2, 4.4–4.6 and 5.1, the implicit-elastic inner iteration procedure outlined in this section generates* $\{(x^{k+1},s(x^{k+1}),y^{k+1},u^{k+1})\}$ *satisfying the inner-iteration stopping conditions (59) for iteration k of Algorithm 5.1 after finitely many steps.*

Fast convergence properties of Algorithm 5.1 may be derived as in Section 4.3.

6 Practical Considerations, Enhancements and Refinements

We might distinguish linear equations $A_E x = b_E$, where A_E has full row rank, from the remaining constraints by including them in the general statement (1). We consider this case since in practice we may aim to find and maintain feasible points for such simple constraints before treating the nonlinear ones, and also to reflect the generality that must be addressed by a practical implementation. Note that explicit treatment of the linear equations preserves the MFCQ.

Explicit linear inequality constraints $A_I x \geq b_I$, including the special case of simple bounds might also be treated directly instead of being penalized. In this case, the objective function of the barrier problem incorporate logarithmic terms to treat the linear inequalities. The Jacobian of the constraints is such that

$$J^{\mathrm{S}}(v_{\mathrm{P}})^T = \begin{bmatrix} J_{\mathscr{E}}(x)^T & J_{\mathscr{I}}(x)^T & A_E^T & A_I^T & 0 & 0 \\ I_{\mathscr{E}} & 0 & 0 & 0 & I_{\mathscr{E}} & 0 \\ 0 & I_{\mathscr{I}} & 0 & 0 & 0 & I_{\mathscr{I}} \end{bmatrix}. \tag{60}$$

Unfortunately, MFCQ is no longer automatically satisfied even in the special case of simple bounds, as it requires that there is a vector d in the nullspace of A_E such that $a_i^T d_i < 0$ for each active inequality $a_i^T x \geq b_i$. A condition such as LICQ on (1) is sufficient for this, and provides a consistent context with Section 4.3.

The convergence theory remains essentially unaltered. The preconditioning matrices P^k used in (23a) and in the trust region must this time be uniformly second-order sufficient, which essentially amounts to uniform positive definiteness on the nullspace of the matrix A_E [10], on which they define uniformly equivalent norms. The seminorms used in (23a) and the trust region are dual of each other and allow for efficient treatment of the linear constraints.

As mentioned earlier, finding an initial strictly feasible estimate (x^0, s^0) for (3) is trivial. Any value $s^0 > \max[0, -c(x^0)]$ is acceptable. In practice, only those s_i (or r_i, depending on the formulation chosen) that are required to be positive because of the initial x need be retained, although it is actually prudent to keep those for which s_i (or r_i) needs to be larger than some "small" positive value (say, 0.1). More generally, it may be beneficial to track each $s_i^{k,j}$ as the iteration progresses and to remove it as soon as the corresponding $c_i(x^{k,j})$ is sufficiently positive. Doing so does not affect the convergence results described in this paper, as there can only be a finite number of these removals.

In the presence of two-sided inequality constraints $c_i^{\mathrm{L}} \leq c_i(x) \leq c_i^{\mathrm{U}}$ the obvious penalty term $\nu \max(c_i^{\mathrm{L}} - c_i(x), c_i(x) - c_i^{\mathrm{U}}, 0)$ may be replaced by νs_i, where s_i is required to satisfy $s_i + c_i^{\mathrm{U}} - c_i(x) \geq 0$, $s_i + c_i(x) - c_i^{\mathrm{L}} \geq 0$ and $s_i \geq 0$. Thus a single elastic variable suffices, rather than the pair that might have been anticipated if $c_i(x) \geq c_i^{\mathrm{L}}$ and $c_i(x) \leq c_i^{\mathrm{U}}$ had been considered separately. If we wish to improve the value of $\phi^{\mathrm{B}}(v_{\mathrm{P}}; \mu, \nu)$ using a magical step, or to use the implicit-elastic approach of Section 5, the defining equation

$$r(x) \equiv \nu(e + e^0_{\mathscr{E}}) - \mu[C(x) - C^{\mathrm{L}} + S(x)]^{-1}e - \mu[C^{\mathrm{U}} - C(x) + S(x)]^{-1}e - \mu S^{-1}(x)e = 0$$

for the $s(x)$ for a two-sided inequality may be reduced to a cubic equation. While it is possible to give an explicit formula for the required root, in practice it is just as easy to use a safeguarded univariate Newton method to find it.

There may be some virtue in adding an upper bound s^{U} on the elastic variables in order to prevent $c(x)$ and s simultaneously diverging to infinity. Of course it is far from obvious what globally a good value for s^{U} might be, but the simple choice of $\max(10, 2s^0)$ has proved to be sufficient in early experiments. The resulting two-sided bound $0 \leq s \leq s^{\mathrm{U}}$ may then be handled exactly as above.

7 Numerical Experience

Algorithm 5.1 has been implemented as a prototype Fortran 95 module in the GALAHAD optimization library of Gould et al. [22]. The inner iteration stopping tolerances are chosen as $\varepsilon^{\mathrm{D}}(\mu) = \varepsilon^{\mathrm{C}}(\mu) = \mu^{1.01}$. The outer iterations stop as soon as the residuals of (23a)–(23c) with $\mu^k = 0$ fall under 1.0e−5. The initial barrier parameter is set to $\mu^0 = 1$ and is updated by simply dividing it by 10 at each outer iteration. The initial penalty parameter is set to $\nu^0 = 1$ and we choose $\tau_1 = 10$ and $\tau_2 = 1$ in (33). The initial guess x_0 specified in the model is honoured and initial elastic variables are chosen so that $r(s(x^0)) = 0$ in (49) and all multipliers are initialized to their primal values. The parameters in the updating rule for the penalty parameter (31)–(32) are $\eta_{\mathscr{E}}^k = \eta_{\mathscr{I}}^k = (\mu^k)^{1.1}$.

Trust-region subproblems (24) are solved by means of the Generalized Lanczos Trust-Region method GLTR of [19] with a preconditioner of the form (28). The block P in (29) is chosen as a band of semi-bandwidth 5 of the Hessian $H(x,\lambda(y,\nu))$. A Cholesky factorization of the coefficient matrix of (29) is then attempted. If it fails, P is replaced by $P+\delta I$ for increasing values of $\delta > 0$. On unsuccessful trust-region steps, a backtracking linesearch is performed along the trust-region step as described by Conn et al. [11]. Prior to solution, problem variables are scaled so they are all $O(1)$ initially, i.e., assuming non-negativity bounds only on the variables for simplicity, the initial (x^0,s^0) is replaced with $(\bar{x}^0,\bar{s}^0)$ where $\bar{x}_i^0 = x_i^0/\max(1,x_i^0)$ and $\bar{s}_i^0 = s_i^0/\max(1,s_i^0)$ for all i. Similarly, $c_i(x)$ is replaced with $c_i(\bar{x})/\max(1,\|\nabla c_i(\bar{x}^0)\|_\infty)$ for all i and $f(x)$ is replaced with $f(\bar{x})/\max(1,\|\nabla f(\bar{x}^0)\|_\infty)$.

Numerical results on the Hock and Schittkowski [27] collection are reported in Table 1. The table headers are, from left to right, the problem name, final objective function value, final primal feasibility, final dual feasibility, final complementarity measure, total number of iterations and running time. The tests were run under OSX on a dual-core Intel Core2 Duo processor and GALAHAD was compiled with the Intel Fortran Compiler version 10.1. A maximum number of 1,000 inner iterations was imposed. Residuals are measured as in Algorithm 5.1. The only failure, on HS87, is indicated by a trailing "F" and is due to the objective function being discontinuous. On HS89, the algorithm stops at a critical point of the ℓ_1 infeasibility measure in the sense of Lemma 3, which is indicated by a trailing "I" in the table. While the results in terms of number of iterations are overall not directly competitive with those of polished production software such as IPOPT [36] or KNITRO [5, 6], they are promising in terms of robustness. Though it is not our goal to conduct a complete comparison here, we note that KNITRO 6.0.0 also terminates at an infeasible point on HS89. IPOPT 3.3 is able to solve HS89 to optimality. Both IPOPT and KNITRO were run with all default settings. Our method takes a rather large number of iterations on a few problems. This behaviour is consistently due to difficulties in reducing dual infeasibility, presumably because of inadequate Lagrange multiplier estimates rather than to degeneracy since the final penalty parameter is never large. We delay extensive benchmarking until we have explored the benefits of all options mentioned in Sections 5 and 6. The full Hessian and banded preconditioners performed almost identically on this problem collection.

Table 1 Results on the Hock and Schittkowski test set

Name	Obj	Pfeas	Dfeas	Comp	Its
HS1	5.182E−08	0.0E+00	6.1E−08	1.8E−07	39
HS2	4.941E+00	0.0E+00	1.3E−09	1.4E−07	18
HS3	1.778E−07	0.0E+00	0.0E+00	1.8E−07	6
HS4	2.666E+00	0.0E+00	6.6E−10	1.8E−07	8
HS5	−1.913E+00	0.0E+00	9.4E−09	1.9E−07	8
HS6	0.000E+00	2.9E−08	1.8E−15	6.4E−06	8
HS7	−1.732E+00	7.4E−08	2.3E−10	3.2E−06	15
HS8	−1.000E+00	2.2E−11	2.2E−11	1.8E−07	8
HS9	−5.000E−01	0.0E+00	7.8E−07	1.8E−08	13
HS10	−9.999E−01	0.0E+00	1.0E−07	1.9E−06	10
HS11	−8.498E+00	0.0E+00	1.2E−06	3.2E−06	11
HS12	−2.999E+01	0.0E+00	1.3E−10	1.8E−07	11
HS13	9.641E−01	5.9E−06	2.6E−09	1.8E−07	57
HS14	1.393E+00	1.0E−08	1.2E−06	8.9E−06	15
HS15	3.065E+02	0.0E+00	7.5E−08	2.0E−07	15
HS16	2.500E−01	0.0E+00	1.9E−07	1.9E−07	14
HS17	1.000E+00	0.0E+00	5.9E−12	2.9E−06	18
HS18	5.000E+00	0.0E+00	5.9E−07	2.0E−07	14
HS19	−6.961E+03	0.0E+00	5.4E−07	1.8E−07	61
HS20	3.819E+01	0.0E+00	6.0E−09	1.8E−07	15
HS21	−9.995E+01	0.0E+00	1.2E−18	1.8E−07	13
HS22	1.000E+00	0.0E+00	2.4E−09	1.9E−07	11
HS23	2.000E+00	0.0E+00	2.7E−07	1.8E−07	31
HS24	−9.999E−01	0.0E+00	6.6E−10	1.8E−07	16
HS25	4.312E−12	0.0E+00	1.5E−08	1.8E−07	30
HS26	1.405E−07	6.6E−06	3.2E−06	6.7E−06	11
HS27	3.999E−02	9.2E−10	1.3E−09	6.3E−06	15
HS28	0.000E+00	1.5E−16	2.5E−13	1.8E−07	6
HS29	−2.262E+01	0.0E+00	1.1E−14	5.6E−06	18
HS30	1.000E+00	0.0E+00	7.4E−09	2.1E−07	9
HS31	6.000E+00	0.0E+00	5.5E−07	2.5E−07	20
HS32	1.000E+00	2.4E−10	9.9E−13	1.8E−07	14
HS33	−4.585E+00	0.0E+00	1.3E−08	1.8E−07	19
HS34	−8.339E−01	0.0E+00	9.8E−06	1.1E−06	21
HS35	1.111E−01	0.0E+00	5.0E−16	1.8E−07	8
HS36	−3.299E+03	0.0E+00	1.9E−08	2.1E−07	21
HS37	−3.455E+03	0.0E+00	6.0E−12	1.8E−07	25
HS38	7.876E+00	0.0E+00	1.0E−07	1.8E−07	9
HS39	−1.000E+00	7.7E−08	1.3E−08	3.2E−06	18
HS40	−2.500E−01	6.1E−09	2.6E−10	3.2E−06	101

(continued)

Table 1 (continued)

Name	Obj	Pfeas	Dfeas	Comp	Its
HS41	1.925E+00	4.0E−10	4.5E−09	2.1E−07	11
HS42	1.385E+01	1.2E−06	1.2E−08	2.0E−07	11
HS43	−4.399E+01	0.0E+00	2.4E−08	2.4E−07	14
HS44	−1.499E+01	0.0E+00	1.8E−15	1.8E−07	13
HS45	1.000E+00	0.0E+00	2.5E−09	1.8E−07	10
HS46	3.746E−08	1.9E−11	6.9E−06	1.8E−08	897
HS47	5.341E−08	4.5E−06	1.2E−06	9.4E−06	23
HS48	0.000E+00	0.0E+00	1.7E−16	1.8E−07	6
HS49	2.315E−05	1.1E−16	5.2E−06	1.8E−07	10
HS50	9.094E−13	5.3E−15	2.3E−09	1.8E−07	8
HS51	0.000E+00	0.0E+00	8.0E−17	1.8E−07	6
HS52	5.326E+00	1.3E−07	1.8E−15	6.5E−06	8
HS53	4.093E+00	5.5E−11	1.9E−12	1.8E−07	15
HS54	−9.080E−01	2.2E−09	2.6E−08	2.1E−07	11
HS55	6.333E+00	3.0E−11	4.5E−07	1.8E−07	27
HS56	−2.362E+00	6.6E−10	2.6E−06	3.2E−06	560
HS57	3.064E−02	0.0E+00	3.6E−06	2.6E−07	15
HS59	−6.749E+00	0.0E+00	3.7E−08	1.8E−07	10
HS60	2.189E+00	5.9E−13	1.0E−11	1.8E−07	18
HS61	−1.436E+02	7.8E−08	2.8E−12	1.8E−07	9
HS62	−2.627E+04	9.0E−09	2.4E−08	1.8E−07	16
HS63	9.617E+02	2.7E−09	2.5E−09	1.8E−07	22
HS64	6.303E+03	1.3E−07	6.6E−06	3.2E−06	18
HS65	9.535E−01	0.0E+00	5.5E−08	3.1E−07	14
HS66	5.181E−01	0.0E+00	7.8E−10	1.8E−07	22
HS67	−1.162E+03	0.0E+00	6.8E−09	2.0E−07	9
HS68	4.650E−05	3.1E−08	7.5E−07	1.1E−06	10
HS69	7.923E−03	3.1E−09	1.3E−07	4.7E−07	13
HS70	1.870E−01	0.0E+00	7.7E−06	2.0E−07	82
HS71	1.701E+01	2.1E−07	1.6E−07	4.1E−07	434
HS72	1.831E+01	6.2E−06	4.5E−07	1.8E−07	30
HS73	2.989E+01	1.6E−09	9.9E−09	1.1E−06	18
HS74	5.126E+03	1.5E−09	1.4E−08	2.2E−07	20
HS75	5.174E+03	4.0E−10	9.4E−09	1.8E−07	20
HS76	−4.681E+00	0.0E+00	2.8E−16	2.0E−07	8
HS77	2.415E−01	1.0E−07	1.7E−06	3.2E−06	19
HS78	−2.919E+00	3.2E−07	1.3E−09	1.8E−07	9
HS79	7.877E−02	8.8E−08	9.2E−10	1.8E−07	7
HS80	5.394E−02	1.6E−09	8.8E−09	1.8E−07	85

(continued)

Table 1 (continued)

Name	Obj	Pfeas	Dfeas	Comp	Its
HS81	5.394E−02	2.2E−07	1.1E−06	4.4E−07	29
HS83	−3.066E+04	0.0E+00	2.5E−10	1.9E−07	14
HS84	−5.279E+06	0.0E+00	6.1E−09	1.8E−07	42
HS85	−2.215E+00	0.0E+00	2.8E−08	1.9E−07	23
HS86	−3.234E+01	0.0E+00	4.0E−09	1.1E−06	16
HS87	4.778E−03	1.0E−09	1.0E+02	1.3E−07	584**f**
HS88	1.362E+00	0.0E+00	9.5E−10	3.2E−06	31
HS89	0.000E+00	1.3E−01	0.0E+00	1.3E−16	25**i**
HS90	1.362E+00	0.0E+00	7.0E−10	3.2E−06	43
HS91	1.362E+00	0.0E+00	3.2E−06	3.2E−06	83
HS92	1.362E+00	0.0E+00	7.5E−10	3.2E−06	44
HS93	1.350E+02	0.0E+00	5.1E−08	2.4E−07	20
HS95	1.562E−02	0.0E+00	9.3E−07	1.1E−06	13
HS96	1.562E−02	0.0E+00	9.3E−07	1.1E−06	13
HS97	4.071E+00	0.0E+00	3.1E−08	1.9E−07	18
HS98	3.135E+00	0.0E+00	4.5E−08	2.6E−07	22
HS99	−8.310E+08	1.9E−07	7.3E−07	1.8E−07	9
HS100	6.806E+02	0.0E+00	3.7E−09	1.9E−07	36
HS101	1.809E+03	0.0E+00	3.1E−07	1.8E−07	36
HS102	9.118E+02	0.0E+00	1.5E−07	1.8E−07	41
HS103	5.436E+02	0.0E+00	2.3E−06	1.8E−07	42
HS104	3.951E+00	0.0E+00	1.9E−08	1.8E−07	24
HS105	1.044E+03	0.0E+00	7.8E−07	1.8E−07	16
HS106	7.049E+03	0.0E+00	1.7E−07	2.4E−07	28
HS107	5.055E+03	6.1E−09	4.5E−09	2.6E−07	24
HS108	−8.660E−01	0.0E+00	8.5E−07	1.8E−07	21
HS109	5.362E+03	6.2E−12	4.1E−07	1.8E−07	68
HS110	−4.577E+01	0.0E+00	2.2E−06	1.8E−07	10
HS111	−4.776E+01	3.4E−11	8.8E−06	1.8E−07	483
HS112	−4.776E+01	3.0E−09	3.8E−07	1.8E−07	24
HS113	2.430E+01	0.0E+00	3.6E−07	7.9E−07	14
HS114	−1.768E+03	9.7E−09	3.0E−06	1.8E−07	235
HS116	9.758E+01	0.0E+00	1.8E−07	1.8E−07	31
HS117	3.235E+01	0.0E+00	2.4E−06	1.8E−06	21
HS118	6.648E+02	0.0E+00	8.4E−15	1.8E−07	25
HS119	2.449E+02	2.2E−10	7.2E−06	2.3E−07	23

8 Concluding Remarks

Clearly, we recognize that the particular approach adopted in this paper is not the only possible one. Another possibility is to use the ℓ_∞ penalty function

$$\phi(x,\nu) = f(x) + \nu \max_{i\in\mathcal{E}} |c_i(x)| + \nu \max_{i\in\mathcal{I}} (-c_i(x), 0) \tag{61}$$

instead of (2). As before, it is easy to show that this may be reformulated as

$$\begin{array}{lll} \underset{x\in\mathbb{R}^n,\ s\in\mathbb{R}}{\text{minimize}} & f(x) + \nu s & \\ \text{subject to} & c_i(x) + s \geq 0, & (i \in \mathcal{E} \cup \mathcal{I}) \\ & s - c_i(x) \geq 0, & (i \in \mathcal{E}) \\ & s \geq 0 & \end{array}$$

involving a single "elastic" variable s. Once again one might apply an interior-point algorithm to such a problem, and again it is trivial to find an initial interior point. The advantage now is clearly this formulation involves significantly fewer surplus variables. The ℓ_∞ approach is also examined in the *elastic mode* in [4].

We believe the method presented in the present paper is appropriate for a variety of degenerate nonlinear programs, and in particular problems for which the MFCQ fails to hold at a solution. At variance with some other methods, the method proposed here is not only able to identify such a solution, but it also delivers a certificate of failure of the MFCQ. This is in line with, e.g., the method proposed in [8].

A substantial advantage of the present approach is that it specializes adequately to the solution of structured degenerate problems, such as mathematical programs with complementarity constraints and mathematical programs with vanishing constraints. Extension of our algorithm to such cases is the subject of current research [12, 13].

Acknowledgements This work was supported in part by the EPSRC grants GR/R46641 and GR/S42170 (Nick I.M. Gould), NSERC Discovery Grant 299010-04 (Dominique Orban), and EPSRC grant GR/S02969 (Philippe L. Toint).

References

1. Armand, P.: A quasi-Newton penalty barrier method for convex minimization problems. Comput. Optim. Appl. **26**(1), 5–34 (2003)
2. Armand, P., Gilbert, J.-Ch., Jan-Jégou, S.: A BFGS-IP algorithm for solving strongly convex optimization problems with feasibility enforced by an exact penalty approach. Math. Program. **92**(3), 393–424 (2000)
3. Bazaraa, M.S., Goode, J.J.: Sufficient conditions for a globally exact penalty-function without convexity. Math. Program. Stud. **19**, 1–15 (1982)

4. Boman, E.G.: Infeasibility and negative curvature in optimization. Ph.D. thesis, Stanford University, Stanford (1999)
5. Byrd, R.H., Gilbert, J.-Ch., Nocedal, J.: A trust region method based on interior point techniques for nonlinear programming. Math. Program. A **89**(1), 149–185 (2000)
6. Byrd, R.H., Nocedal, J., Waltz, R.A.: KNITRO: An integrated package for nonlinear optimization. In: di Pillo, G., Roma, M. (eds.) Large-Scale Nonlinear Optimization, pp. 35–59. Springer, New York (2006)
7. Charalambous, C.: A lower bound for the controlling parameters of the exact penalty functions. Math. Program. **15**(3), 278–290 (1978)
8. Chen, L., Goldfarb, D.: Interior-point ℓ_2-penalty mehods for nonlinear programming with strong global convergence properties. Math. Program. **108**, 1–36 (2006)
9. Coleman, T.F., Conn, A.R.: Second-order conditions for an exact penalty function. Math. Program. **19**(2), 178–185 (1980)
10. Conn, A.R., Gould, N.I.M., Orban, D., Toint, Ph.L.: A primal-dual trust-region algorithm for non-convex nonlinear programming. Math. Program. **87**(2), 215–249 (2000)
11. Conn, A.R., Gould, N.I.M., Toint, Ph.L.: Trust-Region Methods. SIAM, Philadelphia (2000)
12. Coulibaly, Z., Orban, D.: An ℓ_1 elastic interior-point methods for mathematical programs with complementarity constraints. SIAM J. Optim. **22**(1), 187–211 (2011)
13. Curatolo, P.-R., Orban, D.: An elastic penalty method for mathematical programs with vanishing constraints with application to structural optimization. Cahier du GERAD G-2014-xx, GERAD, Montréal, QC (2014, in preparation)
14. Fletcher, R.: Practical Methods of Optimization: Unconstrained Optimization, 2nd edn. Wiley, Chichester (1987)
15. Gauvin, J.: A necessary and sufficient regularity condition to have bounded multipliers in nonconvex programming. Math. Program. **12**, 136–138 (1977)
16. Gill, P.E., Murray, W., Wright, M.H.: Practical Optimization. Academic, London (1981)
17. Gill, P.E., Murray, W., Saunders, M.A.: SNOPT: An SQP algorithm for large-scale constrained optimization. SIAM J. Optim. **12**(4), 979–1006 (2002)
18. Gould, N.I.M.: On practical conditions for the existence and uniqueness of solutions to the general equality quadratic-programming problem. Math. Program. **32**(1), 90–99 (1985)
19. Gould, N.I.M., Lucidi, S., Roma, M., Toint, Ph.L.: Solving the trust-region subproblem using the Lanczos method. SIAM J. Optim. **9**(2), 504–525 (1999)
20. Gould, N.I.M., Orban, D., Sartenaer, A., Toint, Ph.L.: Superlinear convergence of primal-dual interior point algorithms for nonlinear programming. SIAM J. Optim. **11**(4), 974–1002 (2001)
21. Gould, N.I.M., Orban, D., Sartenaer, A., Toint, Ph.L.: Componentwise fast convergence in the solution of full-rank systems of nonlinear equation. Math. Program. Ser. B **92**(3), 481–508 (2002)
22. Gould, N.I.M., Orban, D., Toint, Ph.L.: GALAHAD—a library of thread-safe Fortran 90 packages for large-scale nonlinear optimization. Trans. ACM Math. Softw. **29**(4), 353–372 (2003)
23. Gould, N.I.M., Orban, D., Toint, Ph.L.: An interior-point ℓ_1-penalty method for nonlinear optimization. Technical Report RAL-TR-2003-022, Rutherford Appleton Laboratory, Chilton, Oxfordshire (2003)
24. Han, S.-P., Mangasarian, O.L.: Exact penalty functions in nonlinear programming. Math. Program. **17**(3), 251–269 (1979)
25. Harwell Subroutine Library. A collection of Fortran codes for large-scale scientific computation. AERE Harwell Laboratory, www.cse.clrc.ac.uk/nag/hsl (2007)
26. Herskovits, J.: A two-stage feasible directions algorithm for nonlinear constrained optimization. Math. Program. **36**(1), 19–38 (1986)
27. Hock, W., Schittkowski, K.: Test Examples for Nonlinear Programming Codes. Lectures Notes in Economics and Mathematical Systems, vol. 187. Springer, Berlin (1981)
28. Huang, L.R., Ng, K.F.: 2nd-order necessary and sufficient conditions in nonsmooth optimization. Math. Program. **66**(3), 379–402 (1994)

29. Lawrence, C.T., Tits, A.L.: Nonlinear equality constraints in feasible sequential quadratic programming. Optim. Methods Softw. **6**(4), 265–282 (1996)
30. Mangasarian, O.L., Fromovitz, S.: The Fritz John necessary optimality conditions in the presence of equality and inequality constraints. J. Math. Anal. Appl. **17**, 37–47 (1967)
31. Mayne, D.Q., Polak, E.: Feasible directions algorithms for optimisation problems with equality and inequality constraints. Math. Program. **11**(1), 67–80 (1976)
32. Ortega, J.M., Rheinboldt, W.C.: Iterative Solution of Nonlinear Equations in Several Variables. Academic, New York (1970)
33. Pietrzykowski, T.: An exact penalty method for constrained maxima. SIAM J. Numer. Anal. **6**, 299–304 (1969)
34. Tits, A.L., Wächter, A., Bakhtiari, S., Urban, T.J., Lawrence, C.T.: A primal-dual interior-point method for nonlinear programming with strong global and local convergence properties. SIAM J. Optim. **14**(1), 173–199 (2003)
35. Vanderbei, R.J., Shanno, D.F.: An interior point algorithm for nonconvex nonlinear programming. Comput. Optim. Appl. **13**, 231–252 (1999)
36. Wächter, A., Biegler, L.T.: On the implementation of an interior-point filter line-search algorithm for large-scale nonlinear programming. Math. Program. Ser. A **106**(1), 25–57 (2006)
37. Waltz, R.A., Morales, J.L., Nocedal, J., Orban, D.: An interior algorithm for nonlinear optimization that combines line search and trust region steps. Math. Program. Ser. A **107** (3, Ser. A), 391–408 (2006)
38. Wright, M.H.: Interior methods for constrained optimization. Acta Numerica **1**, 341–407 (1992)
39. Wright, S.J., Orban, D.: Local convergence of the newton/log-barrier method for degenerate problems. Math. Oper. Res. **27**(3), 585–613 (2002)

An ℓ_1-Penalty Scheme for the Optimal Control of Elliptic Variational Inequalities

M. Hintermüller, C. Löbhard, and M.H. Tber

Abstract An ℓ_1-penalty scheme in function space for the optimal control of elliptic variational inequalities is proposed. In an L^2-tracking context, an iterative algorithm is proven to generate a sequence which converges to some weakly C-stationary point and, under certain conditions, even to a strongly stationary point of the original problem. In the case of point tracking control, where the objective contains pointwise function evaluations of the state variable, a modified model problem with constraints on the dual variable associated with the variational inequality constraint is introduced and an auxiliary problem that penalizes not only the complementarity, but also the state constraint, is analyzed. Passing to the limit with the penalty parameter in the stationarity system of the auxiliary problem yields some weak form of a C-stationarity system for the original problem if the additional dual constraints are not active. Finally, numerical results obtained by the new algorithms are documented.

Keywords Optimal Control of Variational Inequalities • ℓ_1-Penalty Methods in Function Space • Point Tracking • Stationarity Conditions for MPECs in Function Space

1 Introduction

In this work we analyze a penalty scheme for the optimal control of variational inequalities, i.e. for problems of the type

M. Hintermüller (✉) • C. Löbhard
Department of Mathematics, Humboldt-Universität zu Berlin, Unter den Linden 6, 10099 Berlin, Germany
e-mail: hint@math.hu-berlin.de; loebhard@math.hu-berlin.de

M.H. Tber
Faculté des Sciences et Technique, Université Sultan Moulay Slimane, Béni-Mellal, Morocco
e-mail: hicham.tber@gmail.com

M. Al-Baali et al. (eds.), *Numerical Analysis and Optimization*, Springer Proceedings in Mathematics & Statistics 134, DOI 10.1007/978-3-319-17689-5_7

$$\text{Minimize } J(y,u) = j(y) + \frac{\nu}{2}\|u\|^2_{L^2(\Omega)} \tag{1a}$$

$$\text{over } (y,u) \in Y \times U_{ad} \tag{1b}$$

$$\text{subject to} \quad \forall z \in K, \quad \langle Ay - u - f, z - y\rangle_{H^{-1}(\Omega)} \geq 0, \tag{1c}$$

where $U_{ad} \subset L^2(\Omega)$ is the non-empty, convex, and closed set of feasible controls u, $\Omega \subset \mathbb{R}^n$ is an (open) domain, $n \in \mathbb{N}$, and $L^2(\Omega)$ is the usual Lebesgue space of square integrable functions (cf. [2]). The cost of the control action is $\nu > 0$, the set $K \neq \emptyset$ is convex and closed, A is linear, bounded, and coercive such that the variational inequality (1c) admits a unique solution for every $u \in U_{ad}$ and given $f \in L^2(\Omega)$. Here, Y is a suitable Banach space for the state variable y such that the solution operator of (1c) maps from U_{ad} into Y, and the summand $j(y)$ in the objective may, for instance, implement the difference of the state variable y to a given desired state, either in the L^2-norm, or in a finite set of tracking points. A more precise definition of the problem data is given in Assumptions 1 and 2 below, respectively.

The constraint (1c) is prototypical for a broad range of problems, such as energy minimization problems or free boundary problems, that can be modeled by variational inequalities. In rather abstract form they have been analyzed since the 1960s. A comprehensive study and a survey on the literature on this subject can be found, for instance, in [10, 20, 29].

In optimal control problems, resulting, e.g., from engineering sciences, one typically can influence a system by a control mechanism, which aims to minimize an objective depending on the state of the system as well as on the control action. Similarly, in inverse problems, a parameter that plays the role of a control variable, has to be determined from (defective) measurements associated with the state y, either on the (whole) domain (L^2 tracking), or in certain predefined locations (point tracking). A classical work on optimal control of systems governed by partial differential equations is [21], but, in recent years, numerous extensions on the analysis and numerical treatment of such problems have been contributed to the literature.

When one aims to control variational inequalities, the inherent non-smoothness of the solution operator associated with the variational inequality (VI) results in constraint degeneracy. Moreover, considering the reduced control problem (written in u), one is confronted with a challenging non-smooth and non-convex minimization problem. In finite dimensions, problems of this structure are termed mathematical programs with equilibrium constraints (MPECs); compare [23, 27]. For characterizing stationarity of a feasible point of the MPEC, the classical Karush-Kuhn-Tucker theory cannot be applied due to the aforementioned constraint degeneracy. Depending on the problem instance, but also depending on the utilized mathematical tools alternative stationarity systems have been derived. A hierarchy of such stationarity conditions in finite dimensions can, e.g., be found in [30]. These concepts were transferred to function space in [14] and later also in [12, 13, 19]. Concerning the choice of mathematical tools, we mention that penalty and smoothing methods as well as concepts from convex or set-valued analysis and generalized differentiation have been used in order to derive stationarity conditions for MPECs in function space; see, e.g., [5, 24–26] in addition to the aforementioned references.

In the literature one often finds classical L^2-tracking type control problems. In this paper, however, inspired by point measurements (e.g., obtained by sensors mounted at fixed locations within a region of interest), we study point tracking of solutions of variational inequalities. We mention that point measurements naturally occur, for instance, in mathematical finance; see, e.g., [1]. Mathematically this in general requires more than $H_0^1(\Omega)$-regularity of the VI-solution, in order to justify evaluations of the state at isolated points within a domain of interest. Such an increased regularity then has various analytical implications on the regularity of associated dual quantities and requires new analytical as well as numerical considerations. For VIs involving second-order elliptic partial differential operators, a stationarity system for control problems with point tracking objectives has been derived in [8].

Since stationarity systems of MPECs are typically non-smooth and, thus, hard to solve numerically, the algorithmic treatment of MPECs is delicate. In function space, penalty and smoothing approaches have been analyzed in [8, 15, 31], a relaxation method can be found in [14], and a descent method has been implemented in [16]. All of these approaches apply some type of relaxation and/or smoothing of the control problem or the VI constraint. Consequently, the solution process depends on parameters which need to be taken to their limits in order to approach some type of stationary point of the original problem. Only, [16] may operate without smoothing, depending on properties of the iterates generated by the associated algorithm. Here, smoothing is only applied when the solution y of (1c) is non-differentiable as a function of u.

In this paper, we extend the elastic mode algorithm of Anitescu et al. [4] to the function space setting. One of the interesting aspects of this algorithm is related to the fact that it relies on an ℓ_1-type penalty approach which, under appropriate conditions, acts as an exact penalty method. Thus, a finite penalty parameter suffices to obtain a solution of the original problem. As, upon discretization, the condition number of the underlying stationarity system typically scales adversely with respect to increasing penalty parameter, the exactness of the penalization is attractive as it allows to keep this parameter (and hence conditioning) bounded. Here we extend and study this method for L^2-tracking as well as for the point-tracking case, which require separate analyses.

The rest of the text is structured as follows: Section 2 treats the L^2-type control of variational inequalities. In order to develop an efficient solution algorithm for this problem class, we begin with the analysis of a penalty method in function space in Section 2.1. In contrast to other available penalty schemes for MPECs, this method does not smoothen the original problem but directly penalizes the critical complementarity condition in the variational inequality. In particular, we prove solvability of the auxiliary problem and consistency of the penalty scheme. Section 2.2 contains stationarity conditions for the auxiliary problem as well as a limiting stationarity system for the MPEC. In [4], the authors treat MPECs in a finite dimensional context, which is more general than a finite dimensional version of the problem treated here. Under certain conditions, strong stationarity of a solution obtained after a finite number of iterations (i.e., γ updates) is proven. The notion of strong stationarity is hereby based on the definition of active and biactive sets. In contrast to the finite dimensional case, where these are specified according to the set of indices where certain solution vectors are zero, it is not straightforward to define the zero set of objects in $H_0^1(\Omega)$ or $H^{-1}(\Omega)$ (see [2] for the definition of these spaces). We give a definition that allows us to prove strong stationarity of feasible first order points of the auxiliary problem in Section 2.3.

In Section 3, we consider point tracking subject to variational inequalities, i.e., the functional $j : Y \to \mathbb{R}$ is given by $j(y) = \frac{1}{2}\sum_{w\in I}(y(w) - y_w)^2$ where $I \subset \Omega$ is finite and for all $w \in I$, $y_w \in \mathbb{R}$. Although the smoothing method of [8] can be understood as an iterative algorithm, that finds limiting ε-almost C-stationary points (or, in the finite dimensional world, C-stationary points) in the limit, we construct a method that penalized the critical complementarity constraint. In contrast to the analysis in the first part of this paper, we have to account for regularity questions that arise from the function evaluations in the objective functional. In particular, in Section 3.1 we modify the problem class and prove consistency of a penalty scheme. We show that a weak version of C-stationarity holds for limits of first order points of the auxiliary problem in Section 3.2.

Finally, we document numerical results obtained by an algorithm associated with our analytic approach. We start with the description of the solution algorithm in Section 4 and provide two examples in Section 5.

Notation

For a measurable subset $\omega \subset \Omega \subset \mathbb{R}$ we denote the characteristic function $\chi_\omega : \Omega \to \{0,1\}$, $\chi_\omega(w) = 1$ if $w \in \omega$ and else $\chi_\omega(w) = 0$, and the averaged characteristic function by $\bar{\chi}_\omega : \Omega \to \{0,1\}$, $\bar{\chi}_\omega(w) = \frac{1}{|\omega|}$ if $w \in \omega$ and else $\chi_\omega(w) = 0$. Here, we assume that the Lebesgue measure $|\omega|$ of ω is positive. If Ω is a Lipschitz domain, we denote the usual Lebesgue, Hilbert, and Sobolev spaces by $L^2(\Omega), H_0^1(\Omega), W_0^{1,q}(\Omega)$, where $q \geq 1$, and the dual spaces by $H^{-1}(\Omega) = (H_0^1(\Omega))^*$, $W^{-1,q'}(\Omega) = (W_0^{1,q}(\Omega))^*$, where $q' = \frac{q}{q-1}$. For their definition and further properties, we refer to [2]. The scalar product in a Hilbert space X is denoted by round brackets $(\cdot,\cdot)_X$. If X is a Banach space, then the duality pairing of an object $x^* \in X^*$ with an object $x \in X$ is denoted by $\langle x^*, x\rangle_{X^*}$. We set

$$\left(H_0^1(\Omega)\right)_+ := \{v \in H_0^1(\Omega) \,|\, v \geq 0 \text{ almost everywhere (a.e.) on } \Omega\},$$
$$\left(H^{-1}(\Omega)\right)_+ := \{\psi \in H^{-1}(\Omega) \,|\, \forall v \in K : \langle \psi, v\rangle_{H^{-1}(\Omega)} \geq 0\}.$$

For a bounded linear operator $A : X \to Y$, where X and Y are Banach spaces, the corresponding adjoint operator is denoted by $A^* : Y^* \to X^*$. For a non-empty, convex, and closed subset $U \subset L^2(\Omega)$, we denote the projection operator which maps an element $f \in L^2(\Omega)$ to its (uniquely defined) projection onto U by $\mathrm{Proj}_U : L^2(\Omega) \to U$. Throughout the text, $C > 0$ denotes a generic constant that may take different values in different situations.

2 The L^2-Tracking Case

In the present section we analyze the optimal control problem (1) under the following assumptions:

Assumption 1. For $n \in \mathbb{N}$, $\Omega \subset \mathbb{R}^n$ is an open bounded domain; the functional $j : L^2(\Omega) \to \mathbb{R}$ is weakly lower semi-continuous, bounded from below and continuously Fréchet differentiable; the set of feasible controls is given by the box constraint

$$U_{ad} = \{v \in L^2(\Omega) \,|\, \underline{u} \leq v \leq \overline{u}\},$$

where $\underline{u}, \overline{u} \in L^2(\Omega) \cup \{-\infty, \infty\}$ satisfy $\underline{u} < \overline{u}$ a.e. in Ω; $\nu \geq 0$ and, if U_{ad} is not bounded in $L^2(\Omega)$, then $\nu > 0$. The feasible set K in the variational inequality is given by $K = \left(H_0^1(\Omega)\right)_+$; the operator $A : H_0^1(\Omega) \to H_0^1(\Omega)$ is linear, bounded, and coercive; and $f \in L^2(\Omega)$.

The canonical example for j is the L^2-tracking objective $j(y) = \frac{1}{2}\|y - y_d\|^2_{L^2(\Omega)}$ where $y_d \in L^2(\Omega)$ is a given desired state. We restate problem (1) with the complementarity formulation of the variational inequality (see, e.g., [29, Prop. 4:5.6]) in the constraint, as follows:

$$\text{Minimize } J(y,u) := j(y) + \frac{\nu}{2}\|u\|^2_{L^2(\Omega)} \tag{2a}$$
$$\text{over } (y,u,\xi) \in H_0^1(\Omega) \times U_{ad} \times H^{-1}(\Omega) \tag{2b}$$
$$\text{subject to} \quad Ay - u - \xi = f \quad \text{in } H^{-1}(\Omega), \tag{2c}$$
$$y \geq 0 \quad \text{in } H_0^1(\Omega), \quad \xi \geq 0 \quad \text{in } H^{-1}(\Omega), \text{ and } \langle \xi, y\rangle_{H^{-1}(\Omega)} = 0. \tag{2d}$$

Below, whenever it is clear from the context, we leave off $H_0^1(\Omega)$ and $H^{-1}(\Omega)$ within the inequalities in (2d).

2.1 Solvability and Consistency of the Penalization Scheme

For a penalty parameter $\gamma > 0$, we define the following auxiliary problem:

$$\text{Minimize } J_\gamma(y,u,\xi) := j(y) + \frac{\nu}{2}\|u\|^2_{L^2(\Omega)} + \gamma\langle \xi, y\rangle_{H^{-1}(\Omega)} \tag{3a}$$

$$\text{over } (y,u,\xi) \in H^1_0(\Omega) \times U_{ad} \times H^{-1}(\Omega) \tag{3b}$$

$$\text{subject to } Ay - u - \xi = f,\, y \geq 0,\, \xi \geq 0. \tag{3c}$$

Given the non-negativity of y and ξ, the term $\gamma\langle \xi, y\rangle_{H^{-1}(\Omega)}$ penalizes the ℓ_1-norm (i.e., the absolute value) of the constraint $\langle \xi, y\rangle_{H^{-1}(\Omega)} = 0$. Note that the auxiliary problem is in general non-convex. We also mention here that a related penalty approach in finite dimensions was considered in [4].

The following lemma is needed to prove solvability and consistency of the penalty scheme.

Lemma 1. *We consider a bounded sequence $(u_k)_{k\in\mathbb{N}} \subset L^2(\Omega)$ and $(y_k,\xi_k)_{k\in\mathbb{N}} \subset H^1_0(\Omega) \times H^{-1}(\Omega)$ such that for all $k \in \mathbb{N}$,*

$$Ay_k - \xi_k = u_k + f \quad \text{in } H^{-1}(\Omega), \quad y_k \geq 0, \quad \xi_k \geq 0, \quad \langle \xi_k, y_k\rangle_{H^{-1}(\Omega)} \leq C, \tag{4}$$

where Ω, A, and f satisfy Assumption 1. Then, there exists a subsequence (also denoted by $(y_k,\xi_k)_{k\in\mathbb{N}}$) such that

$$u_k \rightharpoonup u \quad \text{in } L^2(\Omega), \quad \xi_k \rightharpoonup \xi \quad \text{in } H^{-1}(\Omega), \quad y_k \rightharpoonup y \quad \text{in } H^1_0(\Omega),$$

and the limit $(y,u,\xi) \in H^1_0(\Omega) \times L^2(\Omega) \times H^{-1}(\Omega)$ satisfies

$$Ay - \xi = u + f \quad \text{in } H^{-1}(\Omega), \quad y \geq 0, \quad \xi \geq 0,$$
$$\liminf\{\langle \xi_k, y_k\rangle_{H^{-1}(\Omega)} \,|\, k \in \mathbb{N}\} \geq \langle \xi, y\rangle_{H^{-1}(\Omega)} \geq 0.$$

If in particular $\lim_{k\to\infty}\langle \xi_k, y_k\rangle_{H^{-1}(\Omega)} = 0$, then (y,u,ξ) is the solution of the complementarity problem (2c), (2d) *and we have the following strong convergences,*

$$y_k \to y \quad \text{in } H^1_0(\Omega), \quad \xi_k \to \xi \quad \text{in } H^{-1}(\Omega).$$

Proof. We test the equation in (4) with y_k and use the coercivity of A (see Assumption 1) to obtain the estimate

$$\begin{aligned} C\|y_k\|^2_{H^1_0(\Omega)} \leq& \langle Ay_k, y_k\rangle_{H^{-1}(\Omega)} = (u_k + f, y_k)_{L^2(\Omega)} + \langle \xi_k, y_k\rangle_{H^{-1}(\Omega)} \\ \leq& (\|u_k\|_{L^2(\Omega)} + \|f\|_{L^2(\Omega)})\,\|y_k\|_{H^1_0(\Omega)} + \langle \xi_k, y_k\rangle_{H^{-1}(\Omega)}. \end{aligned} \tag{5}$$

The bounds on $\langle \xi_k, y_k \rangle_{H^{-1}(\Omega)}$ and on u_k according to the assumptions thus yield a uniform bound on $\|y_k\|_{H_0^1(\Omega)}$, and further, on

$$\|\xi_k\|_{H^{-1}(\Omega)} = \|Ay_k - u_k - f\|_{H^{-1}(\Omega)} \leq C. \tag{6}$$

We now consider a subsequence still denoted by (y_k, u_k, ξ_k) with weak limit (y, u, ξ) in $H_0^1(\Omega) \times L^2(\Omega) \times H^{-1}(\Omega)$. The limit satisfies $y \geq 0$, $\xi \geq 0$ and it holds that

$$0 = Ay_k - \xi_k - u_k - f \rightharpoonup Ay - \xi - u - f \text{ in } H^{-1}(\Omega). \tag{7}$$

We thus have $Ay - \xi - u - f = 0$ in $H^{-1}(\Omega)$. The compact embedding of $H_0^1(\Omega)$ into $L^2(\Omega)$ yields strong convergence of $(y_k)_{k\in\mathbb{N}}$ to its limit y in $L^2(\Omega)$. Hence, the product $(u_k, y_k)_{L^2(\Omega)}$ converges to $(u, y)_{L^2(\Omega)}$ and the weak lower semi-continuity of $z \mapsto \langle Az, z \rangle_{H^{-1}(\Omega)}$ in $H_0^1(\Omega)$ yields that

$$\begin{aligned}
\liminf_{k\in\mathbb{N}} \langle \xi_k, y_k \rangle_{H^{-1}(\Omega)} &= \liminf_{k\in\mathbb{N}} \langle Ay_k, y_k \rangle_{H^{-1}(\Omega)} - (u_k, y_k)_{L^2(\Omega)} - (f, y_k)_{L^2(\Omega)} \\
&\geq \langle Ay, y \rangle_{H^{-1}(\Omega)} - (u, y)_{L^2(\Omega)} - (f, y)_{L^2(\Omega)} = \langle \xi, y \rangle_{H^{-1}(\Omega)} \geq 0.
\end{aligned}$$

This proves the first part of the assertion. We know from (4) and (7) that

$$0 = Ay - u - \xi - f - (Ay_k - u_k - \xi_k - f) = A(y - y_k) - (u - u_k) - (\xi - \xi_k).$$

Using $y_k - y$ as test function, and the coercivity of A, we obtain that

$$C\|y - y_k\|^2_{H_0^1(\Omega)} \leq (u - u_k, y - y_k)_{L^2(\Omega)} + \langle \xi - \xi_k, y - y_k \rangle_{H^{-1}(\Omega)}.$$

The first product in the last term converges to zero as $k \to \infty$ owing to the weak convergence of $u - u_k$ to zero, and the strong convergence of $y - y_k$ to zero in $L^2(\Omega)$. The second product can be expressed as

$$\langle \xi - \xi_k, y - y_k \rangle_{H^{-1}(\Omega)} = \langle \xi, y \rangle_{H^{-1}(\Omega)} - \langle \xi_k, y \rangle_{H^{-1}(\Omega)} - \langle \xi, y_k \rangle_{H^{-1}(\Omega)} + \langle \xi_k, y_k \rangle_{H^{-1}(\Omega)}.$$

The mixed terms in the middle both converge to $\langle \xi, y \rangle_{H^{-1}(\Omega)}$. If we have the additional assumption that $\langle \xi_k, y_k \rangle_{H^{-1}(\Omega)} \to 0$ for $k \to \infty$, then we have $\langle \xi, y \rangle_{H^{-1}(\Omega)} = 0$ by the first part of the lemma, and thus, $\langle \xi - \xi_k, y - y_k \rangle_{H^{-1}(\Omega)} \to 0$ for $k \to \infty$. This shows that $\|y - y_k\|^2_{H_0^1(\Omega)} \to 0$, i.e. the strong convergence of $(y_k)_{k\in\mathbb{N}}$ to y in $H_0^1(\Omega)$ and thus also the convergence of $(\xi_k)_{k\in\mathbb{N}}$ to $\xi = Ay - u - f$ in $H^{-1}(\Omega)$.

Note that although in the solution of the variational inequality the slack variable ξ satisfies $\xi \in L^2(\Omega)$, we cannot guarantee this regularity for ξ_k in the auxiliary problem. An attempt to prove convergence of the slack variables in $L^2(\Omega)$ thus fails in our setting.

Given Ω, A, and f due to Assumption 1 we denote the feasible set of (3) by

$$\mathscr{F} := \{(y,u,\xi) \in H_0^1(\Omega) \times L^2(\Omega) \times H^{-1}(\Omega) \,|\, Ay - u - \xi = f, \quad y \geq 0, \quad \xi \geq 0\}.$$

Note that $\mathscr{F}$ does not depend on the penalty parameter γ.

With this preparation, we can state the following existence result:

Proposition 1. *If Assumption 1 holds, then for every penalty parameter $\gamma > 0$, problem* (3) *has a solution $(y_\gamma, u_\gamma, \xi_\gamma)$.*

Proof. The objective J_γ is bounded from below on the feasible set $\mathscr{F}$. Assume that $(y_k, u_k, \xi_k)_{k\in\mathbb{N}}$ is an infimizing sequence, i.e. for all $k \in \mathbb{N}$, $(y_k, u_k, \xi_k) \in \mathscr{F}$ is feasible and

$$\lim_{k\to\infty} J_\gamma(y_k, u_k, \xi_k) = \inf\{J_\gamma(y,u,\xi) \,|\, (y,u,\xi) \in \mathscr{F}\} =: M.$$

Since the sequence of objective values $\big(J_\gamma(y_k,u_k,\xi_k)\big)_{k\in\mathbb{N}}$ is bounded from above owing to its convergence and since the first summands $j(y_k) + \frac{\nu}{2}\|u_k\|^2_{L^2(\Omega)}$ are bounded from below, we infer that $\Big(\gamma\langle \xi_k, y_k\rangle_{H^{-1}(\Omega)}\Big)_{k\in\mathbb{N}}$ is bounded from above. The sequence $(u_k)_{k\in\mathbb{N}}$ is then bounded by Assumption 1. This follows immediately, when U_{ad} is bounded, or follows from the uniform bound on $\Big(\frac{\nu}{2}\|u_k\|^2_{L^2(\Omega)}\Big)_{k\in\mathbb{N}}$, when $\nu > 0$. The feasibility of $(y_k, u_k, \xi_k) \in \mathscr{F}$ then guarantees that (4) is satisfied such that we can apply the first part of Lemma 1. This yields a subsequence with weak limit (y,u,ξ) which is feasible and ensures that $\liminf\{\langle \xi_k, y_k\rangle_{H^{-1}(\Omega)} \,|\, k \in \mathbb{N}\} \geq \langle \xi, y\rangle_{H^{-1}(\Omega)}$. The weak lower semi-continuity of j and the norm $\|\cdot\|_{L^2(\Omega)}$ imply the optimality of the limit as follows:

$$M \geq \liminf_{k\in\mathbb{N}} j(y_k) + \liminf_{k\in\mathbb{N}} \frac{\nu}{2}\|u_k\|^2_{L^2(\Omega)} + \liminf_{k\in\mathbb{N}} \gamma\langle \xi_k, y_k\rangle_{H^{-1}(\Omega)} \geq J_\gamma(y,u,\xi).$$

The next proposition states consistency of the penalty scheme.

Proposition 2. *Assume that $(\gamma_k)_{k\in\mathbb{N}} \subset \mathbb{R}$ with $\gamma_k > 0$ and $\gamma_k \to \infty$ for $k \to \infty$, and that for every $k \in \mathbb{N}$, (y_k, u_k, ξ_k) solves* (3) *with $\gamma = \gamma_k$ and Ω, A and f from Assumption 1. Then there exists a subsequence, still denoted by $(y_k, u_k, \xi_k)_{k\in\mathbb{N}}$ such that*

$$y_k \to y^\star \quad \text{in } H_0^1(\Omega), \qquad u_k \rightharpoonup u^\star \quad \text{in } L^2(\Omega), \qquad \xi_k \to \xi^\star \quad \text{in } H^{-1}(\Omega)$$

and $(y^\star, u^\star, \xi^\star)$ solves the optimal control problem (2).

Proof. For all $k \in \mathbb{N}$ and a tuple (y,u,ξ) satisfying the complementarity problem (2d), optimality of (y_k, u_k, ξ_k), feasibility of $(y,u,\xi) \in \mathscr{F}$ and the definition of J_{γ_k} imply that

$$J_{\gamma_k}(y_k, u_k, \xi_k) \leq J_{\gamma_k}(y,u,\xi) = J(y,u).$$

Given the boundedness of j and of $\frac{\nu}{2}\|\cdot\|^2_{L^2(\Omega)}$ from below, one derives the uniform boundedness of $\gamma_k\langle\xi_k,y_k\rangle_{H^{-1}(\Omega)}$. Then, in the same way as in the proof of Proposition 1, we may infer that $\|u_k\|_{L^2(\Omega)} \leq C$. Lemma 1 thus yields a subsequence with weak limit $(y^\star,u^\star,\xi^\star)$. Since U_{ad} is closed and convex, it is weakly closed and thus contains the weak limit $u^\star \in U_{ad}$. The convergence of $\gamma_k \to \infty$ implies that $\langle\xi_k,y_k\rangle_{H^{-1}(\Omega)} \to 0$ and the second part of Lemma 1 proves feasibility of $(y^\star,u^\star,\xi^\star)$ in (2) and the strong convergence of y_k and ξ_k. Moreover, using the weak lower semi-continuity of J, the non-negativity of the penalty term and optimality of (y_k,u_k,ξ_k) for the auxiliary problem, we obtain that for any feasible (y,u,ξ) it holds that

$$\begin{aligned} J(y^\star,u^\star) &\leq \liminf\{J(y_k,u_k)\,|\,k\in\mathbb{N}\} \leq \liminf\{J_{\gamma_k}(y_k,u_k,\xi_k)\,|\,k\in\mathbb{N}\} \\ &\leq \liminf\{J_{\gamma_k}(y,u,\xi)\,|\,k\in\mathbb{N}\} = J(y,u). \end{aligned}$$

Therefore $(y^\star,u^\star)$ is feasible and optimal and thus a solution of (2).

2.2 *First Order Stationarity*

In order to derive first order stationarity conditions for the penalized problem (3) we aim to apply [34, Thm. 3.1] and thus have to guarantee the respective constraint qualification (regularity of solutions in the sense of [34, p. 51]).

We define the following spaces and mappings:

$$\begin{aligned} X &= H_0^1(\Omega)\times L^2(\Omega)\times H^{-1}(\Omega), & Y &= H^{-1}(\Omega)\times H_0^1(\Omega)\times H^{-1}(\Omega), \\ \mathscr{C} &= H_0^1(\Omega)\times U_{ad}\times H^{-1}(\Omega), & \mathscr{K} &= \{0\}\times K\times \left(H^{-1}(\Omega)\right)_+, \\ g &: X\to Y, & g(y,u,\xi) &= (Ay-u-\xi-f,y,\xi). \end{aligned}$$

For a subset S of a vector space V and an element $z\in V$, $S(z)$ denotes the conical hull of $S-\{z\}$; see, e.g. [34]. With these denotations, we can formulate the following statement on regularity of feasible points (and in particular of solutions) of the auxiliary problem.

Proposition 3. *Under Assumption 1, every feasible point $\bar{x}=(y_\gamma,u_\gamma,\xi_\gamma)$ of problem* (3) *with penalty parameter $\gamma>0$ is regular in the sense of [34], i.e. it holds that $g'(\bar{x})\mathscr{C}(\bar{x})-\mathscr{K}(g(\bar{x}))=Y$.*

Proof. The inclusion $g'(\bar{x})\mathscr{C}(\bar{x})-\mathscr{K}(g(\bar{x}))\subset Y$ is generically satisfied. Assume that $z=(z_1,z_2,z_3)\in Y$. We aim to show that there exist $c\in\mathscr{C}(\bar{x})$ and $k\in\mathscr{K}(g(\bar{x}))$ such that $g'(\bar{x})c-k=z$. Note that $\mathscr{C}(\bar{x})=H_0^1(\Omega)\times U_{ad}(u_\gamma)\times H^{-1}(\Omega)$, where $U_{ad}(u_\gamma)=\{\beta(\tilde{c}_u-u_\gamma)\,|\,\beta\geq 0,\tilde{c}_u\in U_{ad}\}$. Due to the feasibility of $\bar{x}$ for problem (3), it holds that $g(\bar{x})=(0,y_\gamma,\xi_\gamma)$. We therefore have

$$\mathscr{K}(g(\bar{x}))=\left\{(0,k_y-\beta y_\gamma,k_\xi-\beta\xi_\gamma)\,|\,k_y\in K,k_\xi\in\left(H^{-1}(\Omega)\right)_+,\beta\geq 0\right\}. \tag{8}$$

The Fréchet derivative of g in $\bar{x}$ applied to $c = (c_y, c_u, c_\xi) \in C(\bar{x})$ reads

$$g'(y_\gamma, u_\gamma, \xi_\gamma)(c_y, c_u, c_\xi) = (Ac_y - c_u - c_\xi, c_y, c_\xi). \tag{9}$$

We choose $c_u = 0 \in U_{ad}(u_\gamma)$ and define $(k_y, k_\xi) \in K \times \left(H^{-1}(\Omega)\right)_+$ as the solution to the complementarity problem

$$Ak_y - k_\xi = z_1 + c_u - Az_2 + z_3 \in H^{-1}(\Omega), \quad k_y \geq 0, \quad k_\xi \geq 0, \quad \langle k_\xi, k_y \rangle_{H^{-1}(\Omega)} = 0.$$

Then, with $\beta = 0$, $c_y = z_2 + k_y$ and $c_\xi = z_3 + k_\xi$, it holds that

$$g'(\bar{x})\begin{pmatrix} c_y \\ c_u \\ c_\xi \end{pmatrix} - \begin{pmatrix} 0 \\ k_y - \beta y_\gamma \\ k_\xi - \beta \xi_\gamma \end{pmatrix} = \begin{pmatrix} Ak_y + Az_2 - c_u - k_\xi - z_3 \\ k_y + z_2 - k_y \\ k_\xi + z_3 - k_\xi \end{pmatrix} = \begin{pmatrix} z_1 \\ z_2 \\ z_3 \end{pmatrix}.$$

Proposition 3 yields the following proposition on the existence of multipliers p_γ, ϑ_γ and τ_γ corresponding to the equality constraint, and the nonnegativity constraints on y and on ξ in (3c), respectively, and on necessary first order conditions for optimal points of the auxiliary problem.

Proposition 4. *Every solution $(y_\gamma, u_\gamma, \xi_\gamma)$ of problem (3) with penalty parameter $\gamma > 0$ and Assumption 1 is a first order point for problem (3), i.e., there exists a multiplier tuple $(p_\gamma, \vartheta_\gamma, \tau_\gamma) \in Y^\star$ such that the following conditions hold:*

$$Ay_\gamma - u_\gamma - \xi_\gamma - f = 0 \quad \text{in } H^{-1}(\Omega), \tag{10a}$$

$$y_\gamma \geq 0, \quad \xi_\gamma \geq 0, \tag{10b}$$

$$A^\star p_\gamma + j'(y_\gamma) + \gamma \xi_\gamma - \vartheta_\gamma = 0 \quad \text{in } H^{-1}(\Omega), \tag{10c}$$

$$u_\gamma - \mathrm{Proj}_{U_{ad}}\left(\frac{1}{\nu} p_\gamma\right) = 0 \quad \text{in } L^2(\Omega), \tag{10d}$$

$$\gamma y_\gamma - p_\gamma - \tau_\gamma = 0 \quad \text{in } H_0^1(\Omega), \tag{10e}$$

$$\vartheta_\gamma \geq 0, \ \langle \vartheta_\gamma, y_\gamma \rangle_{H^{-1}(\Omega)} = 0, \ \tau_\gamma \geq 0, \ \langle \xi_\gamma, \tau_\gamma \rangle_{H^{-1}(\Omega)} = 0. \tag{10f}$$

For a first order point $(y_\gamma, u_\gamma, \xi_\gamma)$ of the auxiliary problem (3) with the multiplier vector $(p_\gamma, \vartheta_\gamma, \tau_\gamma)$ we define

$$\lambda_\gamma := \vartheta_\gamma - \gamma \xi_\gamma \quad \text{and} \quad \mu_\gamma := \tau_\gamma - \gamma y_\gamma. \tag{11}$$

We will show that for $\gamma_k \to \infty$, there exist accumulation points of $(\lambda_{\gamma_k})_{k \in \mathbb{N}}$ and $(\mu_{\gamma_k})_{k \in \mathbb{N}}$ which play the role of the multipliers λ and μ in the C-stationarity system for the MPEC (1). The following lemma provides the required uniform bounds and thus prepares the proof of this convergence result.

Proposition 5. *Assume that besides the standard Assumption 1, $\gamma > 0$ is given and that $(y_\gamma, u_\gamma, \xi_\gamma) \in X$ is a first order point of problem* (3) *with multiplier tuple $(p_\gamma, \vartheta_\gamma, \tau_\gamma) \in Y^\star$. If $\|u_\gamma\|_{L^2(\Omega)} \leq C$ and $\langle \xi_\gamma, y_\gamma \rangle_{H^{-1}(\Omega)} \leq C$, then the following estimates hold with a constant $C > 0$ that does not depend on γ:*

$$\langle \lambda_\gamma, \mu_\gamma \rangle_{H^{-1}(\Omega)} \geq 0, \quad \|\mu_\gamma\|_{H_0^1(\Omega)} \leq C, \quad 0 \leq \gamma^2 \langle \xi_\gamma, y_\gamma \rangle_{H^{-1}(\Omega)} \leq C, \tag{12a}$$

$$\|\lambda_\gamma\|_{H^{-1}(\Omega)} \leq C, \quad -C \leq \gamma \langle \xi_\gamma, \mu_\gamma \rangle_{H^{-1}(\Omega)} \leq 0. \tag{12b}$$

Proof. Given the feasibility of first order points and the uniform bound on $\|u_\gamma\|_{L^2(\Omega)}$ and on $\langle \xi_\gamma, y_\gamma \rangle_{H^{-1}(\Omega)}$, we can utilize (5) and (6) in the proof of Lemma 1 and derive uniform bounds on $\|y_\gamma\|_{H_0^1(\Omega)}$ and on $\|\xi_\gamma\|_{H^{-1}(\Omega)}$. Multiply the adjoint equation (10c) by p_γ and use the coercivity of A^* and the definition of λ_γ to obtain the estimate

$$c\,\|p_\gamma\|^2_{H_0^1(\Omega)} \leq \langle A^* p_\gamma, p_\gamma \rangle_{H^{-1}(\Omega)} = \langle \lambda_\gamma, p_\gamma \rangle_{H^{-1}(\Omega)} - \left(j'(y_\gamma), p_\gamma\right)_{L^2(\Omega)}.$$

The fact that $j : L^2(\Omega) \to \mathbb{R}$ is continuously Fréchet-differentiable by Assumption 1 yields that

$$|\left(j'(y_\gamma), p_\gamma\right)_{L^2(\Omega)}| \leq \|j'(y_\gamma)\|_{\mathscr{L}(L^2(\Omega),\mathbb{R})} \|p_\gamma\|_{L^2(\Omega)} \leq C\,\|p_\gamma\|_{L^2(\Omega)}.$$

This yields the estimate

$$c\,\|p_\gamma\|^2_{H_0^1(\Omega)} - \langle \lambda_\gamma, p_\gamma \rangle_{H^{-1}(\Omega)} \leq C\,\|p_\gamma\|_{H_0^1(\Omega)}.$$

Since $p_\gamma = -\mu_\gamma$ by (10e) and the definition of μ_γ, we can write

$$c\,\|\mu_\gamma\|^2_{H_0^1(\Omega)} + \langle \lambda_\gamma, \mu_\gamma \rangle_{H^{-1}(\Omega)} \leq C\,\|\mu_\gamma\|_{H_0^1(\Omega)}. \tag{13}$$

The definition of λ_γ and μ_γ in (11) yields, together with the complementarity and sign conditions in (10f), that

$$\langle \lambda_\gamma, \mu_\gamma \rangle_{H^{-1}(\Omega)} = \gamma^2 \langle \xi_\gamma, y_\gamma \rangle_{H^{-1}(\Omega)} + \langle \vartheta_\gamma, \tau_\gamma \rangle_{H^{-1}(\Omega)} \geq 0.$$

This is the first estimate in (12a). Together with (13) it additionally guarantees the second bound in (12a). We plug the expression for the dual pairing $\langle \lambda_\gamma, \mu_\gamma \rangle_{H^{-1}(\Omega)}$ into (13) and obtain

$$c\,\|\mu_\gamma\|^2_{H_0^1(\Omega)} + \gamma^2 \langle \xi_\gamma, y_\gamma \rangle_{H^{-1}(\Omega)} + \langle \vartheta_\gamma, \tau_\gamma \rangle_{H^{-1}(\Omega)} \leq C\,\|\mu_\gamma\|_{H_0^1(\Omega)} \leq C,$$

and in particular, the last bound in (12a). We once again employ the adjoint equation (10c) to bound $\|\lambda_\gamma\|_{H^{-1}(\Omega)} = \|A^* p_\gamma + y_\gamma - y_d\|_{H^{-1}(\Omega)} \leq C$. Finally, making use of (10e) and the complementarity $\langle \tau_\gamma, \xi_\gamma \rangle_{H_0^1(\Omega)} = 0$, we have

$$\gamma\langle \xi_\gamma, \mu_\gamma\rangle_{H^{-1}(\Omega)} = \gamma\langle \xi_\gamma, -p_\gamma\rangle_{H^{-1}(\Omega)} = \gamma\langle \xi_\gamma, \tau_\gamma - \gamma y_\gamma\rangle_{H^{-1}(\Omega)} = -\gamma^2\langle \xi_\gamma, y_\gamma\rangle_{H^{-1}(\Omega)}$$

and thus the same estimates hold for both terms and we proved (12b).

Assume that $(\gamma_k)_{k\in\mathbb{N}} \subset \mathbb{R}$ with $\gamma_k > 0$ and $\gamma_k \to \infty$ for $k \to \infty$, and that for every $k \in \mathbb{N}$, $(y_k, u_k, \xi_k, p_k, \vartheta_k, \tau_k)$ satisfies (10) with $\gamma = \gamma_k$. If $(\|u_k\|_{L^2(\Omega)})_{k\in\mathbb{N}}$ and $\langle \xi_k, y_k\rangle_{H^{-1}(\Omega)}$ are bounded, then we can apply Proposition 5 to obtain the uniform boundedness of the sequences

$$(\|p_k\|_{H_0^1(\Omega)})_{k\in\mathbb{N}}, \quad (\|\lambda_k\|_{H^{-1}(\Omega)})_{k\in\mathbb{N}}, \quad (\|\mu_k\|_{H_0^1(\Omega)})_{k\in\mathbb{N}} \quad \text{and} \quad \gamma_k^2\langle \xi_k, y_k\rangle_{H^{-1}(\Omega)}.$$

Then, one can extract subsequences with weak limits $p^\star$, $\lambda^\star$ and $\mu^\star$. Additionally, the last bound in (12a) implies that $\langle \xi_k, y_k\rangle_{H^{-1}(\Omega)} \to 0$ for $k \to \infty$. The second part of Lemma 1 thus yields the strong convergences of $y_k \to y^\star$ in $H_0^1(\Omega)$ and $\xi_k \to \xi^\star$ in $H^{-1}(\Omega)$. Using the compact embedding of $H^{-1}(\Omega)$ into $L^2(\Omega)$ we find that $p_k \to p^\star$ in $L^2(\Omega)$ and so, the continuity of the projection operator $\mathrm{Proj}_{U_{ad}} : L^2(\Omega) \to L^2(\Omega)$ gives that $u_k = \mathrm{Proj}_{U_{ad}}\left(\frac{1}{\nu}p_k\right) \to \mathrm{Proj}_{U_{ad}}\left(\frac{1}{\nu}p^\star\right) = u^\star$ in $L^2(\Omega)$. This yields the following corollary:

Corollary 1. *With the notation and under the assumptions of the previous paragraph and in particular, under the condition that $(u_k)_{k\in\mathbb{N}}$ is bounded in $L^2(\Omega)$, there exists a subsequence of $(y_k, u_k, \xi_k, p_k, \vartheta_k, \tau_k)_{k\in\mathbb{N}}$ (denoted the same) such that*

$$y_k \to y^\star \text{ in } H_0^1(\Omega), \qquad u_k \to u^\star \text{ in } L^2(\Omega), \qquad \xi_k \to \xi^\star \text{ in } H^{-1}(\Omega)$$
$$p_k \rightharpoonup p^\star \text{ in } H_0^1(\Omega), \quad \vartheta_k - \gamma_k\xi_k \rightharpoonup \lambda^\star \text{ in } H^{-1}(\Omega), \quad \tau_k - \gamma_k y_k \rightharpoonup \mu^\star \text{ in } H_0^1(\Omega).$$

We are now ready to prove limiting stationarity conditions.

Theorem 1. *With the notation and under the assumptions of Corollary 1, the limit point $(y^\star, u^\star, \xi^\star, p^\star, \lambda^\star, \mu^\star)$ satisfies the following conditions:*

$$Ay^\star - u^\star - \xi^\star = f \qquad \text{in } H^{-1}(\Omega), \tag{14a}$$

$$\xi^\star \geq 0, \quad y^\star \geq 0, \quad \langle \xi^\star, y^\star\rangle_{H^{-1}(\Omega)} = 0, \tag{14b}$$

$$A^* p^\star + j'(y^\star) - \lambda^\star = 0 \qquad \text{in } H^{-1}(\Omega), \tag{14c}$$

$$u^\star - \mathrm{Proj}_{U_{ad}}\left(\frac{1}{\nu}p^\star\right) = 0 \qquad in\ H_0^1(\Omega), \tag{14d}$$

$$\mu^\star + p^\star = 0 \qquad in\ H_0^1(\Omega), \tag{14e}$$

$$\langle \lambda^\star, y^\star \rangle_{H^{-1}(\Omega)} = 0,\ \langle \mu^\star, \xi^\star \rangle_{H^{-1}(\Omega)} = 0,\ \langle \lambda^\star, \mu^\star \rangle_{H^{-1}(\Omega)} \geq 0. \tag{14f}$$

Proof. The second part of Lemma 1 yields the feasibility of $(y^\star, u^\star, \xi^\star, p^\star, \lambda^\star, \mu^\star)$ for problem (2) as stated in (14a)–(14b). Moreover the limit point satisfies the adjoint equation (14c) owing to weak continuity of A^*, the continuity of j' and the definition of λ_k. The continuity of the projection operator and the strong convergence of p_k to p in $L^2(\Omega)$ as well as the definition of μ_k and (10e) directly resolve to (14d) and (14e). The dual pairing $\langle \xi_k, p_k \rangle_{H^{-1}(\Omega)}$ converges to $-\langle \xi^\star, \mu^\star \rangle_{H^{-1}(\Omega)} = \langle \xi^\star, p^\star \rangle_{H^{-1}(\Omega)}$. On the other hand, the uniform bound on $\left(\gamma_k \langle \xi_k, p_k \rangle_{H^{-1}(\Omega)}\right)_{k\in\mathbb{N}}$ from (12b) in Proposition 5 implies that

$$-\langle \xi^\star, \mu^\star \rangle_{H^{-1}(\Omega)} = \lim_{k\to\infty} \langle \xi_k, p_k \rangle_{H^{-1}(\Omega)} = 0.$$

Similarly, we infer from the complementarity of ϑ_k and y_k in (10f) and from the uniform bound on $(\gamma_k^2 \langle \xi_k, y_k \rangle_{H^{-1}(\Omega)})_{k\in\mathbb{N}}$ in (12b) that

$$\langle \lambda^\star, y^\star \rangle_{H^{-1}(\Omega)} = \lim_{k\to\infty} \langle \vartheta_k, y_k \rangle_{H^{-1}(\Omega)} - \gamma_k \langle \xi_k, y_k \rangle_{H^{-1}(\Omega)} = 0.$$

Finally, (12a) in Proposition 5 implies that $\liminf\{\langle \lambda_k, \mu_k \rangle_{H^{-1}(\Omega)} \,|\, k \in \mathbb{N}\} \geq 0$ and we employ the adjoint equation in order to show that

$$\langle \lambda^\star, \mu^\star \rangle_{H^{-1}(\Omega)} \geq \liminf\{\langle \lambda_k, \mu_k \rangle_{H^{-1}(\Omega)} \,|\, k \in \mathbb{N}\} \geq 0 \tag{15}$$

as it is postulated in (14f): Firstly, basic considerations ensure the following estimate,

$$\begin{aligned}\liminf\{-\langle Ap_k, p_k \rangle_{H^{-1}(\Omega)} \,|\, k \in \mathbb{N}\} &= -\limsup\{\langle Ap_k, p_k \rangle_{H^{-1}(\Omega)} \,|\, k \in \mathbb{N}\}\\ &\leq -\liminf\{\langle Ap_k, p_k \rangle_{H^{-1}(\Omega)} \,|\, k \in \mathbb{N}\}.\end{aligned}$$

The weak lower semi-continuity of the mapping $v \mapsto \langle Av, v \rangle_{H^{-1}(\Omega)}$ then results in

$$-\liminf\{\langle Ap_k, p_k \rangle_{H^{-1}(\Omega)} \,|\, k \in \mathbb{N}\} \leq -\langle Ap^\star, p^\star \rangle_{H^{-1}(\Omega)}.$$

Secondly we infer the convergence $(j'(y_k),p_k)_{L^2(\Omega)} \to (j'(y^\star),p^\star)_{L^2(\Omega)}$ from the continuity of j' and the strong convergence of y_k to y in $L^2(\Omega)$. Now we examine the limit inferior in (15):

$$\begin{aligned}
&\liminf\{\langle\lambda_k,\mu_k\rangle_{H^{-1}(\Omega)}\,|\,k\in\mathbb{N}\}\\
&\quad=\liminf\{-\langle Ap_k,p_k\rangle_{H^{-1}(\Omega)}-\big(j'(y_k),p_k\big)_{L^2(\Omega)}\,|\,k\in\mathbb{N}\}\\
&\quad=\liminf\{-\langle Ap_k,p_k\rangle_{H^{-1}(\Omega)}\,|\,k\in\mathbb{N}\}-\big(j'(y^\star),p^\star\big)_{L^2(\Omega)}\\
&\quad\le-\langle Ap^\star,p^\star\rangle_{H^{-1}(\Omega)}-\big(j'(y^\star),p^\star\big)_{L^2(\Omega)}=\langle\lambda^\star,\mu^\star\rangle_{H^{-1}(\Omega)}.
\end{aligned}$$

This yields the sign condition in (14f).

The conditions stated in Theorem 1 are weaker than the C- or strong stationarity conditions that are known from the literature, cf. [15, 24, 33]. But, as, e.g., in [15], the analysis here is constructive in the sense that it suggests an iterative solution algorithm for the MPEC. We show in the subsequent section that under certain conditions, such solutions are strongly stationary.

We end this section by comparing the system in Theorem 1 with the ones obtained in [15], or [19] where several approaches to deriving stationarity have been investigated, respectively. In fact, in [15] the complementarity relations (14f) are augmented by relations guaranteeing that $\lambda^\star$ is zero in a dual sense on an inner ε-approximation of the inactive set with respect to $y^\star$ and $p^\star = 0$ on the corresponding active set. These conditions thus yield a "sharper" system when compared to the one in Theorem 1. The system in [19, Sec. 4] is either equivalent to the one in Theorem 1 or even stronger. The latter ambiguity is due to alternatives with respect to the additional conditions on $\lambda^\star$. We note, however, that the numerical realization of the approach in [19, Sec. 4] would require to solve a sequence of MPECs, which is computationally potentially very demanding. In [19, Sec. 3] Mordukhovich's limiting calculus is applied for deriving the stationarity system. The resulting conditions are weaker than the ones in Theorem 1 in the sense that the last relation in (14f) is replaced by a limsup-condition along certain approximating sequences, but it may be stronger depending on which alternative, i.e. (42), (43), or (44) of [19] is relevant. Finally, we point out that in our case—and this is perhaps one of the appealing aspects of the ℓ_1-penalty technique—strong stationarity (which represents a sharper system than all systems mentioned before) may be achieved. This is the subject of the next section.

2.3 Remarks on Exactness of the Penalty Scheme and on Strong Stationarity

In [4] the authors assume an algorithm that finds second order points of the penalized problem for a sequence $(\gamma_k)_{k\in\mathbb{N}}$ of penalty parameters with $\lim_{k\to\infty}\gamma_k=\infty$. It terminates as soon as the penalized complementarity is satisfied exactly, i.e., if $\langle \xi_k, y_k \rangle_{H^{-1}(\Omega)} = 0$. Theorem 4 in [4] says that if this algorithm does not terminate after a finite number of steps then every accumulation point of the generated sequence of solutions $(x_k)_{k\in\mathbb{N}}$ is either infeasible for the original MPEC or fails to satisfy MPEC-LICQ (which requires that an MPEC satisfies the linear independence constraint qualification (LICQ) when the product condition $\langle \xi, y \rangle_{H^{-1}(\Omega)} = 0$ is omitted, cf. [4, Def. 2]). If no control constraints are active on the biactive set where $y = 0$ and $\xi = 0$ simultaneously, then an analogue to the MPEC-LICQ in function space is satisfied in every feasible point. In particular, when there are no control constraints, this constraint qualification is satisfied. This means that in the special case discussed here, and if the elastic mode penalty method in the sense of Anitescu et al. [4] computes second order points of the auxiliary problem, then after a finite number of steps the iterate is feasible for the original problem. Theorem 2 in [4] proves that any first order point of the auxiliary problem which is feasible for the MPEC is in fact strongly stationary. In total, the two theorems thus indicate exactness of the penalty method (if second order points are computed), which means that for every finite-dimensional restriction of the MPEC (2), it computes a strongly stationary point after a finite number of iterations.

This fact provokes the question whether it is possible to prove exactness of the penalty scheme also in function space and, indeed, we prove here the counterpart of Anitescu et al. [4, Thm. 2] in function space. It is not clear how the second ingredient, namely the convergence of a sequence of second order points, can be utilized to prove a result that is analog to [4, Thm. 4]. Moreover, the non-convexity in the objective of the auxiliary problem seems to preclude second order conditions.

We define the zero set of a function y in $H_0^1(\Omega)$ in the same way as, e.g., in [28]: Utilizing a quasi-continuous representative $\tilde{y}$ of y, see, e.g., [7], we set

$$A^f(y) := \{\tilde{y} = 0\} = \{x \in \Omega \,|\, \tilde{y}(x) = 0\}. \tag{16}$$

Since the quasi-continuous representative is unique up to capacity zero, cf. [7, Lemma 6.55], this definition is also unique up to a set of capacity zero. In general we abbreviate $\{y =^f 0\} := \{x \in \Omega \,|\, \tilde{y}(x) = 0\}$, $\{y >^f 0\} := \{x \in \Omega \,|\, \tilde{y}(x) > 0\}$, etc. as sets defined up to set of capacity zero. We may thus understand $A^f(y)$ without specifying $\tilde{y}$, as the set where any representative is zero quasi everywhere (q.e.), such that in contrast to the set $\{y = 0\}$ which is defined in the sense of almost everywhere, the set $A^f(y)$ is defined in the sense of quasi everywhere.

Since any feasible ξ is non-negative, it can be interpreted as a positive measure, [3] and we define the *finely active set* utilizing the fine topology [3, Sec. 6.3] as it is done in the preprint [33]:

$$A^f(\xi) := \{\xi =^f 0\} := \bigcup \{\omega \subset \Omega \,|\, \omega \text{ finely open}, \xi(\omega) = 0\}. \tag{17}$$

This set is the union of finely open sets, and thus finely open.

Its complement f-supp(ξ), the *fine support of* ξ, is used to define the *fine strongly active set* A_s^f and the *finely biactive set* B^f up to a set of capacity zero as follows:

$$A_s^f(y,\xi) := \text{f-supp}(\xi) \cap A^f(y) = (\Omega \setminus A^f(\xi)) \cap A^f(y), \tag{18}$$

$$B^f(y,\xi) := A^f(y) \cap A^f(\xi). \tag{19}$$

At first, we prove two lemmas which appear in a similar form in [33].

Lemma 2. *Assume that* $v \in H_0^1(\Omega)$ *and* $\zeta \in H^{-1}(\Omega)$ *are both non-negative, and that the dual pairing* $\langle \zeta, v \rangle_{H^{-1}(\Omega)} = 0$. *Then, it holds that* $v = 0$ ζ*-almost everywhere on* Ω *and in particular,*

$$\zeta(\{x \in \Omega \,|\, v(x) > 0\}) = 0.$$

Proof. The complementarity conditions $v \geq 0$, $\zeta \geq 0$, $\langle \zeta, v \rangle_{H^{-1}(\Omega)} = 0$ are equivalent to the variational inequality

$$\forall z \in H_0^1(\Omega), \quad z \geq 0: \qquad \langle \zeta, z - v \rangle_{H^{-1}(\Omega)} \geq 0.$$

Consider a compact subset C of Ω and a smooth function $\chi_C \in C_0^\infty(\Omega)$ with compact support in Ω which takes values in the interval $[0,1]$ and is equal to 1 on C. We set $z = (1 - \chi_C)v \in H_0^1(\Omega)$. Since $z \geq 0$, the assumptions yield $\langle \zeta, z - v \rangle_{H^{-1}(\Omega)} \geq 0$. On the other hand, we can write $z - v = -\chi_C v \leq 0$ and infer that $\langle \zeta, z - v \rangle_{H^{-1}(\Omega)} \leq 0$ from the signs of ζ and $-\chi_C v$. These two inequalities imply $\langle \zeta, \chi_C v \rangle_{H^{-1}(\Omega)} = 0$. We write the dual pairing as a finite integral with respect to the measure ζ,

$$0 = \langle \zeta, \chi_C v \rangle_{H^{-1}(\Omega)} = \int_\Omega \chi_C v \, d\zeta,$$

and employ the non-negativity of $\chi_C v$ to obtain that $\chi_C v = 0$ ζ-almost-everywhere, which in turn means that $v = 0$ ζ-almost-everywhere on arbitrary compact sets $C \subset \Omega$. Finally, Ω is the countable union of compact sets, e.g. of all closed balls with rational midpoints and rational radii in Ω, and the σ-additivity of $\langle \zeta, \cdot \rangle_{H^{-1}(\Omega)}$ yields the assertion.

Lemma 3. *For* $\xi \in \left(H^{-1}(\Omega)\right)_+$ *and* $y \in K$ *with* $y = 0$ ξ*-a.e. it holds that*

$$\{v \in H_0^1(\Omega) \,|\, v = 0 \ \xi\text{-a.e.}\} = \{v \in H_0^1(\Omega) \,|\, v = 0 \ \text{q.e. on } A_s^f(y,\xi)\}.$$

Proof. Assume that $z \in \{v \in H_0^1(\Omega) \,|\, v = 0 \ \xi\text{-a.e.}\}$. We thus know that $\xi(\{z \neq^f 0\}) = 0$, and the set $\{z \neq^f 0\}$ can be supposed to be finely open because it is quasi-open and differs only by a set of capacity zero from its fine interior (cf. [3, Thm. 6.4.13]). Hence, (17) guarantees that $\{z \neq^f 0\} \subset \{\xi =^f 0\} = \Omega \setminus \text{f-supp}(\xi)$ and one can infer from (18) that

$$\operatorname{cap}\left(\{z \neq^f 0\} \cap A_s^f(y,\xi)\right) \leq \operatorname{cap}\left(\{z \neq^f 0\} \cap \text{f-supp}(\xi)\right) = 0.$$

This proves that the first set is included in the second one in the assertion. Now consider $z \in \{v \in H_0^1(\Omega) \,|\, v = 0 \text{ q.e. on } A_s^f(y,\xi)\}$. It holds that

$$\operatorname{cap}\left(\{z \neq^f 0\} \cap \text{f-supp}(\xi) \cap A(y)\right) = 0$$

and thus, $\xi\left(\{z \neq^f 0\} \cap \text{f-supp}(\xi) \cap A(y)\right) = 0$. Since

$$\{z \neq^f 0\} \subset \left(\{z \neq^f 0\} \cap \text{f-supp}(\xi) \cap A(y)\right) \cup \{\xi =^f 0\} \cup \{y >^f 0\},$$

and $\xi(\{\xi =^f 0\}) = 0$ as well as $\xi(\{y >^f 0\}) = 0$ we can infer that $\xi(\{z \neq^f 0\}) = 0$.

The next theorem is the counterpart of Anitescu et al. [4, Thm. 2].

Theorem 2. *Assume that* $(y_\gamma, u_\gamma, \xi_\gamma)$ *is a first order point for* (3) *with multipliers* $(p_\gamma, \vartheta_\gamma, \tau_\gamma)$ *and that* $(y_\gamma, u_\gamma, \xi_\gamma)$ *is feasible for problem* (2) *with Assumption 1. Then* $(y_\gamma, u_\gamma, \xi_\gamma)$ *is strongly stationary for problem* (2) *in the sense that for*

$$(y,u,\xi,p) = (y_\gamma, u_\gamma, \xi_\gamma, p_\gamma), \quad \lambda = \vartheta_\gamma - \gamma\xi_\gamma, \quad \mu = \tau_\gamma - \gamma y_\gamma,$$

the assertions (14a)–(14e) *and the following complementarity and sign conditions hold:*

$$\forall \phi \in H_0^1(\Omega),\ \phi = 0 \text{ q.e. on } A^f(y): \qquad \langle \lambda, \phi \rangle_{H^{-1}(\Omega)} = 0, \tag{20}$$

$$\forall \phi \in H_0^1(\Omega),\ \phi \geq 0 \text{ q.e. on } B^f(y,\xi),\ \phi = 0 \text{ q.e. on } A_s^f(y,\xi): \qquad \langle \lambda, \phi \rangle_{H^{-1}(\Omega)} \geq 0, \tag{21}$$

$$\mu = 0 \text{ q.e. on } A_s^f(y,\xi), \tag{22}$$

$$\mu \geq 0 \text{ q.e. on } B^f(y,\xi). \tag{23}$$

Proof. The condition on feasibility of (y,u,ξ) for the original problem (2) directly implies (14a), (14b). Equations (14c)–(14e) result from the definition of λ and μ and the first order stationarity conditions (10c)–(10e). For $\phi \in H_0^1(\Omega)$ with $\phi = 0$ q.e. on $A^f(y)$ we have

$$\langle \lambda, \phi \rangle_{H^{-1}(\Omega)} = \langle \vartheta_\gamma, \phi \rangle_{H^{-1}(\Omega)} - \gamma \langle \xi_\gamma, \phi \rangle_{H^{-1}(\Omega)}. \tag{24}$$

The non-negativity of ϑ_γ permits us to interpret it as a measure and we can split the dual pairing $\langle \vartheta_\gamma, \phi \rangle_{H^{-1}(\Omega)}$ into the following integrals,

$$\langle \vartheta_\gamma, \phi \rangle_{H^{-1}(\Omega)} = \int_\Omega \phi \, d\vartheta_\gamma = \int_{A^f(y)} \phi \, d\vartheta_\gamma + \int_{\{y >^f 0\}} \phi \, d\vartheta_\gamma.$$

The first integral vanishes because $\phi = 0$ q.e. on $A^f(y)$. By Lemma 2, the complementarity and sign conditions on y and ϑ_γ imply that $\vartheta_\gamma(\{y >^f 0\})) = 0$, and so the second integral also vanishes. In the same way, the dual pairing $\langle \xi_\gamma, \phi \rangle_{H^{-1}(\Omega)}$ is split into

$$\langle \xi_\gamma, \phi \rangle_{H^{-1}(\Omega)} = \int_\Omega \phi d\xi_\gamma = \int_{A^f(y)} \phi d\xi_\gamma + \int_{\{y >^f 0\}} \phi d\xi_\gamma,$$

and with the same arguments as above we observe that, together with $\langle \xi_\gamma, \phi \rangle_{H^{-1}(\Omega)}$, $\langle \lambda, \phi \rangle_{H^{-1}(\Omega)}$ vanishes. This proves (20). Consider $\phi \in H_0^1(\Omega)$ with $\phi \geq 0$ q.e. on $B^f(y,\xi)$ and $\phi = 0$ q.e. on $A_s^f(y,\xi)$. We reuse (24) and again analyze the two summands separately. With the disjoint decomposition $A^f(y) = \left(A^f(y) \cap A^f(\xi)\right) \dot\cup \left(A^f(y) \cap (\Omega \setminus A^f(\xi))\right)$ and the definition of $A_s^f(y,\xi)$ and $B^f(y,\xi)$ in (18) and (19) it is possible to split the first summand into

$$\langle \vartheta_\gamma, \phi \rangle_{H^{-1}(\Omega)} = \int_{\{y >^f 0\}} \phi d\vartheta_\gamma + \int_{A_s^f(y,\xi)} \phi d\vartheta_\gamma + \int_{B^f(y,\xi)} \phi d\vartheta_\gamma.$$

The first integral vanishes with the same argument as above. The conditions on ϕ imply that the second one is zero and the third one is non-negative. Replacing ϑ_γ by ξ_γ in the representation above we immediately obtain that $\langle \xi_\gamma, \phi \rangle_{H^{-1}(\Omega)} = 0$ because

$$\xi_\gamma(\{y >^f 0\}) + \xi_\gamma(B^f(y,\xi)) = \xi_\gamma(\{y >^f 0\} \dot\cup B^f(y,\xi)) = \xi_\gamma(A^f(y)) = 0.$$

This proves (21). We now turn our attention to $\mu = \tau_\gamma - \gamma y_\gamma$. To begin with, it is clear that $y_\gamma = y$ vanishes q.e. on the fine strongly active set $A_s^f(y)$ as well as on the finely biactive set $B^f(y,\xi)$ both of which are a subset of $A^f(y)$, the fine zero set of y. Lemma 2 guarantees that $\mu_\gamma = 0 \xi_\gamma$-a.e. which by Lemma 3 implies that $\mu_\gamma = 0$ q.e. on $A_s^f(y)$. This yields (22). On the finely biactive set $B^f(y,\xi)$, we now know that $\mu = \tau_\gamma$ q.e. which is non-negative q.e. on Ω. We thus have the claimed sign of μ on the biactive set from (23).

3 Point Tracking Control Problem

In this section we consider point tracking subject to a variational inequality. In fact, we assume that a finite set of tracking points I and desired values $y_w \in R$ (for $w \in I$) are given and that the mapping $j : Y \to \mathbb{R}$ in the objective is defined by

$$j : C(\bar\Omega) \to \mathbb{R}, \quad j(y) = \frac{1}{2} \sum_{w \in I} (y(w) - y_w)^2.$$

In the lower level problem, we consider an additional constraint on the slack variable ξ which restricts, for instance in an application where the elastic deformation of a membrane is modeled by the variational inequality, the force that the elastic membrane exerts on the obstacle in the contact region. We choose the state space $W_0^{1,q}(\Omega)$ which embeds into $C(\bar{\Omega})$ if q is larger than the dimension of the computational domain Ω.

3.1 *Model Problem, Penalty Scheme: Solvability and Consistency*

Consider the problem

$$\text{Minimize } J(y,u) = \frac{1}{2}\sum_{w\in I}(y(w)-y_w)^2 + \frac{\nu}{2}\|u\|^2_{L^2(\Omega)} \tag{25a}$$

$$\text{over } (y,u,\xi) \in W_0^{1,q}(\Omega)\times U_{ad}\times \Xi_{ad} \tag{25b}$$

$$\text{subject to } Ay-u-\xi=f, \quad y\geq 0, \quad (y,\xi)_{L^2(\Omega)}=0, \tag{25c}$$

with the following data: For an open bounded domain $\Omega\subset\mathbb{R}^n$, $n\in\mathbb{N}$, we consider $a_{ij}\in L^\infty(\Omega)$ $(i,j\in\{1,\dots,n\})$ collected in the matrix $(a_{ij})\in L^\infty(\Omega)^{n\times n}$ such that for all $\zeta\in\mathbb{R}^n$ and $x\in\Omega$,

$$\zeta^\top(a_{ij}(x))\zeta \geq \Sigma_A|\zeta|^2, \qquad |(a_{ij}(x))\zeta|\leq C_A|\zeta|, \tag{26}$$

where $\Sigma_A, C_A > 0$, and we define the bounded and uniformly elliptic differential operator $A: H_0^1(\Omega)\to H^{-1}(\Omega)$ by

$$A = -\sum_{i,j=1}^n \frac{\partial}{\partial x_i}a_{ij}\frac{\partial}{\partial x_j} = -\operatorname{div}((a_{ij})\nabla\cdot). \tag{27}$$

If the dimension of the problem is at most 2, i.e., $\Omega\subset\mathbb{R}^n$, $n\in\{1,2\}$, then we assume that Ω is a Lipschitz domain and $q\in(2,Q)$ with $Q>2$ from [8, 22]. If the dimension of the problem is larger than 2, i.e. $\Omega\subset\mathbb{R}^n$, $n\in\mathbb{N}$, $n\geq 3$, then we assume that $\partial\Omega\in C^1$, that the coefficients (a_{ij}) of the operator A defined in (27) satisfy additionally $a_{ij}\in C(\bar{\Omega})$ and that $q>n$ is given due to the regularity result in [29, Thm. 5:2.5 (i)]. Then, by Bensoussan et al. [6, Thm. 4.2] and Gröger [11, Thm. 3], respectively, the operator $A: W_0^{1,q}(\Omega)\to W^{-1,q}(\Omega)$ is invertible with continuous inverse operator $A^{-1}: W^{-1,q}(\Omega)\to W_0^{1,q}(\Omega)$. The set of feasible controls is given by the box constraint

$$U_{ad} = \{v\in L^2(\Omega)\,|\,\underline{u}\leq v\leq\overline{u}\},$$

where $\underline{u}, \overline{u} \in L^2(\Omega) \cup \{-\infty, \infty\}$ satisfy $\underline{u} < \overline{u}$ a.e. on Ω, $\nu \geq 0$ and, if U_{ad} is not bounded in $L^2(\Omega)$, then $\nu > 0$, and $f \in L^2(\Omega)$.

We collect these definitions in the following assumption.

Assumption 2. The quantities $\Omega, A, U_{ad}, I, (y_w)_{w \in I}, q > n$ are given as specified in the previous paragraph and in the beginning of Section 3, and it holds that

$$-f \in U_{ad}, \; \Xi_{ad} = \{v \in L^2(\Omega) \,|\, 0 \leq v \leq \phi\} \text{ with } \phi \in L^2(\Omega), \, \phi > 0 \text{ a.e. on } \Omega. \tag{28}$$

By use of an infimizing sequence argument, the solvability of the problem class stated above can be argued: Given the weak closedness of the (non-empty) set $U_{ad} \times \Xi_{ad}$ in $L^2(\Omega) \times L^2(\Omega)$ and the compact embedding of $L^2(\Omega)$ into $W^{-1,q}(\Omega)$, the continuity of the solution operator $A^{-1} : W^{-1,q}(\Omega) \to W_0^{1,q}(\Omega)$ yields the feasibility of an accumulation point of an infimizing sequence of J over the feasible set. The fact that the point evaluation mapping $\delta_w : W_0^{1,q}(\Omega) \to \mathbb{R}, \delta_w(z) := z(w)$ in the first term of J is linear and bounded and the weak lower semi-continuity of the norm mapping in the second part gives the optimality of such accumulation points. This proves the following proposition.

Proposition 6. *Under Assumption 2, problem* (25) *has a solution.*

We propose the following regularized and penalized version of problem (25):

$$\text{Minimize } \tilde{J}_{\gamma,r}(y,u,\xi) = \frac{1}{2} \sum_{w \in I} \fint_{B_r(w)} (y - y_w)^2 \, dx + \frac{\nu}{2} \|u\|^2_{L^2(\Omega)} \tag{29a}$$

$$+ \frac{\gamma}{2} \|(y)_-\|^2_{L^2(\Omega)} + \delta(\gamma) \, (y, \xi)_{L^2(\Omega)} \tag{29b}$$

$$\text{over } (y,u,\xi) \in H_0^1(\Omega) \times U_{ad} \times \Xi_{ad} \tag{29c}$$

$$\text{subject to } Ay - u - \xi = f. \tag{29d}$$

Here, $r > 0$ is an averaging parameter that serves in the same way as in [8]: We define $B_r(w) = \{x \in \Omega \,|\, |x - w| < r\}$ and approximate the point tracking term in the original problem by the integrals in the first summand of $\tilde{J}_{\gamma,r}$. In difference to the ℓ_1-penalty from Section 2, we also penalize the constraint $y \geq 0$ here to avoid a constraint degeneracy. A lack of complementarity in $(y, \xi)_{L^2(\Omega)}$ contributes in the term $\delta(\gamma) \, (y, \xi)_{L^2(\Omega)}$ to the objective, involving a mapping $\delta : \mathbb{R}^{>0} \to \mathbb{R}^{>0}$ with $\delta(\gamma) \to \infty$ for $\gamma \to \infty$.

The following two lemmas guarantee boundedness and convergence properties that are necessary to prove solvability and consistency of the auxiliary problem (29).

Lemma 4. *For all $\gamma, \delta = \delta(\gamma), r > 0$ and $y \in W_0^{1,q}(\Omega)$, $u \in L^2(\Omega)$ and $\xi \in \Xi_{ad}$ from Assumption 2, the objective functional $\tilde{J}_{\gamma,r}$ satisfies*

$$\tilde{J}_{\gamma,r}(y,u,\xi) \geq -\frac{\delta(\gamma)^2}{2\gamma} \|\phi\|^2_{L^2(\Omega)}.$$

Proof. The lower boundedness of $\tilde{J}_{\gamma,r}$ is non-trivial only on the account of its last term. Since $\xi \in \Xi_{ad}$ is non-negative, the product $(y,\xi)_{L^2(\Omega)}$ satisfies

$$(y,\xi)_{L^2(\Omega)} \geq \int_{\{y<0\}} y\xi \, dx \geq -\left\|(y)_-\right\|_{L^2(\Omega)} \|\xi\|_{L^2(\Omega)} \geq -\left\|(y)_-\right\|_{L^2(\Omega)} \|\phi\|_{L^2(\Omega)}. \tag{30}$$

For all feasible (y,u,ξ) we thus have

$$\tilde{J}_{\gamma,r}(y,u,\xi) \geq \frac{\gamma}{2} \left\|(y)_-\right\|^2_{L^2(\Omega)} - \delta(\gamma) \left\|(y)_-\right\|_{L^2(\Omega)} \|\phi\|_{L^2(\Omega)}. \tag{31}$$

The analysis of the right-hand side, which is a quadratic function in $\left\|(y)_-\right\|_{L^2(\Omega)}$, then yields the assertion: For $v \geq 0$ it holds that

$$\frac{\gamma}{2} v^2 - \delta(\gamma) \|\phi\|_{L^2(\Omega)} v \geq -\frac{\delta(\gamma)^2}{2\gamma} \|\phi\|^2_{L^2(\Omega)}. \tag{32}$$

Lemma 5. *Let $(u_k)_{k\in\mathbb{N}}$ and $(\xi_k)_{k\in\mathbb{N}}$ be sequences in $L^2(\Omega)$ that converge weakly to u and $\xi \in L^2(\Omega)$, respectively. Then, the sequence $(y_k)_{k\in\mathbb{N}}$ defined by $y_k = A^{-1}(u_k + \xi_k + f)$ converges strongly in $W_0^{1,q}(\Omega)$ to $y = A^{-1}(u + \xi + f)$ and for every sequence $(r_k)_{k\in\mathbb{N}} \in \mathbb{R}^{>0}$ with $r_k \to 0$, we have that*

$$\sum_{w\in I} \fint_{B_{r_k}(w)} (y_k(x) - y_w)^2 \, dx \to \sum_{w\in I} (y_k(w) - y_w)^2.$$

Proof. The compact embedding of $L^2(\Omega)$ into $W^{-1,q}(\Omega)$ and the continuity of A^{-1} as a mapping from $W^{-1,q}(\Omega)$ to $W_0^{1,q}(\Omega)$ shows the first assumption. We then use the embedding of $W_0^{1,q}(\Omega)$ into $C(\bar{\Omega})$ and [8, L. 2.10] to prove the second assertion.

Lemma 4 guarantees the existence of a feasible infimizing sequence of $\tilde{J}_{\gamma,r}$, and Lemma 5 allows to derive the existence of a solution for the auxiliary problem by use of the typical Weierstraß-argument. This yields the following proposition.

Proposition 7. *Under Assumption 2, problem* (29) *has a solution for any set of parameters $r, \gamma, \delta(\gamma) > 0$.*

The following assumption on the parameters for the auxiliary problem is motivated by the dependence of the lower bound on the objective from Lemma 4.

Assumption 3. In our penalty scheme we use positive sequences $(r_k)_{k\in\mathbb{N}}$, $(\gamma_k)_{k\in\mathbb{N}} \subset \mathbb{R}$ with $r_k \to 0$, $\gamma_k \to \infty$ for $k \to \infty$, and a mapping $\delta : \mathbb{R}^{>0} \to \mathbb{R}^{>0}$ which satisfies $\lim_{\gamma\to\infty} \delta(\gamma) = \infty$ and $\lim_{\gamma\to\infty} \frac{\delta(\gamma)^2}{\gamma} = 0$.

The following lemma on a strong-weak lower semi-continuity property will be helpful in the proof of consistency of the penalty scheme.

Lemma 6. *Let* $(r_k)_{k\in\mathbb{N}}$, $(\gamma_k)_{k\in\mathbb{N}} \subset \mathbb{R}$ *and* $\delta : \mathbb{R}^{>0} \to \mathbb{R}^{>0}$ *satisfy Assumption 3 and for each* $k \in \mathbb{N}$ *assume that* (y_k, u_k, ξ_k) *solves the auxiliary problem* (29) *where Assumption 2 holds true. If* $y_k \to y^\star$ *in* $W_0^{1,q}(\Omega)$ *and* $u_k \rightharpoonup u^\star$ *in* $L^2(\Omega)$, *then*

$$J(y^\star, u^\star) \le \liminf\{\tilde{J}_{\gamma_k, r_k}(y_k, u_k, \xi_k) \,|\, k \in \mathbb{N}\}. \tag{33}$$

Proof. We examine the summands of $\tilde{J}_{\gamma_k, r_k}(y_k, u_k, \xi_k)$ separately. Lemma 5 yields that

$$\lim_{k\to\infty} \frac{1}{2} \sum_{w\in I} ⨍_{B_{r_k}(w)} (y_k - y_w)^2 \, dx = \frac{1}{2} \sum_{w\in I} (y^\star(w) - y_w)^2.$$

From the weak lower semi-continuity of the norm $\|\cdot\|_{L^2(\Omega)} : L^2(\Omega) \to \mathbb{R}$ we infer that

$$\liminf\{\frac{\nu}{2} \|u_k\|^2_{L^2(\Omega)} \,|\, k \in \mathbb{N}\} \ge \frac{\nu}{2} \|u^\star\|^2_{L^2(\Omega)}.$$

Now we use the estimate (32) in the proof of Lemma 4 to see that for all $k \in \mathbb{N}$,

$$\frac{\gamma_k}{2} \|(y_k)_-\|^2_{L^2(\Omega)} + \delta(\gamma_k)\,(y_k, \xi_k)_{L^2(\Omega)} \ge -\frac{\delta(\gamma_k)^2}{2\gamma_k} \|\phi\|^2_{L^2(\Omega)}.$$

Hence, Assumption 3 leads to

$$\begin{aligned}
&\liminf\left\{\frac{\gamma_k}{2} \|(y_k)_-\|^2_{L^2(\Omega)} + \delta(\gamma_k)\,(y_k, \xi_k)_{L^2(\Omega)} \,|\, k \in \mathbb{N}\right\} \\
&\ge \liminf\{-\frac{\delta(\gamma_k)^2}{2\gamma_k} \|\phi\|^2_{L^2(\Omega)} \,|\, k \in \mathbb{N}\} = \lim_{k\to\infty} -\frac{\delta(\gamma_k)^2}{2\gamma_k} \|\phi\|^2_{L^2(\Omega)} = 0.
\end{aligned}$$

Summing up the terms yields the assertion.

Note that for $s, t > 0$ and the function $h : \mathbb{R} \to \mathbb{R}$ defined by $h(v) = sv^2 - tv$ it holds that if $v, \kappa > 0$ are given such that $h(v) \le \kappa$, then

$$v \le \sqrt{\frac{\kappa}{s} + \frac{t^2}{4s^2}} + \frac{t}{2s}. \tag{34}$$

Now we prove consistency of the penalty scheme.

Proposition 8. *Let* $(r_k)_{k\in\mathbb{N}}, (\gamma_k)_{k\in\mathbb{N}} \subset \mathbb{R}$ *and* $\delta : \mathbb{R}^{>0} \to \mathbb{R}^{>0}$ *satisfy Assumption 3, and for each* $k \in \mathbb{N}$ *assume that* (y_k, u_k, ξ_k) *solves the auxiliary problem* (29) *with Assumption 2. Then, there exists a subsequence such that for* $k \to \infty$*, it holds that* $y_k \to y^\star$ *in* $W_0^{1,q}(\Omega)$*,* $u_k \rightharpoonup u^\star$ *in* $L^2(\Omega)$ *and* $\xi_k \rightharpoonup \xi^\star$ *in* $L^2(\Omega)$ *and the limit* $(y^\star, u^\star, \xi^\star) \in W_0^{1,q}(\Omega) \times U_{ad} \times \Xi_{ad}$ *solves problem* (25).

Proof. The weak convergence of a subsequence of $(u_k)_{k\in\mathbb{N}}$ and $(\xi_k)_{k\in\mathbb{N}}$ is due to the boundedness of U_{ad} (resp. the term $\frac{\nu}{2}\|u_k\|^2_{L^2(\Omega)}$ in the objective $\tilde{J}_{\gamma_k,r_k}$, cf. (35) below) and Ξ_{ad}. We denote the weak limits by $u^\star$ and $\xi^\star$, respectively, and note that $y_k := A^{-1}(u_k + \xi_k + f)$ converges to $y^\star = A^{-1}(u^\star + \xi^\star + f)$ in $W_0^{1,q}(\Omega)$ due to the compact embedding of $L^2(\Omega)$ into $W^{-1,q}(\Omega)$ and the continuity of $A^{-1} : W^{-1,q}(\Omega) \to W_0^{1,q}(\Omega)$. The weak limits $u^\star$ and $\xi^\star$ are feasible owing to the weak closedness of U_{ad} and Ξ_{ad} in $L^2(\Omega)$ and $y^\star$ satisfies the partial differential equation in (25c). For all $k \in \mathbb{N}$ it holds that

$$\frac{\gamma_k}{2}\left\|(y_k)_-\right\|^2_{L^2(\Omega)} - \delta(\gamma_k)\left\|(y_k)_-\right\|_{L^2(\Omega)}\|\phi\|_{L^2(\Omega)} \le \tilde{J}_{\gamma_k,r_k}(y_k, u_k, \xi_k).$$

The fact that $(y, u, \xi) = (0, -f, 0)$ is feasible for the auxiliary problem for all penalty and averaging parameters yields the uniform bound

$$\tilde{J}_{\gamma_k,r_k}(y_k, u_k, \xi_k) \le \tilde{J}_{\gamma_k,r_k}(0, -f, 0) = \frac{1}{2}\sum_{w\in I} y_w^2 + \frac{\nu}{2}\|f\|^2_{L^2(\Omega)} =: \kappa. \tag{35}$$

We apply formula (34) for $s = \frac{\gamma_k}{2}$, $t = \delta(\gamma_k)\|\phi\|_{L^2(\Omega)}$, $v = \left\|(-y_k)_-\right\|_{L^2(\Omega)}$ and κ as defined above to derive that

$$\left\|(y_k)_-\right\|_{L^2(\Omega)} \le \sqrt{\frac{2\kappa}{\gamma_k} + \frac{\delta(\gamma_k)^2}{\gamma_k^2}\|\phi\|^2_{L^2(\Omega)}} + \frac{\delta(\gamma_k)}{\gamma_k}\|\phi\|_{L^2(\Omega)}.$$

By Assumption 3 we thus have $\lim_{k\to\infty}\left\|(y_k)_-\right\|_{L^2(\Omega)} \le 0$, i.e., $y^\star \ge 0$. To complete feasibility of the limit point we prove the complementarity of $y^\star$ and $\xi^\star$. The strong convergence of $y_k \to y^\star$ in $L^2(\Omega)$ and the weak convergence of $\xi_k \rightharpoonup \xi^\star$ as well as feasibility of $y^\star$ and $\xi^\star$ yields that

$$\lim_{k\to\infty}(\xi_k, y_k)_{L^2(\Omega)} = (\xi^\star, y^\star)_{L^2(\Omega)} \ge 0.$$

Since all other terms in the auxiliary objective $\tilde{J}_{\gamma_k,r_k}(y_k, u_k, \xi_k)$ are non-negative, we derive the uniform bound

$$\delta(\gamma_k)\,(\xi_k, y_k)_{L^2(\Omega)} \le \tilde{J}_{\gamma_k,r_k}(y_k, u_k, \xi_k) \le \frac{1}{2}\sum_{w\in I} y_w^2 + \frac{\nu}{2}\|f\|^2_{L^2(\Omega)}$$

from (35) and so, $\delta(\gamma_k) \to \infty$ implies that $\lim_{k\to\infty}(\xi_k, y_k)_{L^2(\Omega)} \leq 0$. In order to prove optimality of the limiting element, we assume that (y,u,ξ) is feasible for problem (25). Then, for every $k \in \mathbb{N}$, (y,u,ξ) is feasible for the respective penalized auxiliary problem (29). Utilizing Lemma 5 for the (constant) sequence $(y)_{k\in\mathbb{N}}$ we infer from the fact that the penalization terms vanish in (y,u,ξ) that

$$J(y,u) = \lim_{k\to\infty} \tilde{J}_{\gamma_k,r_k}(y,u,\xi) = \liminf\{\tilde{J}_{\gamma_k,r_k}(y,u,\xi) \,|\, k \in \mathbb{N}\}.$$

The optimality of (y_k,u_k,ξ_k) additionally yields that $\tilde{J}_{\gamma_k,r_k}(y,u,\xi) \geq \tilde{J}_{\gamma_k,r_k}(y_k,u_k,\xi_k)$ and we hence have

$$J(y,u) = \liminf\{\tilde{J}_{\gamma_k,r_k}(y,u,\xi) \,|\, k \in \mathbb{N}\} \geq \liminf\{\tilde{J}_{\gamma_k,r_k}(y_k,u_k,\xi_k) \,|\, k \in \mathbb{N}\}.$$

We finally utilize Lemma 6 to obtain

$$J(y,u) \geq \liminf\{\tilde{J}_{\gamma_k,r_k}(y_k,u_k,\xi_k) \,|\, k \in \mathbb{N}\} \geq J(y^\star,u^\star).$$

3.2 *First Order Stationarity Conditions*

In the same way as in Section 2 we use [34, Thm. 3.1] to derive a system of first order conditions for the auxiliary problem (29) and perform a limiting analysis to derive necessary first order conditions for the original problem.

For

$$x = (y,u,\xi) \in \mathscr{C} = H_0^1(\Omega) \times U_{ad} \times \Xi_{ad} \subset X = H_0^1(\Omega) \times L^2(\Omega) \times L^2(\Omega)$$

define $g : X \to Y = H^{-1}(\Omega)$ by $g(y,u,\xi) = Ay - u - \xi - f$. Then, the constraint set in problem (29), which does not depend on the parameters γ_k, ε_k, is described by $\{x \in \mathscr{C} \,|\, g(x) \in \{0\} \subset H^{-1}(\Omega)\}$. In order to show that $Y \subset g'(x_k)\mathscr{C}(x_k)$ (where $x_k = (y_k,u_k,\xi_k)$, and $\mathscr{C}(x_k)$ is the conical hull of $\mathscr{C} - x_k$) assume that $z \in Y$. Choose $c_u \in U_{ad}$, $c_\xi = 0$, and $c_y = A^{-1}(f + c_u + z)$ to obtain that

$$g'(y_k,u_k,\xi_k)(\beta(c_y - y_k), \beta(c_u - u_k), \beta(c_\xi - \xi_k)) = \beta(Ac_y - c_u - c_\xi - f) = z, \tag{36}$$

and thus, every feasible point $(y_k,u_k,\xi_k) \in H_0^1(\Omega) \times U_{ad} \times \Xi_{ad}$ of (29) is regular in the sense of [34, Eq. (1.4)]. This yields the following result.

Proposition 9. *If* $(y,u,\xi) \in H_0^1(\Omega) \times U_{ad} \times \Xi_{ad}$ *is optimal for the auxiliary problem* (29) *with Assumption 2 and parameters* $(\gamma,r,\delta) \in (\mathbb{R}^{>0})^3$*, then there exist*

$p \in H_0^1(\Omega)$ and $\tau = (\tau)_+ + (\tau)_- \in H_0^1(\Omega)$ such that the following first order conditions hold:

$$A^\star p + \sum_{w\in I}(y-y_w)\bar{\chi}_{B_{r_k}(w)} + \gamma(y)_- + \delta\xi = 0 \qquad \text{in } H^{-1}(\Omega), \tag{37a}$$

$$u - \mathrm{Proj}_{U_{ad}}\left(\frac{1}{\nu}p\right) = 0 \qquad \text{in } L^2(\Omega), \tag{37b}$$

$$p + \tau - \delta y = 0 \qquad \text{in } H_0^1(\Omega), \tag{37c}$$

$$\langle \xi, (\tau)_+\rangle_{H^{-1}(\Omega)} = 0, \qquad \langle \xi - \phi, (\tau)_-\rangle_{H^{-1}(\Omega)} = 0. \tag{37d}$$

Note that τ is split into its positive part, which is the multiplier for the non-negativity condition on ξ, and its negative part, which corresponds to the upper bound on ξ.

We now turn our attention to limiting first order conditions. Let $(r_k)_{k\in\mathbb{N}}$, $(\gamma_k)_{k\in\mathbb{N}} \subset \mathbb{R}$ and $\delta : \mathbb{R}^{>0} \to \mathbb{R}^{>0}$ satisfy Assumption 3 and assume that for every $k \in \mathbb{N}$, $(y_k, u_k, \xi_k) \in H_0^1(\Omega) \times U_{ad} \times \Xi_{ad}$ is feasible for problem (29) with parameters $(\gamma_k, r_k, \delta(\gamma_k))$ and under Assumption 2. Moreover, let $p_k \in H_0^1(\Omega)$ and $\tau_k \in H_0^1(\Omega)$ be given such that the first order conditions (37) hold. We define

$$\lambda_k = -\gamma_k (y_k)_- - \delta(\gamma_k)\xi_k, \qquad \mu_k = (\tau_k)_+ - \delta(\gamma_k)y_k. \tag{38}$$

Proposition 10 below yields uniform bounds and hence the existence of accumulation points of the sequence $(y_k, u_k, \xi_k, p_k, \lambda_k, \mu_k)_{k\in\mathbb{N}}$. In its proof, we need the following assumption.

Assumption 4. We assume the following uniform upper bounds:

$$\frac{\delta(\gamma_k)^3}{\gamma_k} \le C, \tag{39a}$$

$$\left(\delta(\gamma_k)\phi + \gamma_k (y_k)_-, (\tau_k)_-\right)_{L^2(\Omega)} \le C, \tag{39b}$$

$$\|u_k\|_{L^2(\Omega)} \le C. \tag{39c}$$

Note that since we are free to choose $\delta : \mathbb{R}^{>0} \to \mathbb{R}^{>0}$, the first assumption (39a) can be guaranteed a priori. The second bound can be understood as an implication of a convergence rate of $\left\|(-y_k)_-\right\|_{L^\infty(\Omega)}$ and (39a) (see [22]). Or, if the upper constraint on ξ is chosen such that it is not active in the iterates (and in the solution), then $(\tau_k)_- = 0$ guarantees (39b). The third bound is satisfied if U_{ad} is bounded in $L^2(\Omega)$, and apart from that, a typical assumption that becomes important in the analysis of merely stationary points.

We observe that

$$\langle (y_k)_-, y_k\rangle_{H^{-1}(\Omega)} = \left((y_k)_-, y_k\right)_{L^2(\Omega)} = \left\|(y_k)_-\right\|^2_{L^2(\Omega)}. \tag{40}$$

Proposition 10. *With the notation from above, let Assumption 4 hold true. Then, we have the following uniform bounds:*

$$\|y_k\|_{W_0^{1,q}(\Omega)} \leq C, \quad \|\xi_k\|_{L^2(\Omega)} \leq C, \quad \|p_k\|_{H_0^1(\Omega)} \leq C, \quad \|\lambda_k\|_{H^{-1}(\Omega)} \leq C,$$

for some constant $C \geq 0$ which does not depend on k.

Proof. The admissible set Ξ_{ad} directly yields the uniform bound on $(\xi_k)_{k\in\mathbb{N}}$ in $L^2(\Omega)$. Then, the primal equation (29d) and the embedding of $L^2(\Omega)$ into $W^{-1,q}(\Omega)$ provides a bound for y_k in $W_0^{1,q}(\Omega)$. Utilizing the definition of λ_k in (38), the adjoint equation (37a) reads

$$A^\star p_k - \lambda_k = -\sum_{w\in I}(y_k - y_w)\bar{\chi}_{B_{r_k}(w)}.$$

Testing with p_k yields the estimate

$$c\,\|p_k\|^2_{H_0^1(\Omega)} \leq \langle A^\star p_k, p_k\rangle_{H^{-1}(\Omega)} = \langle \lambda_k, p_k\rangle_{H^{-1}(\Omega)} - \sum_{w\in I} \fint_{B_{r_k}(w)} (y_k - y_w) p_k \, dx,$$

with $0 < c$. The uniform bound on y_k in $W_0^{1,q}(\Omega)$, which embeds continuously into $L^\infty(\Omega)$, then gives the estimate

$$c\,\|p_k\|^2_{H_0^1(\Omega)} - \langle \lambda_k, p_k\rangle_{H^{-1}(\Omega)} \leq C\,\|p_k\|_{H_0^1(\Omega)}, \tag{41}$$

where $0 < c \leq C$. In order to obtain a bound on $\|p_k\|_{H_0^1(\Omega)}$, we provide an upper bound for the dual pairing $\langle \lambda_k, p_k\rangle_{H^{-1}(\Omega)}$. The definition of λ_k in (38) and (37c) yield that

$$\begin{aligned}\langle \lambda_k, p_k\rangle_{H^{-1}(\Omega)} =& \gamma_k\langle (y_k)_-, \tau_k\rangle_{H^{-1}(\Omega)} - \gamma_k\delta(\gamma_k)\langle (y_k)_-, y_k\rangle_{H^{-1}(\Omega)} \\ &+ \delta(\gamma_k)\langle \xi_k, \tau_k\rangle_{H^{-1}(\Omega)} - \delta(\gamma_k)^2\langle \xi_k, y_k\rangle_{H^{-1}(\Omega)}.\end{aligned}$$

The second term on the right-hand side can be simplified using (40). Furthermore we split $\tau_k = (\tau_k)_+ + (\tau_k)_-$ and use the complementarities in (37d) to obtain that

$$\begin{aligned}\langle \lambda_k, p_k\rangle_{H^{-1}(\Omega)} =& \gamma_k\langle (y_k)_-, (\tau_k)_+\rangle_{H^{-1}(\Omega)} + \gamma_k\langle (y_k)_-, (\tau_k)_-\rangle_{H^{-1}(\Omega)} \\ &- \gamma_k\delta(\gamma_k)\,\|(y_k)_-\|^2_{L^2(\Omega)} + \delta(\gamma_k)\langle \phi, (\tau_k)_-\rangle_{H^{-1}(\Omega)} - \delta(\gamma_k)^2\langle \xi_k, y_k\rangle_{H^{-1}(\Omega)}.\end{aligned} \tag{42}$$

We drop the first product as it is certainly not positive and estimate

$$\begin{aligned}\langle\lambda_k,p_k\rangle_{H^{-1}(\Omega)} \leq \langle\gamma_k(y_k)_- + \delta(\gamma_k)\phi,(\tau_k)_-\rangle_{H^{-1}(\Omega)}\\ -\gamma_k\delta(\gamma_k)\left\|(y_k)_-\right\|^2_{L^2(\Omega)} - \delta(\gamma_k)^2\langle\xi_k,y_k\rangle_{H^{-1}(\Omega)}.\end{aligned} \tag{43}$$

The first term is bounded by (39b) in Assumption 4. Note that although the term $-\gamma_k\delta(\gamma_k)\left\|(y_k)_-\right\|^2_{L^2(\Omega)}$ is clearly not positive, we rather keep it in the estimate and further analyze the sum

$$\gamma_k\delta(\gamma_k)\left\|(y_k)_-\right\|^2_{L^2(\Omega)} + \delta(\gamma_k)^2\,(\xi_k,y_k)_{L^2(\Omega)},$$

which we need to bound from below. Similarly as in the proof of Proposition 6, we estimate

$$\delta(\gamma_k)^2\,(\xi_k,y_k)_{L^2(\Omega)} \geq -\delta(\gamma_k)^2\left\|\phi\right\|_{L^2(\Omega)}\left\|(y_k)_-\right\|_{L^2(\Omega)}$$

and study a quadratic function $\bar{h}_{\gamma_k}(v) = \gamma_k\delta(\gamma_k)v^2 - \delta(\gamma_k)^2\left\|\phi\right\|_{L^2(\Omega)}v$. We hence find that $\bar{h}_{\gamma_k}(v) \geq \bar{h}_{\gamma_k}\left(\frac{\delta(\gamma_k)}{2\gamma_k}\left\|\phi\right\|_{L^2(\Omega)}\right)$ and so

$$\gamma_k\delta(\gamma_k)\left\|(y_k)_-\right\|^2_{L^2(\Omega)} + \delta(\gamma_k)^2\,(\xi_k,y_k)_{L^2(\Omega)} \geq -\frac{\delta(\gamma_k)^3}{4\gamma_k}\left\|\phi\right\|^2_{L^2(\Omega)}. \tag{44}$$

The last term is bounded from below by (39a) in Assumption 4, and so, the dual pairing $\langle\lambda_k,p_k\rangle_{H^{-1}(\Omega)}$ from (43) is bounded from above. Plugging this into (41), we have

$$c\left\|p_k\right\|^2_{H^1_0(\Omega)} - \tilde{C} \leq C\left\|p_k\right\|_{H^1_0(\Omega)},$$

i.e., a uniform bound on $\left\|p_k\right\|_{H^1_0(\Omega)}$. One can then derive the boundedness of the sequence $(\left\|\lambda_k\right\|_{H^{-1}(\Omega)})_{k\in\mathbb{N}}$ from the adjoint equation (37a).

The next lemma prepares the limiting analysis of a sequence of first order points.

Lemma 7. *Under the conditions of Proposition 10 it holds that*

$$\delta(\gamma_k)\left|\gamma_k\left\|(y_k)_-\right\|^2_{L^2(\Omega)} + \delta(\gamma_k)\,(\xi_k,y_k)_{L^2(\Omega)}\right| \leq C, \tag{45}$$

$$\lim_{k\to\infty}\delta(\gamma_k)\left\|(y_k)_-\right\|_{L^2(\Omega)} = 0, \tag{46}$$

$$\lim_{k\to\infty}\delta(\gamma_k)\,(\xi_k,y_k)_{L^2(\Omega)} = 0. \tag{47}$$

Proof. The estimates (44) and (39a) guarantee that

$$-\delta(\gamma_k)\left(\gamma_k \left\|(y_k)_-\right\|^2_{L^2(\Omega)} + \delta(\gamma_k)\,(\xi_k, y_k)_{L^2(\Omega)}\right) \le \frac{\delta(\gamma_k)^3}{4\gamma_k}\,\|\phi\|^2_{L^2(\Omega)} \le C. \qquad (48)$$

Using the expression (42) for $\langle \lambda_k, p_k\rangle_{H^{-1}(\Omega)}$ in (41) and the uniform boundedness of $\|p_k\|_{H_0^1(\Omega)}$ one derives that

$$-\gamma_k\langle (y_k)_-, (\tau_k)_+\rangle_{H^{-1}(\Omega)} + \delta(\gamma_k)\left(\gamma_k \left\|(y_k)_-\right\|^2_{L^2(\Omega)} + \delta(\gamma_k)\langle \xi_k, y_k\rangle_{H^{-1}(\Omega)}\right)$$
$$\le C + \langle \gamma_k (y_k)_- + \delta(\gamma_k)\phi, (\tau_k)_-\rangle_{H^{-1}(\Omega)}.$$

The non-negativity of the first term on the left-hand side, and the uniform bound on the last term on the right-hand side from Assumption 4 thus yield that

$$\delta(\gamma_k)\left(\gamma_k \left\|(y_k)_-\right\|^2_{L^2(\Omega)} + \delta(\gamma_k)\langle \xi_k, y_k\rangle_{H^{-1}(\Omega)}\right) \le C.$$

Combined with (48) this proves (45). Estimating

$$\langle \xi_k, y_k\rangle_{H^{-1}(\Omega)} \ge -\|\phi\|_{L^2(\Omega)} \left\|(y_k)_-\right\|_{L^2(\Omega)}$$

we obtain that

$$\delta(\gamma_k)\left(\gamma_k \left\|(y_k)_-\right\|^2_{L^2(\Omega)} - \delta(\gamma_k)\,\|\phi\|_{L^2(\Omega)} \left\|(y_k)_-\right\|_{L^2(\Omega)}\right) \le C.$$

Then Equation (34) for $s = \gamma_k$, $t = \delta(\gamma_k)\,\|\phi\|_{L^2(\Omega)}$, $\kappa = \frac{C}{\delta(\gamma_k)}$ and $v = \left\|(y_k)_-\right\|_{L^2(\Omega)}$ gives the bound

$$\left\|(y_k)_-\right\|_{L^2(\Omega)} \le \sqrt{\frac{C}{\delta(\gamma_k)\gamma_k} + \frac{\delta(\gamma_k)^2}{4\gamma_k^2}\,\|\phi\|^2_{L^2(\Omega)}} + \frac{\delta(\gamma_k)}{\gamma_k}\,\|\phi\|_{L^2(\Omega)}.$$

Therefore, $\left\|(y_k)_-\right\|_{L^2(\Omega)} \to 0$ for $k \to \infty$, and, owing to the convergence $\frac{\delta(\gamma_k)^2}{\gamma_k} \to 0$, we even have $\delta(\gamma_k)\left\|(y_k)_-\right\|_{L^2(\Omega)} \to 0$. We utilize this convergence to see that from

$$\delta(\gamma_k)\langle \xi_k, y_k\rangle_{H^{-1}(\Omega)} \ge -\delta(\gamma_k)\,\|\phi\|_{L^2(\Omega)} \left\|(y_k)_-\right\|_{L^2(\Omega)}$$

it follows that

$$\liminf_{k\to\infty} \delta(\gamma_k)\langle \xi_k, y_k\rangle_{H^{-1}(\Omega)} \ge \liminf_{k\to\infty} -\delta(\gamma_k)\,\|\phi\|_{L^2(\Omega)} \left\|(y_k)_-\right\|_{L^2(\Omega)} = 0. \qquad (49)$$

Assume, on the other hand, that $\limsup\{\delta(\gamma_k)\langle\xi_k, y_k\rangle_{H^{-1}(\Omega)} \,|\, k \in \mathbb{N}\} = 2\varepsilon > 0$. Then we have a subsequence denoted the same and a natural number $K \in \mathbb{N}$ such that for all $k \geq K$ it holds that $\delta(\gamma_k)\langle\xi_k, y_k\rangle_{H^{-1}(\Omega)} > \varepsilon$. This implies

$$\delta(\gamma_k)\left(\gamma_k \left\|(y_k)_-\right\|^2_{L^2(\Omega)} + \delta(\gamma_k)\langle\xi_k, y_k\rangle_{H^{-1}(\Omega)}\right) > \delta(\gamma_k)\gamma_k \left\|(y_k)_-\right\|^2_{L^2(\Omega)} + \delta(\gamma_k)\varepsilon$$

which is a contradiction to the boundedness of the term on the left-hand side. We thus have $\limsup\{\delta(\gamma_k)\langle\xi_k, y_k\rangle_{H^{-1}(\Omega)} \,|\, k \in \mathbb{N}\} \leq 0$, which yields, together with (49), the convergence in (47).

Theorem 3. *With the notation of Assumptions 2–4 let $(y_k, u_k, \xi_k) \in H_0^1(\Omega) \times U_{ad} \times \Xi_{ad}$ be a first order point for problem* (29) *with parameters $(\gamma_k, r_k, \delta(\gamma_k))$ and let $p_k \in H_0^1(\Omega)$, $\tau_k \in H_0^1(\Omega)$ be the respective multipliers for every $k \in \mathbb{N}$. Moreover, let λ_k, μ_k be given by* (38). *Then there exists a subsequence of first order points (denoted the same) with*

$$y_k \to y \text{ in } W_0^{1,q}(\Omega), \qquad u_k \to u \text{ in } L^2(\Omega), \qquad \xi_k \rightharpoonup \xi \text{ in } L^2(\Omega),$$
$$p_k \rightharpoonup p \text{ in } H_0^1(\Omega), \qquad \lambda_k \rightharpoonup \lambda \text{ in } W^{-1,q'}(\Omega).$$

The limit point $(y, u, \xi, p, \lambda, \mu)$ satisfies the following conditions:

$$Ay - u - \xi = f, \tag{50a}$$
$$u \in U_{ad}, \qquad \xi \in \Xi_{ad}, \tag{50b}$$
$$y \geq 0, \qquad \langle\xi, y\rangle_{H^{-1}(\Omega)} = 0, \tag{50c}$$
$$A^*p + \sum_{w\in I}(y(w) - y_w)\delta_w - \lambda = 0, \tag{50d}$$
$$\langle\lambda, y\rangle_{W^{-1,q'}(\Omega)} = 0. \tag{50e}$$

If $(\xi_k)_{k\in\mathbb{N}}$ is a (sub-)sequence such that $\xi_k < \phi$ a.e. on Ω for all $k \in \mathbb{N}$ and $\lim_{k\to\infty} \frac{\delta(\gamma)^3}{\gamma_k} = 0$, then it holds additionally that

$$\mu_k \rightharpoonup \mu \text{ in } H_0^1(\Omega), \quad (\xi, \mu)_{L^2(\Omega)} = 0, \quad \liminf\{\langle\lambda_k, \mu_k\rangle_{H^{-1}(\Omega)} \,|\, k \in \mathbb{N}\} \geq 0. \tag{50f}$$

Proof. We start with similar arguments as in the proof of Proposition 8 on consistency of the penalty scheme. The bound on $(\|p_k\|_{H_0^1(\Omega)})_{k\in\mathbb{N}}$ from Proposition 10 yields a weak limit $p \in H_0^1(\Omega)$ of a subsequence of $(p_k)_{k\in\mathbb{N}}$, and the strong convergence of this subsequence in $L^2(\Omega)$ guarantees that

$$u_k = \mathrm{Proj}_{U_{ad}}\left(\frac{1}{\nu}p_k\right) \to \mathrm{Proj}_{U_{ad}}\left(\frac{1}{\nu}p\right) = u \in U_{ad}.$$

The slack constraint set Ξ_{ad} is bounded and weakly closed, which means that $(\xi_k)_{k\in\mathbb{N}}$ contains a subsequence with weak limit $\xi \in \Xi_{ad}$ and

$$y_k = A^{-1}(u_k + \xi_k + f) \to A^{-1}(u + \xi + f) =: y \qquad \text{in } W_0^{1,q}(\Omega)$$

for $k \to \infty$. We employ the adjoint equation (37a) and the definition of λ_k in (38) to derive the convergence of $(\lambda_k)_{k\in\mathbb{N}}$ as follows. Firstly, $A^* : W_0^{1,q'}(\Omega) \to W^{-1,q'}(\Omega)$ is a bounded linear operator and as such weakly continuous such that $A^*p_k \rightharpoonup A^*p$ in $W^{-1,q'}(\Omega)$. Owing to the convergence of $(y_k)_{k\in\mathbb{N}}$ to y in $W_0^{1,q}(\Omega)$ we can apply Lemma 5 to obtain that

$$\lambda_k = A^*p_k + \sum_{w\in I}(y_k - y_w)\bar{\chi}_{B_{r_k}(w)} \rightharpoonup A^*p + \sum_{w\in I}(y(w) - y_w)\delta_w = \lambda \quad \text{in } W^{-1,q'}(\Omega).$$

We hence showed (50a), (50b) and (50d). The non-negativity of the limit state y in (50c) follows from (46) in Lemma 7: Since $y_k \to y$ for $k \to \infty$ we have

$$\|(-y)_+\|_{L^2(\Omega)} = \lim_{k\to\infty} \|(-y_k)_+\|_{L^2(\Omega)} = 0.$$

The convergence (47) in Lemma 7 implies that

$$\langle \xi, y\rangle_{H^{-1}(\Omega)} = \lim_{k\to\infty} \langle \xi_k, y_k\rangle_{H^{-1}(\Omega)} = 0.$$

This proves (50c). We next analyze the dual pairing

$$\langle \lambda, y\rangle_{W^{-1,q'}(\Omega)} = \lim_{k\to\infty} \langle \lambda_k, y_k\rangle_{W^{-1,q'}(\Omega)}.$$

The definition of λ_k in resolves to

$$|\langle \lambda_k, y_k\rangle_{W^{-1,q'}(\Omega)}| = |\gamma_k \|(-y_k)_+\|^2_{L^2(\Omega)} + \delta(\gamma_k)\,(\xi_k, y_k)_{L^2(\Omega)}|.$$

The term on the right-hand side converges to zero by (45) in Lemma 7 and we thus proved (50e). In the second part of the assertion, it holds for all $k \in \mathbb{N}$ that $\xi_k - \phi < 0$ and $(\xi_k - \phi, (\tau_k)_-)_{L^2(\Omega)} = 0$, which indicates that $(\tau_k)_- = 0$. This implies that

$$\mu_k = (\tau_k)_+ - \delta(\gamma_k)y_k = \tau_k - \delta(\gamma_k)y_k = -p_k$$

and we infer the weak convergence of μ_k to $\mu = -p$ in $H_0^1(\Omega)$ from the respective convergence of the adjoint state variables p_k. We write $(\xi, \mu)_{L^2(\Omega)} = \lim_{k\to\infty} (\xi_k, \mu_k)_{L^2(\Omega)}$ and compute

$$(\xi_k, \mu_k)_{L^2(\Omega)} = (\xi_k, (\tau_k)_+ - \delta(\gamma_k)y_k)_{L^2(\Omega)} = -\delta(\gamma_k)\,(\xi_k, y_k)_{L^2(\Omega)}.$$

Using (47) in Lemma 7 we directly obtain that $(\xi, \mu)_{L^2(\Omega)} = 0$. Finally, for all $k \in \mathbb{N}$, the definition of λ_k and μ_k in (38) yields

$$\begin{aligned}
\langle \lambda_k, \mu_k \rangle_{H^{-1}(\Omega)} &= \left(-\gamma_k (y_k)_- - \delta(\gamma_k)\xi_k, (\tau_k)_+ - \delta(\gamma_k) y_k \right)_{L^2(\Omega)} \\
&= -\gamma_k \left((y_k)_-, (\tau_k)_+ \right)_{L^2(\Omega)} + \gamma_k \delta(\gamma_k) \left\| (y_k)_- \right\|^2_{L^2(\Omega)} + \delta(\gamma_k)^2 \left(\xi_k, y_k \right)_{L^2(\Omega)} \\
&\geq \delta(\gamma_k) \left(\gamma_k \left\| (y_k)_- \right\|^2_{L^2(\Omega)} + \delta(\gamma_k) \left(\xi_k, y_k \right)_{L^2(\Omega)} \right).
\end{aligned}$$

The bound from Lemma 7 resolves to

$$\liminf \{ \langle \lambda_k, \mu_k \rangle_{H^{-1}(\Omega)} \,|\, k \in \mathbb{N} \} \geq -C,$$

but one can consider (48) to refine this estimate and derive (50f) by the assumption that $\lim_{k \to \infty} \frac{\delta(\gamma)^3}{\gamma_k} = 0$.

4 Algorithm

In this section, we set up a function space algorithm according to the penalty schemes discussed above. In fact, we provide a path-following method for the solution of optimal control problems of the type (1) in function space, cf. Algorithm 1. In this connection, the so-called *path* is a sequence of first order points $x_k = (y_k, u_k, \xi_k)$ of the auxiliary problem (3) or (25), with respective multiplier vectors Λ_k for a sequence of penalty parameters $(\gamma_k)_{k \in \mathbb{N}} \subset \mathbb{R}^{>0}$ with $\gamma_k \to \infty$ for $k \to \infty$ and, in the point tracking case, $\delta : \mathbb{R}^{>0} \to \mathbb{R}^{>0}$ and $(r_k)_{k \in \mathbb{N}} \subset \mathbb{R}^{>0}$. We choose $(\gamma_k)_{k \in \mathbb{N}} = (\gamma \cdot (\delta\gamma)^k)_{k \in \mathbb{N}}$ for an initial penalty parameter $\gamma > 0$ and a factor $\delta\gamma > 1$. The sequence $(x_k, \Lambda_k)_{k \in \mathbb{N}}$ is computed in steps 6 or 10 of the outer loop (Algorithm 1) by the subroutine `solvePenMPEC` which will be referred to as the inner loop. In each iteration, the multipliers λ_k, μ_k that occur in the stationarity systems (14) and (50) are reconstructed from (y_k, u_k, ξ_k, p_k), see steps 7, 8/11, 12. Then, the norm of the residual pertinent to the conditions in (14) and (50) is computed in step 14.

The break criterion for the outer loop relies on a sufficient decrease of the residual by means of a prescribed $\bar{r} > 0$. In case that the break criterion is not satisfied, the penalty parameter is increased and the corresponding auxiliary problem is solved. Since we proved the convergence only on a subsequence, it may happen that the chosen sequence of penalty parameters is not suitable. We indicate this by defining a maximum number of iterations M, and act on the assumption that the sequence does not converge if M is reached and the residual pertinent to C-stationarity of the suggested solution is not satisfactory small. In this case we reset γ to a fixed value (here γ_0) and decrease $\delta\gamma$ in step 20. Note that the algorithm thus selects an alternative subsequence in line 20 in the infinite loop from line 3 to 21. This step is in practice performed only very rarely.

Algorithm 1 solveMPEC

Input: Either data for problem (2) or,
if point tracking problem (PT=true), data for problem (25),
initial values for y, u, ξ

Choose $0 < \bar{r} \ll 1$, $\gamma = \gamma_0 > 0$, $\delta\gamma > 1$ and $M \in \mathbb{N}$, set $i := 1$.
If PT, set $\delta = \gamma^{1/3}$, $r = \gamma^{-1}$.
loop
  while $i \leq M$ **do**
    if PT **then**
      $(y, u, \xi, p, \tau) =$ `solvePenPT`$(\text{DATA}, \gamma, \delta, r, y, u, \xi)$
      Set $\lambda := A^* p + \sum_{w \in I} (y(w) - y_w) \delta_w$ according to (50d).
      Set $\mu := (\tau)_+ - \delta y$ according to (38).
    else
      $(y, u, \xi, p) =$ `solvePen`$(\text{DATA}, \gamma, y, u, \xi)$
      Set $\lambda := A^* p + j'(y)$ according to (14c).
      Set $\mu := -p$ according to (14e).
    end if
    Compute $r_\gamma =$`residual`$(y, u, \xi, p, \lambda(, \mu, \tau))$ due to (14) or (50)
    if $r_\gamma \leq \bar{r}$ **then**
      return y, u, ξ.
    end if
    Set $\gamma = \delta\gamma \cdot \gamma$, $i = i + 1$, and, if PT, set $\delta = \gamma^{1/3}$, $r = \gamma^{-1}$.
  end while
  Reset $\gamma = \gamma_0$ and if PT, set $\delta\gamma = \frac{8\delta\gamma + 1}{9}$, $\delta = \gamma^{1/3}$, $r = \gamma^{-1}$.
end loop

The auxiliary problems (3) and (29) can be solved with standard optimization tools for problems with smooth objective and linear constraints. In our numerical test computations, we discretize the auxiliary stationarity systems and solve the resulting finite dimensional complementarity system by a damped semi-smooth Newton method (cf. [9, 17, 32]).

5 Numerical Tests

5.1 Numerical Results for an L^2-Tracking Problem

We consider an example from [14, Example 6.1].

Example 1. We use the Laplace operator $A = -\Delta = -\nabla \cdot \nabla : H_0^1(\Omega) \to H^{-1}(\Omega)$ on the square domain $\Omega = (0,1) \times (0,1)$ and

$$z_1(x_1) = -4096x_1^6 + 6144x_1^5 - 3072x_1^4 + 512x_1^3,$$

$$z_2(x_2) = -244.140625x_2^6 + 585.9375x_2^5 - 468.75x_2^4 + 125x_2^3,$$

$$y^*(x_1,x_2) = \begin{cases} z_1(x_1)z_2(x_2) & \text{in } (0,0.5)\times(0,0.8), \\ 0 & \text{else,} \end{cases}$$

$$u^*(x_1,x_2) = y^*(x_1,x_2),$$

$$\xi^*(x_1,x_2) = 2\max\{0, -|x_1-0.8| - |(x_2-0.2)x_1-0.3|+0.35\}.$$

The data f, y_d is set to

$$f = -\Delta y^* - u^* - \xi^*, \qquad y_d = y^* + \xi^* - \nu\Delta u^*.$$

The parameter for the cost of the control is $\nu = 1$, there are no constraints on the control, and the objective functional is defined by

$$J(y,u) = \frac{1}{2}\|y-y_d\|^2_{L^2(\Omega)} + \frac{\nu}{2}\|u\|_{L^2(\Omega)}.$$

The solution $y = u = p$, the multiplier $\xi = -\lambda$, and the strongly active and biactive sets are shown in Figure 1.

We discretize the state as well as the control space by use of P^1-finite elements on a (regular) triangulation of the domain Ω. The MPEC solver can be run on a fixed mesh, or, in order to improve the efficiency of the method, it can be run on gradually refined meshes, optionally also with an a posteriori error estimation procedure from [18] which adaptively adjusts the discretization to the solution of the concrete problem. On each refinement level, the optimization routine (Algorithm 1) is employed with $\bar{r} = 10^{-6}$, $\gamma = 10^{-3}$, $\delta\gamma = 1.5$ and $M = 500$.

Figure 2 shows the convergence history of the ℓ_1-penalty scheme for Example 1 on two different meshes. The L^2-error of the control is plotted in black, and the residual is plotted in gray in a logarithmic scale against the value of the penalty parameter γ. The errors and residuals of the respective last steps of the algorithm are not plotted, because the extremely small residual (around 10^{-15}, see also Figure 3) spoils the scale. The three plots correspond to mesh sizes of 2^{-3} (49 free nodes), 2^{-4} (225 free nodes) and 2^{-5} (961 free nodes) on the square domain. Note that we choose a quite small initial value for the penalty parameter γ and avoid a tuning of parameters to fit certain test examples. All graphs in Figure 2 thus start with a sequence of more or less unreasonable solutions until a penalty parameter of around 10^{-1} is reached. When using this penalty scheme on fixed meshes without an adaptive or uniform refinement loop, the initial value should thus be increased. Especially in the first row, the discretization error can be seen: At some point, the L^2-error of u [i.e., the distance of the discrete control to the exact control in $L^2(\Omega)$], denoted by e_γ, does not decrease anymore while the solution of the discrete problem is not yet found, while the residual r_γ still decreases. This indicates that if one aims to compute an approximation of the function space solution, then the break criterion for the penalty method should be linked to the mesh size.

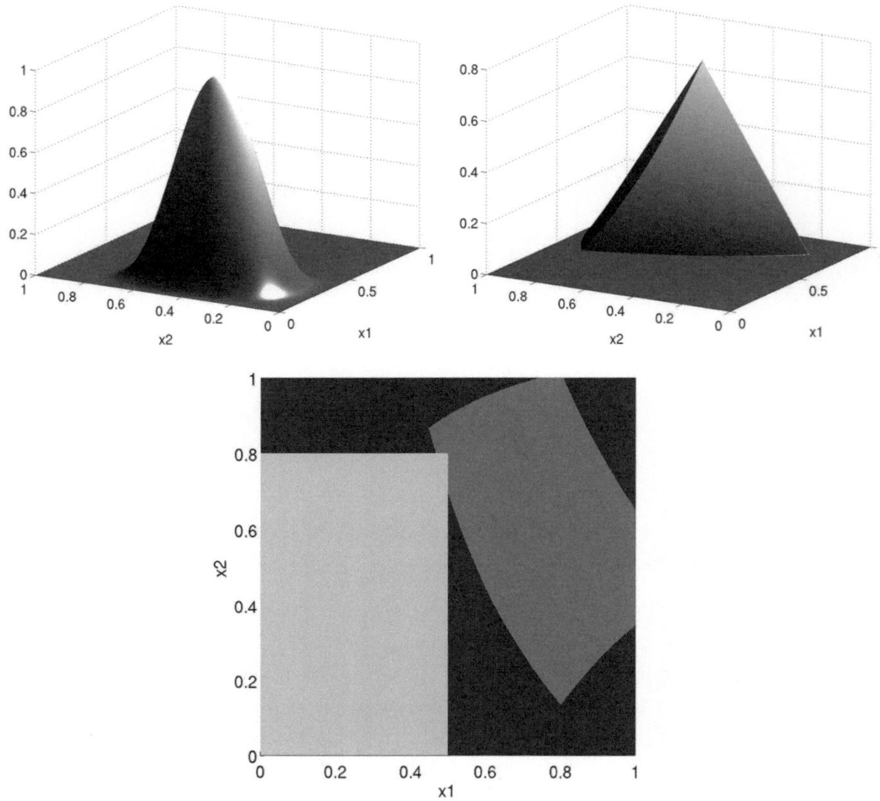

Fig. 1 Solution graphs for Example 1, showing y ($= u$, *upper left*), ξ ($= -\lambda$, *upper right*), and the active sets (*lower plot*). The inactive set is depicted in *light gray*, the strongly active set in *medium gray*, and the biactive set in *dark gray*

Figure 3 shows the convergence history of the algorithm including successive mesh refinement. The two plots show the data pertinent to uniformly refined meshes (left) and adaptive refinement (right). Every plot shows the number of nodes [= number of degrees of freedom (DOF)] in the mesh in black and outer iterations on the horizontal axes. On a fixed mesh (i.e., for a constant number of nodes), the outer loop increases the penalty parameter γ until the residual that belongs to the C-stationarity system $r_\gamma = \texttt{residual}(x,\Lambda)$ in an iterate (x,Λ) (gray graph) is below a level $\bar{r}$ (dotted in gray). Then, the mesh is refined by bisection of every triangle or due to the estimator and thus the number of nodes increases. We cut each plot after 100 outer iterations. The L^2-norm of the error in the control variable $e_u = \|u^\star - u_k\|_{L^2(\Omega)}$ in an iterate u_k is plotted in black. Its value combines the discretization error with the error in optimality. More specifically, it decreases when γ is increased until the discretization error is reached, but when the mesh is refined and the iterate u_k is prolongated (by local interpolation) to the larger finite element

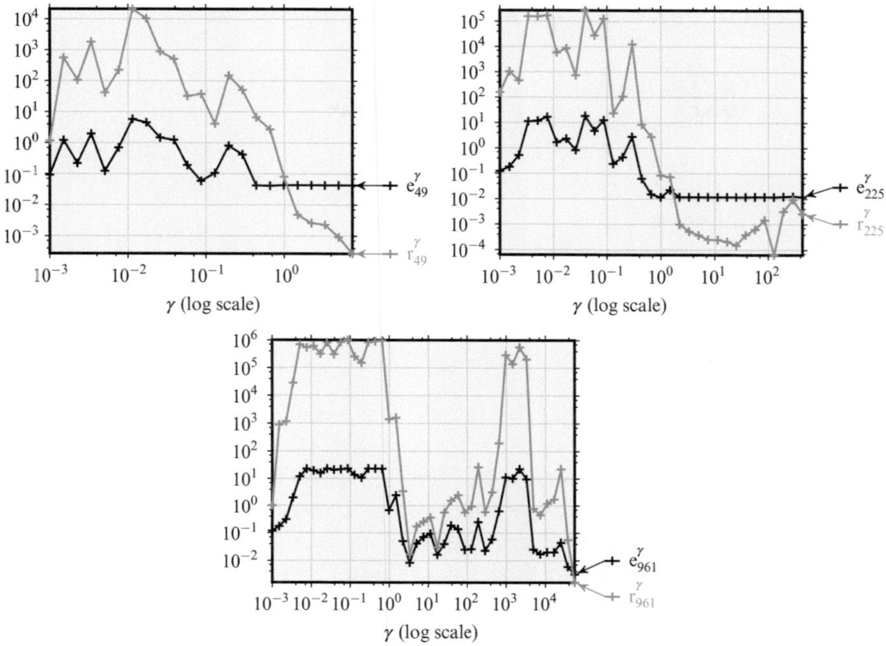

Fig. 2 Convergence history for the elastic mode algorithm 1 without multigrid/adaptive refinement. The convergence of the error (e^γ, *black*) and the residual (r^γ, *gray*) are plotted against γ for Example 1 on three different meshes

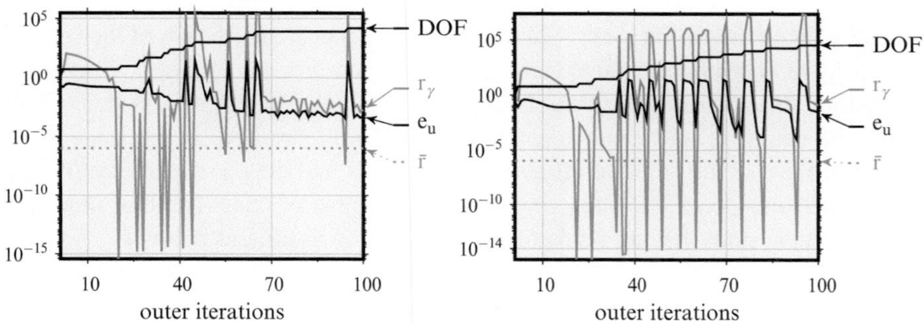

Fig. 3 Convergence history for the elastic mode algorithm 1 with multigrid/adaptive refinement applied to Example 1. The *left plot* shows the convergence of the error (e^γ, *black*) and the residual (r^γ, *gray*) plotted against the outer iterations for the uniform method, the *right plot* refers to the respective data in the adaptive method

space, e_u increases. Moreover, the error graph indicates that together with finer meshes (with increasing DOF), we obtain smaller discretization errors. In this plot we did not cut the last steps and see the sudden decrease of the residual on each mesh. This indicates that the algorithm in fact finds an exact solution to the discrete problem, cf. Section 2.3.

5.2 Numerical Results for a Point Tracking Problem

Example 2. We now consider the L-shaped domain $\Omega = (-1,0) \times (-1,1) \cup (-1,1) \times (0,1)$. The set of tracking points is

$$I = \{(-0.5,-0.5),(-0.5,0),(-0.5,0.5),(0,0.5),(0.5,0.5)\}$$

and we want the state variable to take the values $y_w = 1$ for $w \in \{(-0.5,-0.5), (-0.5,0)\}$, $y_w = -0.1$ for $w \in \{(-0.5,0.5),(0,0.5)\}$ and $y_w = 0$ for $w = (0.5,0.5)$. The parameter belonging to the control costs is set to $\nu = 0.01$, the force acting on the state is defined by

$$f(x_1,x_2) = 0.5 + 0.5(x_1 - x_2).$$

There are no control constraints. The solutions calculated by our algorithm are depicted in Figure 4. The example again admits a biactive set, and the solutions have a low regularity (consider, for instance, λ on the right-hand side, middle row in Figure 4). For the penalty method 1 we choose $\bar{r} = 10^{-5}$, $\gamma_0 = 0.01$, $\delta\gamma = 1.5$, and $M = 300$.

We again start with a test of the algorithm on fixed meshes with different complexity. Figure 5 shows the convergence of the residuals for Example 2 on a mesh with 833 nodes (bright gray graph, $h = 2^{-4}$ on the L-shaped domain), with 3,201 nodes (dark gray graph, $h = 2^{-5}$) and with 12,545 nodes (black graph, $h = 2^{-6}$). The algorithm shows a clearly mesh independent convergence. Note that since we do not know the exact solution, we do not plot $L^2(\Omega)$-errors of the control variable, but instead use the residuals as an error measure.

Figure 6 shows the convergence history of the overall algorithm for both uniform (left) and adaptive mesh refinement (right). The structure here is the same as in Figure 3: We plot the outer iterations of Algorithm 1 and the mesh refinement steps on the horizontal axis. The black line shows the number of nodes and thus indicates the refinement steps that are performed as soon as the residual is below $\bar{r}$.

The residual shows a similar trend as in Figure 3 on the L^2-tracking case. It decreases until, at a certain value of γ, the iterate "falls" into the solution and the outer loop breaks in fact with a residual that is far below its bound $\bar{r}$. The adaptive method (right plot) has an advantage because of its capability of a rather accurate detection of the active sets. The sudden increase of the residual in the left plot comes from a reset of γ to γ_0: The upper bound M in Algorithm 1 is reached, and the solution pertinent to the new (small) penalty parameter is of course a bad candidate for a solution of the C-stationarity system.

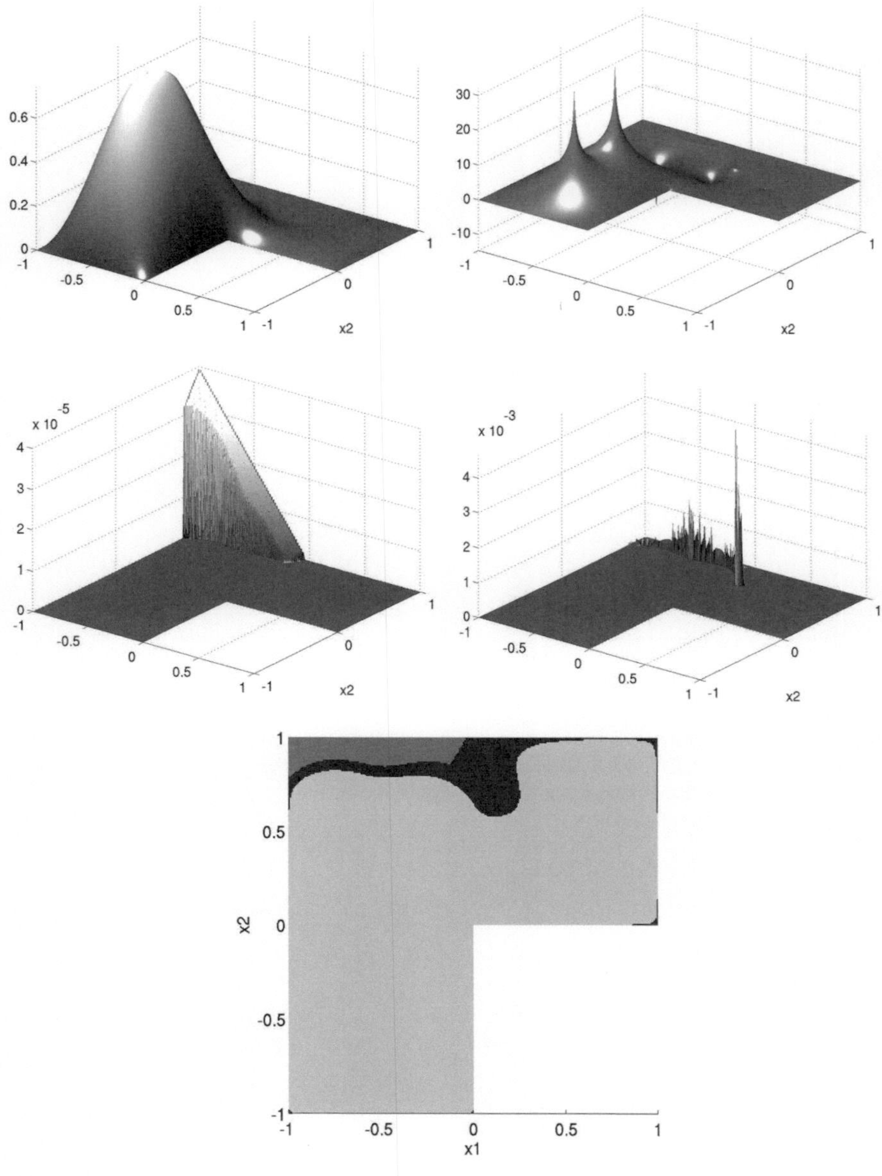

Fig. 4 Solution graphs for Example 2, showing y (*upper left*), u (*upper right*), ξ (*middle left*), λ (*middle right*) and the active sets (*lower plot*). The inactive set is depicted in *light gray*, the strongly active set in *medium gray*, and the biactive set in *dark gray*

References

1. Achdou, Y.: An inverse problem for a parabolic variational inequality arising in volatility calibration with American options. SIAM J. Control Optim. **43**(5), 1583–1615 (2005)

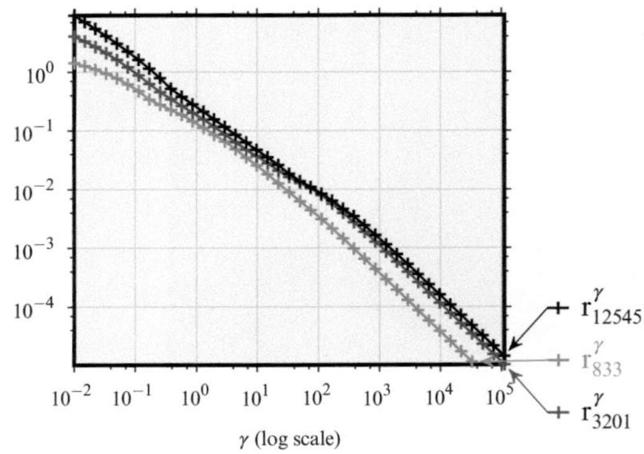

Fig. 5 Convergence history of the residuals of the C-stationarity system for the elastic mode algorithm 1 without multigrid/adaptive refinement plotted against the value of the penalty parameter γ. The *subscript values* in the labels of the graphs give the complexity of the problem which they belong to

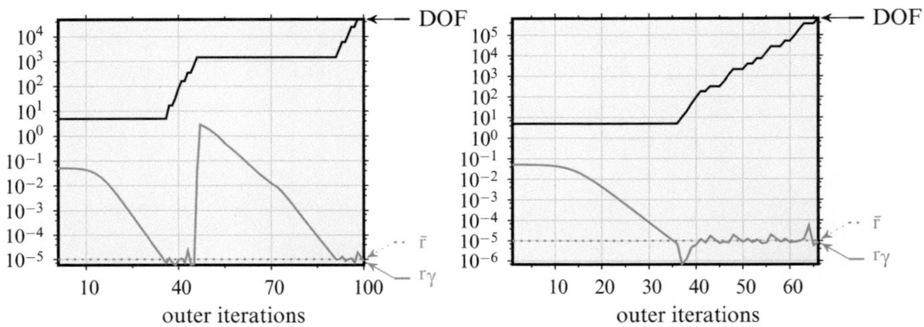

Fig. 6 Convergence history for the elastic mode algorithm 1 with multigrid/adaptive refinement applied to Example 2. The *left column* shows the convergence of the residual r^γ (*gray*) and DOF (*black*) plotted against the outer iterations for the uniform method, the *right column* refers to the residual in the adaptive method

2. Adams, R., Fournier, J.: Sobolev Spaces. Pure and Applied Mathematics, vol. 140. Elsevier, Amsterdam (2003)
3. Adams, R., Hedberg, L.: Function Spaces and Potential Theory. A Series of Comprehensive Studies in Mathematics, vol. 314. Springer, Berlin (1996)
4. Anitescu, M., Tseng, P., Wright, S.: Elastic-mode algorithms for mathematical programs with equilibrium constraints: global convergence and stationarity properties. Math. Program. **110**, 337–371 (2007)
5. Barbu, V.: Optimal Control of Variational Inequalities. Addison-Wesley, Reading (1984)
6. Bensoussan, A., Lions, J., Papanicolaou, G.: Asymptotic Analysis for Periodic Structures. North-Holland, Amsterdam (1978)
7. Bonnans, J., Shapiro, A.: Perturbation Analysis of Optimization Problems. Springer, New York (2000)
8. Brett, C., Elliott, C., Hintermüller, M., Löbhard, C.: Mesh adaptivity in optimal control of elliptic variational inequalities with point-tracking of the state. Start project, IFB-Report No. 67 (09/2013), Institute of Mathematics and Scientific Computing, University of Graz (2013)
9. Facchinei, F., Pang, J.: Finite-Dimensional Variational Inequalities and Complementarity Problems. Springer Series in Operations Research, vol. 1. Springer, New York (2003)
10. Glowinski, R., Lions, J., Trémolières, R.: Numerical Analysis of Variational Inequalities. Studies in Mathematics and its Applications, vol. 8. North-Holland, Amsterdam (1981)

11. Gröger, K.: A $W^{1,p}$-estimate for solutions to mixed boundary value problems for second order elliptic differential equations. Math. Ann. **283**, 679–687 (1989)
12. Herzog, R., Meyer, C., Wachsmuth, G.: C-stationarity for optimal control of static plasticity with linear kinematic hardening. SIAM J. Control Optim. **50**(5), 3052–3082 (2012)
13. Herzog, R., Meyer, C., Wachsmuth, G.: B- and strong stationarity for optimal control of static plasticity with hardening. SIAM J. Optim. **23**(1), 321–352 (2013)
14. Hintermüller, M., Kopacka, I.: Mathematical programs with complementarity constraints in function space: C- and strong stationarity and a path-following algorithm. SIAM J. Optim. **20**, 868–902 (2009)
15. Hintermüller, M., Kopacka, I.: A smooth penalty approach and a nonlinear multigrid algorithm for elliptic MPECs. Comput. Optim. Appl. **50**, 111–145 (2011)
16. Hintermüller, M., Surowiec, T.: A bundle-free implicit programming approach for a class of MPECs in function space. Start Project, IFB-Report No. 60 (09/2012), Institute of Mathematics and Scientific Computing, University of Graz (2013)
17. Hintermüller, M., Ito, K., Kunisch, K.: The primal-dual active set strategy as a semismooth Newton method. SIAM J. Optim. **13**, 865–888 (2002)
18. Hintermüller, M., Hoppe, R., Löbhard, C.: A dual-weighted residual approach to goal-oriented adaptivity for optimal control of elliptic variational inequalities. ESAIM Control Optim. Calc. Var. **20**, 524–546 (2014)
19. Hintermüller, M., Mordukhovich, B., Surowiec, T.: Several approaches for the derivation of stationarity conditions for elliptic MPECs with upper-level control constraints. Math. Program. **146**, 555–582 (2014)
20. Kinderlehrer, D., Stampacchia, G.: An Introduction to Variational Inequalities and Their Applications. Academic, New York (1980)
21. Lions, J.: Optimal Control of Systems Governed by Partial Differential Equations. Grundlehren der mathematischen Wissenschaften. Springer, Heidelberg (1971)
22. Löbhard, C.: Optimal control of variational inequalities: Numerical methods and point tracking. Ph.D. thesis, Humboldt-Universität zu Berlin, Institute of Mathematics (2014)
23. Luo, Z., Pang, J., Ralph, D.: Mathematical Programs with Equilibrium Constraints. Cambridge University Press, Cambridge (1996)
24. Mignot, F., Puel, J.: Optimal control in some variational inequalities. SIAM J. Control Optim. **22**(3), 466–476 (1984)
25. Mordukhovich, B.: Variational Analysis and Generalized Differentiation I: Basic Theory. Grundlehren Der Mathematischen Wissenschaften, vol. 330. Springer, Berlin (2006)
26. Mordukhovich, B.: Variational Analysis and Generalized Differentiation II: Applications. Grundlehren Der Mathematischen Wissenschaften, vol. 331. Springer, Berlin (2006)
27. Outrata, J., Kočvara, M., Zowe, J.: Nonsmooth Approach to Optimization Problems with Equilibrium Constraints: Theory, Applications, and Numerical Results. Nonconvex Optimization and Its Applications, vol. 152. Kluwer Academic, Dordrecht (1998)
28. Outrata, J., Jarušek, J., Stará, J.: On optimality conditions in control of elliptic variational inequalities. Set Valued Anal. **19**, 23–42 (2011)
29. Rodrigues, J.-F.: Obstacle Problems in Mathematical Physics. North-Holland, Amsterdam (1987)
30. Scheel, H., Scholtes, S.: Mathematical programs with complementarity constraints: stationarity, optimality, and sensitivity. Math. Oper. Res. **25**(1), 1–22 (2000)
31. Schiela, A., Wachsmuth, D.: Convergence analysis of smoothing methods for optimal control of stationary variational inequalities with control constraints. ESAIM Math. Model. Numer. Anal. **47**, 771–787 (2013)

32. Ulbrich, M.: Semismooth Newton methods for operator equations in function spaces. SIAM J. Optim. **13**, 805–842 (2003)
33. Wachsmuth, G.: Strong stationarity for optimal control of the obstacle problem with control constraints. SIAM J. Optim. **24**, 1914–1932 (2014)
34. Zowe, J., Kurcyusz, S.: Regularity and stability for the mathematical programming problem in Banach spaces. Appl. Math. Optim. **5**, 49–62 (1979)

Reduced Space Dynamics-Based Geo-Statistical Prior Sampling for Uncertainty Quantification of End Goal Decisions

Lior Horesh, Andrew R. Conn, Eduardo A. Jimenez, and Gijs M. van Essen

Abstract The inversion of large-scale ill-posed problems introduces multiple challenges. These include, identifying appropriate noise model, prescription of suitable prior information, design of an informative experiment, uncertainty quantification, incorporation of heterogeneous sources of data, and definition of an appropriate optimization scheme. In the context of flow in porous media, subsurface parameters are inferred through the inversion of oil production data (a process called history matching). In this study, the inherent uncertainty of the problem is mitigated by devising efficient and comprehensive approaches for prior sampling. Despite meticulous efforts to minimize the variability of the solution space, the distribution of the posterior may remain intractable. In particular, geo-statisticians may often propose large sets of prior samples that regardless of their apparent geological distinction are almost entirely flow equivalent. As an antidote, a reduced space hierarchical clustering of flow relevant indicators is proposed for aggregation of these samples. The effectiveness of the method is demonstrated both with synthetic and field scale data. In addition, numerical linear algebra techniques that exploit the special structure of the underlying problems are elucidated.

Keywords Uncertainty quantification • Reduced space • Dynamic indicator • Prior sampling • Geo-statistics • Hierarchical clustering • Goal-oriented prediction • Dynamic similarity • History matching.

1 Introduction

History matching in the context of multi-phase flow in porous medium is an ill-posed problem as the measured data do not convey sufficient information for complete and stable recovery of the underlying subsurface parameters [21, 58]. Part of the missing information can be supplemented by incorporation of additional

L. Horesh (✉) • A.R. Conn
IBM Thomas J. Watson Research Center, Yorktown Heights, NY, USA
e-mail: lhoresh@us.ibm.com; arconn@us.ibm.com

E.A. Jimenez • G.M. van Essen
Shell, Katy, TX, USA
e-mail: Eduardo.Jimenez@shell.com; Gijs.VanEssen@shell.com

M. Al-Baali et al. (eds.), *Numerical Analysis and Optimization*, Springer Proceedings in Mathematics & Statistics 134, DOI 10.1007/978-3-319-17689-5_8

independent information, for example seismic data, gravity, electromagnetic survey information, and additional log data; or prescribing structure information based upon a-priori knowledge either in the form of regularization or re-parametrization. Complementary, improved experimental design [18–20, 26, 31, 32, 53] and handling of model mis-specification errors [15, 22–24, 44] can improve upon the utility of observable data.

Each of these resolutions introduces various challenges. Information from heterogeneous sources of data (potentially related to different physical entities) may be seemingly conflicting [17, 33, 47, 55, 56]. When it comes to imposition of structure, mathematical interpretation of abstractly described structural notions is far from trivial. This often leads to situations where ad-hoc structural regularization schemes are utilized, which introduce non-desired bias to the solutions [25, 34, 38, 54]. Furthermore, there is a computational cost and increased complexity typically associated with these tasks. Moreover, the supplemental information, even when extremely useful, is unlikely to accurately and completely resolve the ill-posed nature of the problem, so as a result we are led to focus our attention and efforts towards development of computationally efficient means to realize uncertainty. This is of great relevance in the course of reservoir management, as useful decision-making requires comprehensive realization of this uncertainty.

1.1 The Role of the Prior in Bayesian Inference

We begin our discussion with a description of the history matching problem in Bayesian inference terms and then identify the role of the prior and its sampling in realization of the posterior distribution. Let $m = \{\kappa, \phi, \rho, \cdots\}$ represent the model parameters (permeability, porosity, seal factor, etc.), $x = \{p, S, \cdots\}$ stand for the state parameters (pressure, saturation, etc.) and $y = \{p_c, q_c, \cdots\}$ define the controls (controlled point pressure, injected water rate, etc.). Further let us assume that the observed production data, d, is linked to the model parameters and the controls through the following observation operator

$$d = g(m, x, y) + \varepsilon,$$

where ε comprises both model mis-specification error and measurement noise. The link between the model prior distribution and the inference posterior is given by the Bayes theorem

$$\pi(m|d) = \frac{\pi(d|m)\pi(m)}{\pi(d)},$$

where $\pi(m)$ stands for the model m prior probability, $\pi(d|m)$ and $\pi(m|d)$ are conditional probabilities of the likelihood and posterior, respectively, and $\pi(d)$ represents the marginal probability.

Proper realization of the inferred model posterior $\pi(m|d)$ is essential for any consequent decision-making processes, and can be regarded as an intermediate stage that supports the decision/prediction/control process $G(\pi(m|d))$ and its respective distribution $\pi(G(\pi(m|d)))$.

1.2 Prior Sampling

Prior sampling provides means for exploration of uncertainty associated with model parameters under consideration. However, since typically the distribution is a non-trivial one, the decision as to which samples should be considered is critical. In the context of reservoirs, in most cases, these samples are generated by geologists (or geo-statisticians) who embody in those their own personal subjective interpretation of the geology (geological concept). Traditionally, for uncertainty quantification, one considers mainly parameters that are not controlled by the underlying algorithm, i.e. the history matching process in our case. Nonetheless, the history-matching problem is ill-posed and nonlinear, thus, an exact definition of the adjustable variables is a moving target, as the large null space renders many parameters redundant. Furthermore, since the search space is usually characterized by multiple local minima, exploration of the search space can be carried out via consideration of multiple starting points. Without a consistent imposition of structure the features that geologists introduce in the process are typically not retained throughout the history matching process, as the conventional optimization problem formulation is agnostic to such considerations. Consequently, the initial postulated structure (as introduced in the sampled prior) is no longer honored when the data itself is (superficially) well-matched. This is a serious impediment upon the veracity of the results produced, particularly since it is the ensuing predictions that are usually of primary importance. Given that even the most ample efforts to account for the ill-posed nature of the problem may almost always yield solutions that inherently involve uncertainty, we focus our attention towards efficient means for quantification of such uncertainty.

1.3 Flow Equivalence

Based upon structural information and geostatistical reasoning, geologists produce multiple model realizations to account for prior uncertainty. However, many seemingly notable geological (static) variations result in little, if any, measurable impact upon flow patterns (i.e., the dynamics). For large-scale, nonlinear problems of non-trivial probability distributions a large set of geological prior models is seemingly required in order to capture uncertainty. Fortunately, this large set contains much redundancy from the point of view of the flow behavior. We provide below (Figure 1) an example where two distinct geological models of different

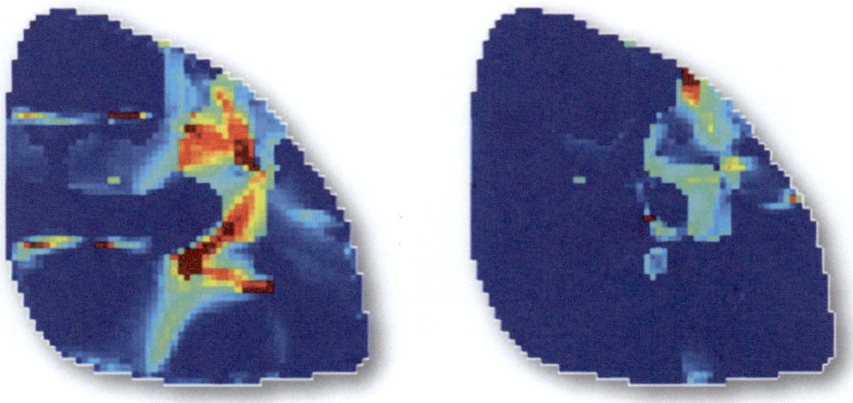

Fig. 1 Two distinct reservoir models that include six producing wells and six injection wells. *Red* indicates high permeability whereas *blue* corresponds to low permeability regions

permeability distributions display nearly identical production rates and pressures (Figure 2), for a particular period in time (500 days) and a given set of controls.[1]

This example demonstrates how multiple distinct reservoir models may correspond to similar production behavior. Yet, we must be careful as to the implications of such a comparison. By virtue of history matching, it is sometimes possible to update two distinct models so that they will equally honor the data for a given time window. The updated models may still differ from each other significantly, either because the inversion forced each of them to move into different local minima, or simply due to the large null space associated with the problem. Unless the models are essentially equivalent and consistent in terms of flow dynamics, their future forecast (i.e., production measures following the history matching period) may differ drastically. An indication of such a situation is provided by the very different production forecasts and predicted net present values that the two models (Figure 1) provide, as can be seen after the first 500 days (Figure 2).

1.4 Proposition

For the above reasons, our overall goal in this study is to provide effective methods for pre-screening geological models prior to the history matching process.

[1]Note that often appropriate integration of structure is not exercised in the course of history matching, consequently, in standard practice the emphasis is on data misfit minimization alone. Such an approach is obviously far from desirable. More generally there are various circumstances whereby the dynamical equivalence of distinct geological models may be observed.

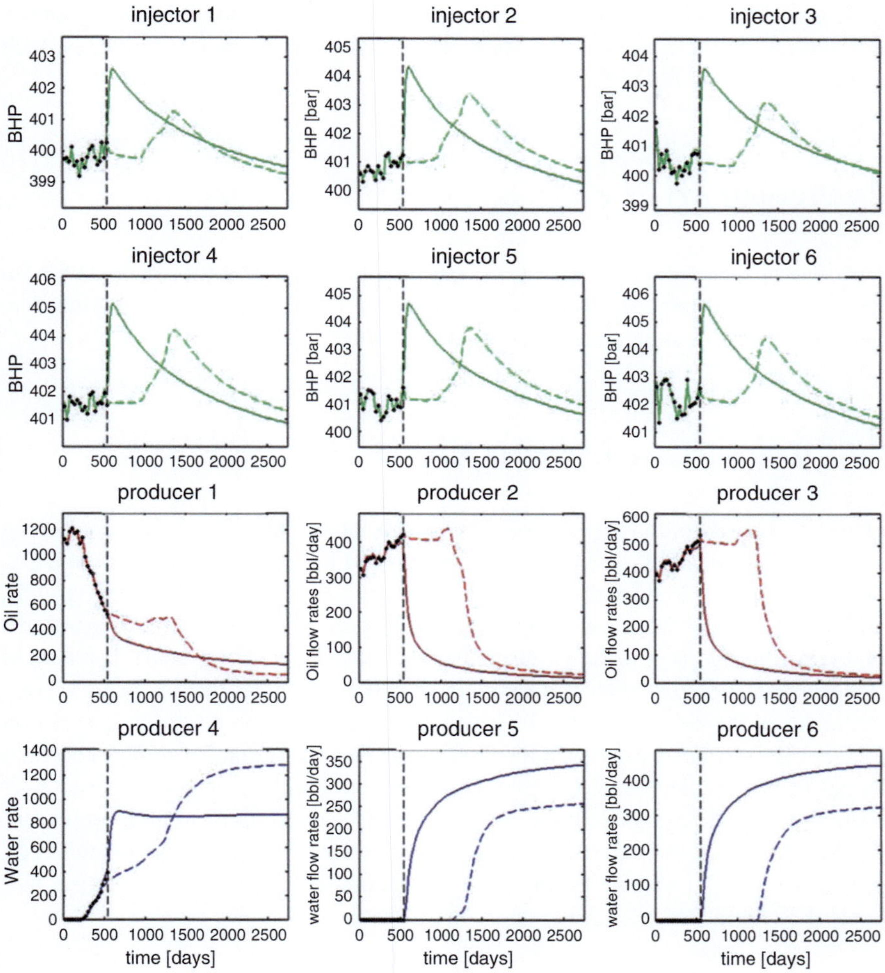

Fig. 2 Extended duration production rates and pressures forecasts for the two distinct reservoir models above. *Continuous lines* represent simulated production measurements of the first model, whereas the *dashed lines* represent the simulated measurements of the second model

Ideally, each geo-statistical prior realization (sample) would undergo a history matching procedure and the variability of their forecasts would be factored for operational decisions. However, the history matching process is computationally intensive; hence, given a large number of possible scenarios, this option is impractical. Fortunately, often many (static) geological scenarios are almost equivalent in some sense, so the challenge is to characterize each distinct scenario.

Our proposition is to cluster prior realizations into flow-equivalent sets, thereby, identifying far fewer representatives whilst still being able to predict the same range of future production, reliably. As a link between the static geology and the dynamics as captured by the geophysics we propose considering dynamic fingerprints, such as mass fluxes and time of flight.

2 Previous Art

Uncertainty could be captured by a joint distribution function [13]; however, in the context of large-scale problems this option is also computationally infeasible. Several alternative strategies can be identified in the literature.

2.1 Heuristics and Ensemble Methods

One approach is to quantify uncertainty by the rather naive means of considering very few extreme cases, such as low, medium, and high, for which few samples of the posterior distribution are computed. While this approach is appealing computationally, it is far from reflecting reliably the posterior distribution of problems of high dimensions. Recently, ensemble methods have also become popular. In these approaches one typically extracts approximations to the first and second moments, in the hope that these will adequately represent the posterior distribution [1, 14, 39, 40]. One of the main appeals of these methods is that the computation can be performed on distributed systems. However, variability of geological structures, such as channels, cannot be captured by first and second moment and, even more fundamentally, the question of the desired sample size dimension to adequately (i.e., providing globally reliable estimates) capture complex posterior distributions is a bone of contention.

2.2 Model Agnostic Approaches

One remedy to the above concern is consideration of multi-point geo-statistics [10, 16]. Some of the more successful applications of the approach was through the utility of a kernel transformation to a higher dimensional space, in which co-distances between aggregated field production were computed [48, 51, 52]. However, this approach introduces several difficulties; first, complete simulation is required for each realization and this requires computation that can be orders of magnitude more expensive than required by the methods proposed in this study. The second difficulty raises a more fundamental issue. While the concept of a kernel transformation is popular in the machine learning community, it is typically used for model-agnostic problems, which is appropriate when one is incapable of deriving concrete links between the data and the desired entity to be classified. For problems already suffering from a deficiency of reliable admissible information in the form of observable data, disregard of available knowledge, in the form of the governing physics, makes little sense. Another limitation of model agnostic approaches is the lack of spatial separation. When only field level production data is considered the information is often too rudimentary for many operational decisions.

2.3 *Physics Based Approaches*

The last category consists of methods that exploit the physics of the problem, such as reduced physics (for example, proper orthogonal decomposition, [5, 6, 42], dynamic mode decomposition [49], streamlines, [3, 7, 11, 37]) and the proposed methodology of this paper. With all these methods a physics-based link between computationally appealing entities and the desired output is usefully exploited. In the context of uncertainty quantification, we shall mention a few approaches. The method of [57] is one in which only a single model is history matched, while multiple realizations are ranked with respect to streamlines properties such as time of flight and flow rate. Idrobo et al. [35] considered an approximated (binary) swept volume measure based on streamlines to characterize the model dynamics. Møyner et al. have developed an interactive fast simulation tool for well placement screening under multiple scenarios, [41]. Of course, other choices for capturing the dynamics can be sought.

3 Methodology

3.1 *Flow Indicators*

An appropriate choice of a flow indicator lays at the core of the proposed workflow for identification of model flow equivalence. There are numerous potential candidates for this task, such as mass fluxes and time of flight. We shall describe the utility of mass fluxes for this purpose. As is well-known, the basis of flow in a porous medium is Darcy's law

$$\mathbf{v}^{\alpha} = -\frac{K\kappa^{\alpha}}{\mu^{\alpha}}\left(\nabla\left(p^{\alpha}+p^{\alpha}_{cap}\right) - g\rho^{\alpha}\nabla z\right),$$

which is essentially a conservation of momentum relation that can be derived through homogenization of the Navier Stokes equations. Here ρ^{α} is the mass density per phase.[2] It provides links between the pressure, p, the pressure difference and the fluid velocity, v. K is an absolute permeability tensor describing how amenable the

[2]The references to α are phase references not power indices.

porous medium is to flow, κ^α is a dimensionless relative permeability parameter indicative of the ease of flow for one phase due to the presence of another phase. This parameter is a major source of nonlinearity. The fluid viscosity is per phase and is given by μ^α. The overall pressure is a composite of the phase pressure, p^α, the capillary pressure, p^α_{cap}, and the vertical pressure drop caused by gravity. For a situation where the friction (K, κ and μ) and the pressure are maintained fixed, the velocity is proportional to pressure difference. For future reference we shall reformulate Darcy law in terms of the phase mass fluxes, F^α.

$$F^\alpha = -K\frac{\rho^\alpha \kappa^\alpha}{\mu^\alpha}\left(\nabla\left(p+p^\alpha_{cap}\right) - g\rho^\alpha \nabla z\right).$$

For a more comprehensive overview covering multi-phase settings, the reader is referred to [2, 43].

A link between these physical entities (mass fluxes) and measurable production can be determined by application of an appropriate observation operator, $\tilde{P}$, upon the mass conservation equation, (1), below, that states that a change in mass in a given domain can be attributed to the total influx and outflux of mass through the volume and to any mass coming from any external sources.

$$\frac{\partial\left(\varphi a^c\right)}{\partial t} - \nabla\cdot F^c = \rho^c q^c \tag{1}$$

Here the component mass fluxes, F^c, link to the phase mass fluxes, F^α through the total mass fraction $\chi^{c\alpha}$ via

$$F^c = \sum_{\alpha=1}^{N_\alpha} \chi^{c\alpha} F^\alpha$$

and the component accumulations relates the link between the phase mass densities, ρ^α, the phase saturations S^α, and the total mass fraction matrix, $\chi^{c\alpha}$ via

$$a^c = \sum_{\alpha=1}^{N_\alpha} \chi^{c\alpha} S^\alpha \rho^\alpha c = 1,\ldots,N_c$$

$$\tilde{P}\left(\frac{\partial\left(\varphi a^c\right)}{\partial t} - \nabla\cdot F^c\right) = \tilde{P}\left(\rho^c q^c\right).$$

As the right-hand side represents volumetric flow rate, the observation operator, $\tilde{P}$, is often linear (or at least approximately linear). Once we have established that mass fluxes and production measurements are closely related, we remark that since mass fluxes provide a detailed (grid resolution) vector map of flow, they offer superior insights for distinguishing between model dynamics compared with production information (which is sparsely distributed spatially). This point is actually subtle. On the one hand, there are situations in which models of similar production output would be decreed unnecessarily different if judged by their mass fluxes alone. On the other hand, if not differentiated, such situations are much more likely to manifest undesired behavior similar to the one illustrated in Figure 2, that is, production measurements would coincide for a giving period, but then later diverge (see, for example, Figure 2 after 500 days). Clearly, we would still want to regard such models as distinct, and thereby, better off relying upon the information this flow indicator is offering, rather than the limited one suggested by production alone. Other virtues that mass fluxes offer as a flow indicator are their insensitivity to flow isolated permeability regions as well as to vorticity effects (which, however, are unlikely for reservoir forces and velocities). And of course, as opposed to permeability (or any other static model parameter), mass fluxes do, naturally, account for the underlying physics.

In (Figure 3) the singular value distribution of water, oil and both mass fluxes for a controlled test case of ten distinct flow patterns is illustrated. It is evident from the charts that mass fluxes have clearly identified the presence of ten distinct modes, and thereafter, variations of these modes. For comparison, we also provide similar singular value analysis of the pressure field and the saturation field (in Figure 4). These results indicate that these flow indicators can potentially provide similar qualitative results, however, distinguishably, they appear to be less discriminative than the mass fluxes.

4 Reduced-Space Representation Eigen-Dynamics

Flow indicators, such as mass fluxes, capture chief characteristics of the dynamics, yet, 4D vector fields are of a large dimension. Clustering in such a large dimensional space is typically impractical. Instead, a reduced order representation is considered. Let us assume that for each realization, i, mass flux (or any other suitable alternative flow indicator) the corresponding vector fields, $\mathbf{F}_i(x,y,z;t)$, are computed.

We perform a singular value decomposition of the vector fields from all realizations to subsequently enable a reduced order representation

$$U\Sigma V^\top = [\mathbf{F}_1(x,y,z;t), \mathbf{F}_2(x,y,z;t), \ldots, \mathbf{F}_n(x,y,z;t)]$$

Representation of each flux realization i, in the singular vector basis is obtained as

$$\alpha_{i,j} = U_j \cdot \mathbf{F}_i(x,y,z;t), \quad (ith\ realization,\ jth\ component)$$

Other than a trivial reduction through the truncation of components, $\alpha_{i,j}$, below a prescribed threshold, the key element in reduction is the fact that the models are

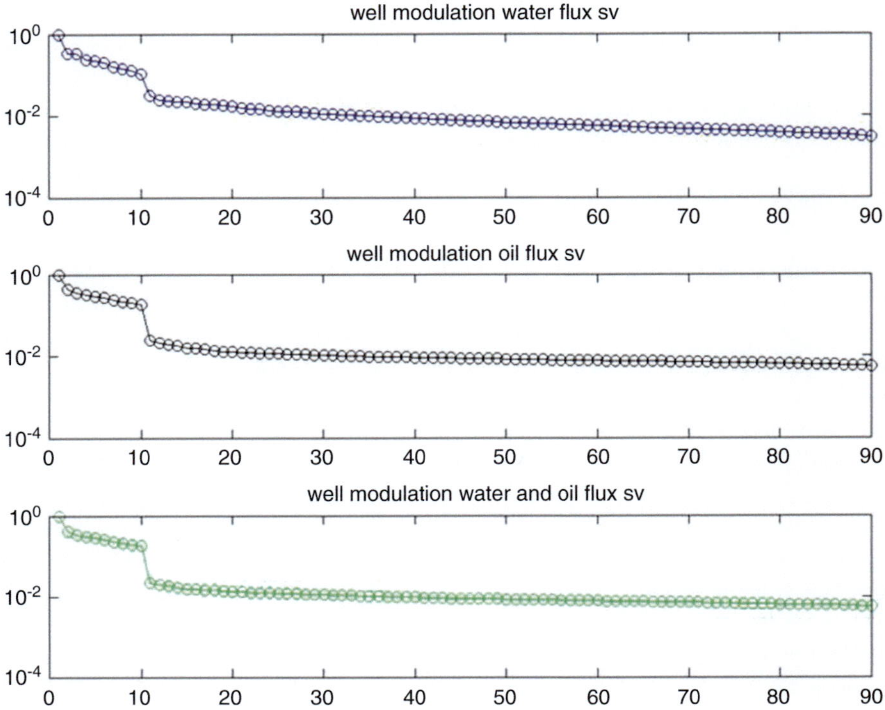

Fig. 3 Singular values spectrum for a set of fluxes for a set of 90 realizations of ten well-distinguished flow patterns. *Top to bottom*: singular values of fluxes of water, fluxes of oil, fluxes of water and oil

now represented by the product of a given (fixed) set of principle mass fluxes via individual sets of coefficients. For most realistic scale problems, the dimension of the coefficient space is of the order of the number of considered realizations (or a truncation of which), in contrast with the dimension of the mass flux instances. Figure 5 gives a simplified illustration if there are only three significant (principal) components in the reduced representation.

Next, the distances between relevant representations of flow indicator characteristics are quantified in the representation coefficient space (Figure 6). Once distances between realizations are computed based on the similarity of flux in the representation coefficient space, clustering in the reduced space can be performed [28].

4.1 Hierarchical Clustering

In this study we consider an agglomerative (i.e., bottom-up) hierarchical clustering construction. Initially, each realization, (i.e. representation coefficients) serves as

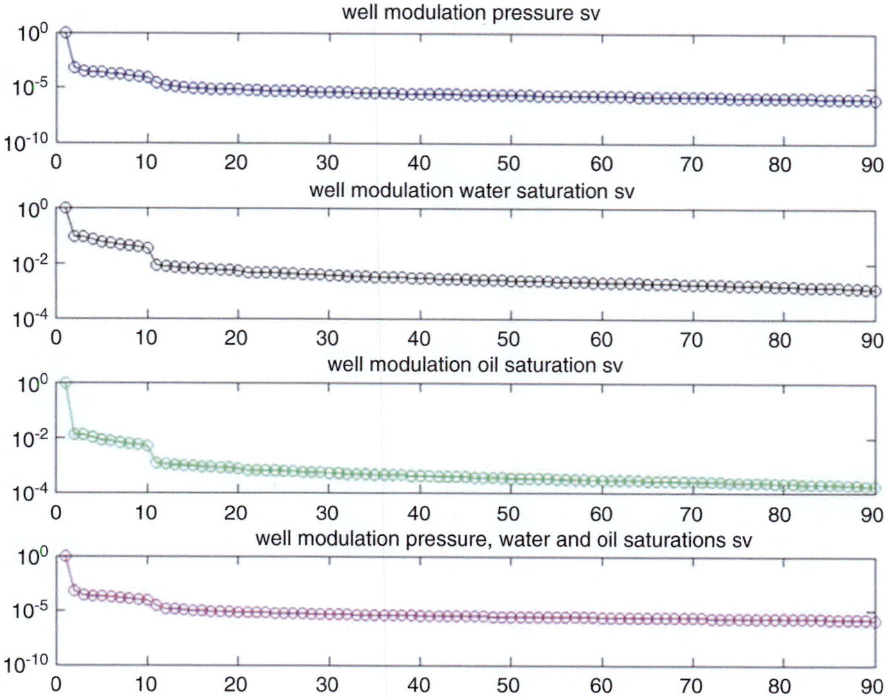

Fig. 4 Singular values spectrum for alternative flow indicators (pressure, saturation) for a set of 90 realizations of ten well-distinguished flow patterns. *Top to bottom*: singular values of pressure, water saturation, oil saturation, water and oil saturation

an independent cluster. Next, clusters are formed when a node and all of its subnodes have consistent values larger than a given threshold. In such a case, all leaves at or below the node are grouped into a cluster. Each link in a cluster tree is characterized by inconsistency coefficients, which compares its height with average height of other links of the same hierarchal levels. The higher the value of this coefficient, the less similar the objects connected by the link. Different distance measures can be considered for quantifying the distances between samples within each cluster, and respectively, the distances between clusters. In some situations, the choice of distance metric can be of cardinal importance in clustering. The choice of hierarchical clustering is by no means exclusive. The advantage of this clustering is that it offers the practitioner some level of control as to the granularity of the obtained clusters. Thus, if a fine level of classification is desired, one can use the output of the lowest level, whereas if only crude classification is required the top levels can be used. Also, in terms of analysis, the tree structure that the clustering process entails, called a dendrogram, offers excellent insight into the

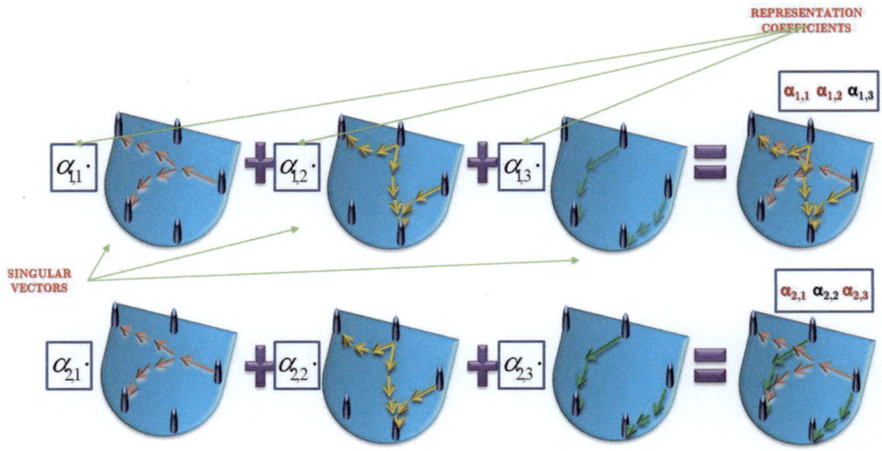

Fig. 5 A simplified illustration of flux representation using the singular basis vectors. The model on the *right top row* is formed as a linear combination (first and second of notable value, while the third is negligible) of the principal flux vectors. The model on the *bottom right* is formed in the same fashion, but with a notable contribution from the first and the third vectors only

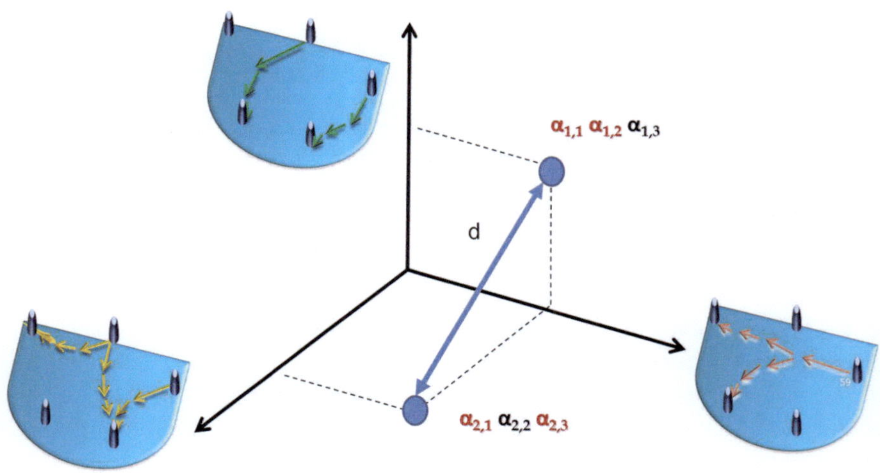

Fig. 6 Simplistic illustration of representation and distances of mass fluxes in a singular vectors coefficient space

relative distances between members of each cluster. The vertical axis (see, for example, Figure 7) is a measure of how similar nodes are in a given cluster level. Once clustering is concluded, ideally, one can choose a single representative of each cluster for further processing without losing any critical information.

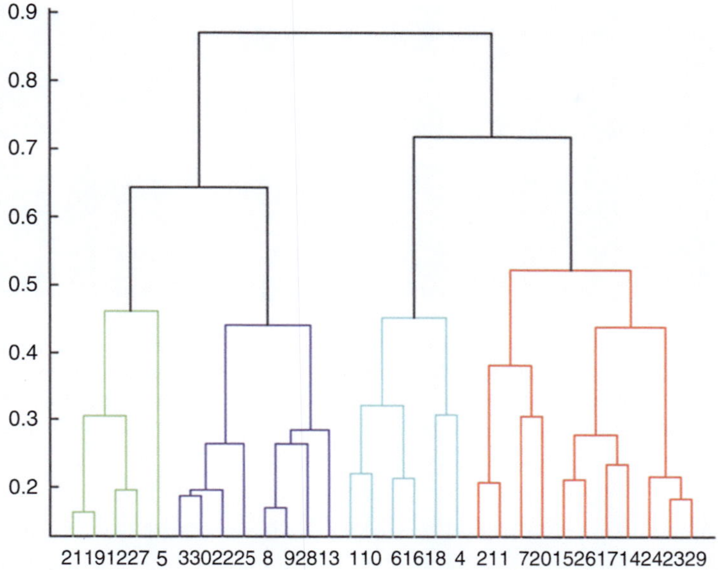

Fig. 7 Schematic illustration of a dendrogram chart. The height of the links corresponds to the level of dissimilarity between the linked members

4.2 *Induced Control Excitation (Modulations) for Enhanced Reservoir Characterization*

In practice, even a complete simulation of a single realization may be computationally intractable, especially when one considers a long simulation time and/or a large number of model samples. Under these circumstances, we would favor devising some efficient manner in which we can approximate flow indicators, such as mass fluxes. More importantly, we would like to be able to predict how models behave (flow-wise) with respect to future control activation, rather than merely historic ones. Data collection is conventionally performed either with fixed controls (e.g., fixed water injection rate or fixed controlled pressures at well sites), or more often, the controls are prescribed in order to serve a direct production objective, such as maximizing net present value, minimizing water production, or maximizing oil production see, for example, [4, 8, 9, 12, 36, 45, 46, 50, 59]. Since the unnecessary perturbation of controls may impair immediate production goals, in practice one typically tends to disturb them as little as possible. The downside of such a strategy is that the sensitivities describing the link between the model and the data are limited to the (typically unknown) effective rank of the varying controls.

However, the sensitivity of the model parameters to the recorded data is an essential component in the recovery of the subsurface attributes and dynamics. The scope of the sensitivities (as can be measured by the corresponding singular value spectrum) allows one to identify how a change in a model parameter influences

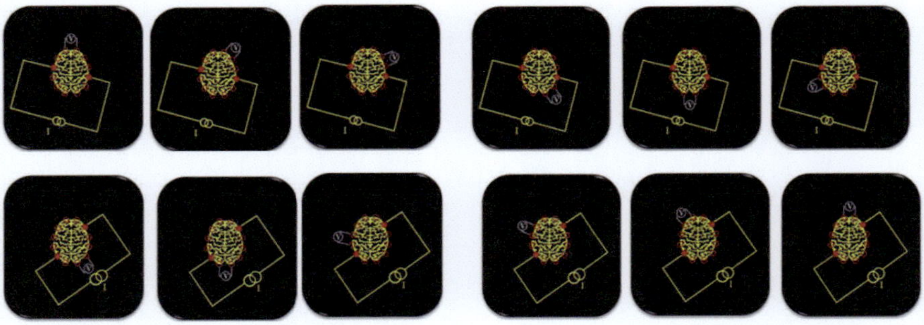

Fig. 8 On the *top row*, the source is static, while the sensors are recording sequentially. On the *bottom row*, a similar sensing sequence is given for a different source location. Through excitation of the source throughout the entire data acquisition sequence, more non-redundant information regarding the interaction between the source and the medium at different configurations can be attained

the measurable data. The more comprehensive these sensitivities, the better our capability to update or determine the model parameters from the available data. In order to enhance sensitivity information, one can probe the medium (reservoir subsurface in our context) via some controlled flow perturbations and its resulting responses. This idea is inspired by medical imaging methodologies, in which the object under investigation is probed through a sequence of source excitations, and the response of the system to a broad range of configurations is examined. A chart sequence that illustrates medical imaging data acquisition is shown in Figure 8.

Since the minimum required excitations for enhancing sensitivity can be rather small (as long as their effect is measurable with respect to measurement noise) their overall impact upon production is negligible and due to the insights they provide into the subsurface, this procedure is generally cost effective. In the context of simulation, for instance, when multiple realizations are evaluated, these induced excitations of the controls cost almost nothing.

In the course of the simulation, rather than maintaining the control values fixed, or contemplating minimal intervention, we propose prescribing a spanning set of control excitations (modulations). The modulations can be performed canonically, that is, while maintaining the complementary controls fixed, one or several controls are changing at a time in a predetermined simple (e.g., off or on at a same fixed level) manner. Each modulation should be short and relatively small in magnitude. Doing so serves two purposes: firstly adverse influence upon production is minimized, and secondly each well control and observables must satisfy clear (typically simple) bounds, or otherwise a reactive measure (such as shutting down a well completely) may result. Thus, invoking small modulations prevents a cascade of well constraint violations. Clearly, to be able to regard the accumulated effects as negligible, our prediction horizon must be finite.

Once the controls are prescribed, simulation can be performed. For sensitivity computations, in addition to the simulation run, additional adjoint simulation is performed. Note that in the simulation context, the overall simulation period for

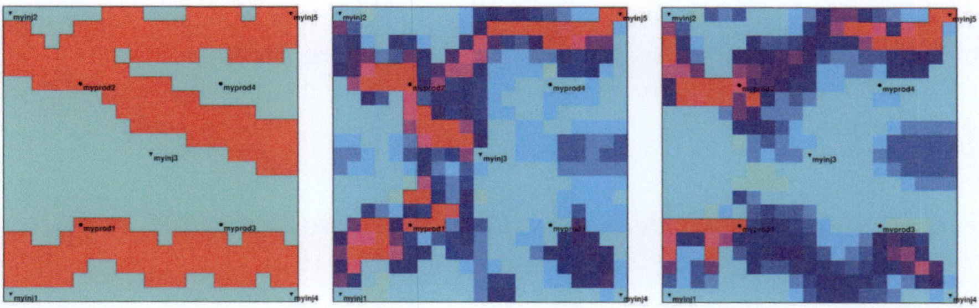

Fig. 9 A comparison of a model (*left*) recovered by conventional means (*center*), as opposed to one which is able to benefit from the modulations (*right*)

the modulations is likely to be significantly shorter than the overall simulation time using a conventional set of controls.

In the image above (Figure 9), an illustration of a simple 2D test model of 21 × 21 grid blocks and wells deployed in a so-called nine spot pattern (five injectors and four producers) is given. The red zones correspond to high permeability channels, whereas the cyan colored area corresponds to less permeable regions.

As expected, due to the ill-posed nature of the problem, none of the recoveries provided a perfect reconstruction of the true model. Given that the initial guess (starting point for the history matching optimization process) was a uniform permeability map (at the value of 1,000 mDarcy), the achieved recoveries are regarded as relatively successful. Comparison of the two recoveries demonstrates a major improvement of the estimated model using the modulations (right) in contrast to the non-modulated model recovery (center). The non-modulated recovered model suggests a high connectivity (high permeability) link between producer $myprod_2$ and $myprod_1$, whereas in reality (as indicated by the true model illustrated above) no such link exists. The true subsurface structure is much more evident for the model estimated using the modulations [27].

5 Algebraic Computational Enhancements

Several computational bottlenecks may arise while dealing with big sample sets for large models. The first is to efficiently handle situations where the size of each realization is augmented. This could happen for a variety of reasons. For instance, another well was placed, a new control setup was required, or simply a consideration of a broader physical domain was desired. In order to address this problem, we have developed a novel singular value decomposition augmentation formulation that allows for the recycled use of previous computations. Next, and complementary, one may ask how additional realizations can be added without the need to recompute the singular value decomposition of the larger set. For this problem, we propose a projection strategy. Lastly, there is the question as to what would be a computationally tractable procedure for the extraction of a spanning set (since a representation only requires a spanning set) in situations where very big sets of realizations of large dimension are considered. We have proposes in the following a multi-level tournament based approach for distributed computation of such a spanning set.

5.1 Singular Value Decompositions Row Augmentation

The singular value decomposition (SVD) is a key factorization form for a broad range of numerical algorithms. Yet, for large-scale matrices its computation can become intensive [$\min(mn^2, m^2n)$ flops for an m by n matrix]. In some situations, one may compute the SVD of a given matrix, and later desire to compute the SVD of its augmentation by rows. An example would be while dealing with indicator realizations[3] for geo-statistical screening. In this situation, computation of the SVD of a given set of realizations has already been invested, yet, we wish to account for their augmentation due to the addition of wells/time-steps/or an extended domain. Rather than conducting a complete computation of the SVD of the resulting augmented matrix we show how to utilize efficiently the information from the original decomposition to derive what is needed.

Let A be an augmentation of the matrix $A_1 \in \mathbb{R}^{m_1 \times n}$ by the matrix $A_2 \in \mathbb{R}^{m_2 \times n}$ so that

$$A = \begin{bmatrix} A_1 \\ A_2 \end{bmatrix} \in \mathbb{R}^{(m_1+m_2)\times n}.$$

In our context n is significantly smaller than $m_1 + m_2$. Let the compact SVD decomposition of these matrices be given by

$$\begin{aligned} A_1 &= U_1 S_1 V_1^\top \\ A_2 &= U_2 S_2 V_2^\top, \end{aligned}$$

where $U_i \in \mathbb{R}^{m_i \times n}, S_i \in \mathbb{R}^{n \times n}, V_i \in \mathbb{R}^{n \times n}$. We wish to determine the compact SVD decomposition of the augmented matrix A

$$A = USV^\top.$$

By definition

$$\underbrace{A^\top A}_{K} V = \left(VS^\top U^\top\right)\left(USV^\top\right) V = VS^2$$

[3]Indicator of the realization could be a function of the realization such as static properties, for example, permeability or porosity, or alternatively, as we have already seen, a dynamic descriptor such as mass flux or time of flight.

and so by substitution

$$\underbrace{A^\top A}_{K} V = \left[A_1^\top \; A_2^\top\right] \begin{bmatrix} A_1 \\ A_2 \end{bmatrix} V = \left[\underbrace{A_1^\top A_1}_{M_1} + \underbrace{A_2^\top A_2}_{M_2} \right] V = VS^2.$$

Note that in our context we are not unduly concerned by the reduced conditioning introduced by forming the normal equations. One can then solve the eigenvalue problem

$$KV = VS^2,$$

which is relatively small in size (run time complexity is only $\mathcal{O}\left(n^3\right)$). Then U is given by

$$U = AVS^{-1} = \left(USV^\top\right) VS^{-1}.$$

Note that here we can save some computation by utilizing the relatively small n by n (number of columns) product when computing

$$A_1^\top A_1 = M_1$$

from the previous computations and therefore we retain it for future use. This process can be repeated giving $M_1, M_2, \ldots, M_k$ so that

$$[M_1 + M_2 + \ldots + M_k]\, V = VS^2$$

can be solved [29].

The superiority of the proposed approach is evident. In particular, in geo-statistical screening, realizations may be available in relatively small batches and a series of SVDs would be required for algorithmic computations. In the above full rank matrices were assumed (often the case because of preprocessing) and compact SVDs. Analogous results when either the matrices are not full rank or not effectively (by virtue of truncation) full rank can be obtained in a straightforward manner.

5.2 *Incremental Addition of Realizations*

In situations where additional realizations are added to the problem, as opposed to maintaining a fixed number of realizations while extending each in length as discussed above, two options can be considered. One (rather inadequate) approach would be to utilize the formulation above in its transposed form. Alternatively, since our goal is to represent our set of realizations in a reduced space, other spanning sets than the singular vector basis are preferable. Considering such settings, instead of recomputing the SVDs of the extended large-scale matrix, we subtract from each new realization a projection of the current spanning set and are thus left with a set of residuals (in addition to the original spanning set). Next the residuals (which are orthogonal to the original set) are sorted according to their norm. As needed, residuals smaller than a prescribed tolerance/threshold can be discarded. As a consequence, it is now possible to sequentially add more and more realizations for processing, without the need to recompute the SVDs of the entire set. It is important to note that the resulting set does not correspond to a singular value decomposition of the extended matrix, yet, it is a compact spanning representation.

5.3 Distributed Spanning Set Computation

Unlike the conventional numerical analysis context, in this section we wish to generate a spanning set rather than a basis, that can be used for compact representation in situations where additional realizations are added. In some situations, the set of realizations may be too large to be stored or processed on a single machine. We shall assume that the effective rank of a matrix comprising of the realizations is far smaller than the number of columns and that we know (at least approximately) what that rank is. We shall assume further that a complete SVD computation of the matrix is prohibitively expensive, and it is therefore desirable to resort to a distributed computation of a small (approximate) spanning set. Thus, rather than computing the SVD of a large matrix, our proposition is to distribute it into sub-matrices, and leverage the (computationally much more tractable) SVDs of the separate parts. More formally, let us assume that there is a set of $n < m$ model realizations, represented by the linear mapping $A \in \mathbb{R}^{m \times n}$. Further assume the effective rank k is relatively small with $k \ll n$ and that A possesses an approximate (typically truncated) singular value decomposition such that

$$\left\| A - U^{(k)} S^{(k)} V^{(k)\top} \right\|_2 \leq \delta_k,$$

where δ_k is a small threshold. Our first task is to partition A into s subsets for which we can effectively compute their SVDs

$$A = [A_1, A_2, \ldots, A_s],$$

while noting that the SVD of each can be computed (ideally in parallel) as

$$U_1 S_1 V_1^\top = A_1,\ U_2 S_2 V_2^\top = A_2,\ \ldots,\ U_s S_s V_s^\top = A_s.$$

Given these singular values, we maintain the top singular entries, making sure that we have the option to include more than the rank. Thus

$$\sum_i k_i = k_s \geq k,$$

where k_i is the effective rank of A_i. Further, we can re-orthogonalize the union of the selected SVDs

$$\left[U_1^{(k_1)} S_1^{(k_1)} V_1^{(k_1)\top}, U_2^{(k_2)} S_2^{(k_2)} V_2^{(k_2)\top}, \ldots, U_s^{(k_s)} S_s^{(k_s)} V_s^{(k_s)\top}\right]$$

and perform a second truncation if needed. The output would be

$$k_T \leq k_s$$

ordered spanning vectors. If required we can randomly mix the remaining vectors for a further distributed processing. The entire process can be repeated until a sufficiently small set is obtained [30].

As an example, we considered a set constructed from 50×20 random vectors. We added variability of an additional 20 entries artificially via a noisy linear combination of the 50×20 set. The resulting 50×40 set was split into two 50×20 sets, see Figure 10. Finally, we ensured that more than 10 SVDs were retained from each set. In Figure 11 the recovery of a truncated set is compared to the one obtained through distributed computation of the truncated set. As can be observed in both cases (bottom row, center and right) the error levels are comparable.

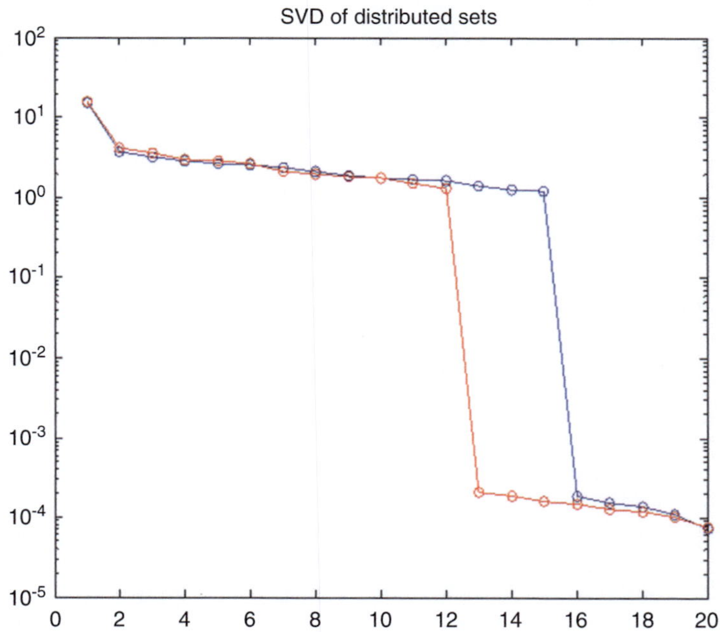

Fig. 10 Singular values of the two sets (*blue* and *red*). Note that although the cumulative number of effective singular values is expected to be 20, the sum of significant singular values of both sets is larger as some level of overlap in both sets is present

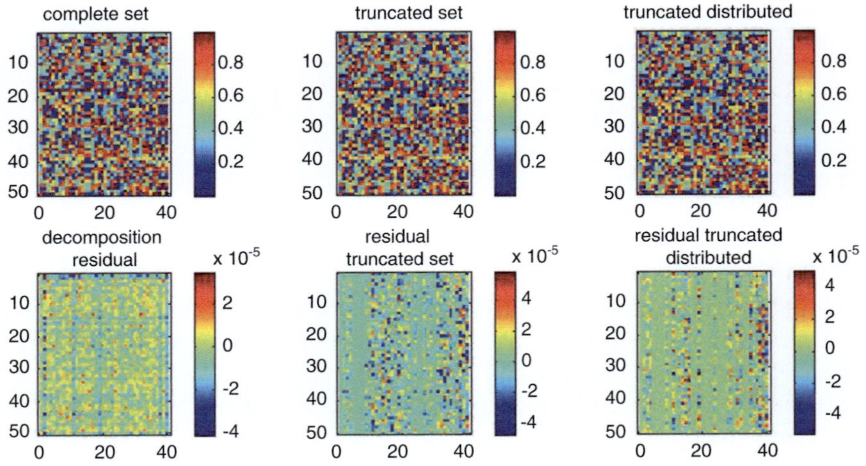

Fig. 11 From *left* to *right*: true set, truncated set, and distributedly computed truncated set. In the *top row*, the sets themselves, on the *bottom*, their error from the true set

6 Results

Having discussed conceptual and computational enhancements, we shall now present some synthetic and field scale results for the geo-statistical prior sampling problem.

6.1 Synthetic Results

6.1.1 Controlled Test Case

In order to test our hypothesis regarding the ability of flow indicators to identify, by means of flow similarity in a reduced space, models that are likely to provide similar production forecasts, we begin with a controlled test case.

For an egg shaped reservoir model (Figure 1) consisting of $60 \times 60 \times 7$ grid cells, eight injectors and four producers, we generated three sets of models, each comprised of high, medium, and small variations of ten facies[4] prototypes. Using Petrel we have intentionally populated a mixed number of variations for each model facie prototype. In order to compute flow indicators, for every realization we used the control modulation strategy elucidated above, in Section 4.2. Since we had $8 + 4 = 12$ well controls to manipulate, we elected to use a canonical

[4]Areas with particular rock characteristics.

excitation pattern in which every 5 simulation days one of the controls is modulated sequentially. The specific choice of 5 days separation between one control change and another was made based upon a systems and control perspective. The duration was chosen to be larger than the rise time (90 % of the step response) of the slowest dynamic response to a step function in the field. Typically this would be an injector-producer pair that are far apart. However, this is not necessarily the case. For the egg shaped model under various facies configurations, that rise time was a little less than 5 days. The total simulation time was therefore the number of controls to be modulated by rise time $= 12 \times 5 = 60$ days. For assessment purposes, we have also conducted a full blown simulation (i.e., 8 years of production profiles). In order to confirm that the production measures of the controlled sets are indeed clustered (i.e., several distinct geological models provide almost similar production output), we first applied the proposed reduced space clustering procedure directly upon the complete (8 years) simulation data. The dendrogram below (Figure 12) confirms the validity of the approach, at least in this instance.

6.1.2 Flow Indicators Clustering Results

Following the reduced space clustering procedure described above, we have clustered each of the three sets of models, based on their dynamics snapshots, as provided by the control-modulated fluxes. The dendrogram for the mass flux clustering of the small permeability variations is provided in Figure 13.

While obtaining a similar number (actually exactly the same in this case) of clusters is indeed reassuring, we would still need to confirm that these clusters correspond to different production scenarios. One way of doing so, which is feasible for a small number of realizations, is to cross-match each leaf (lowest level in the

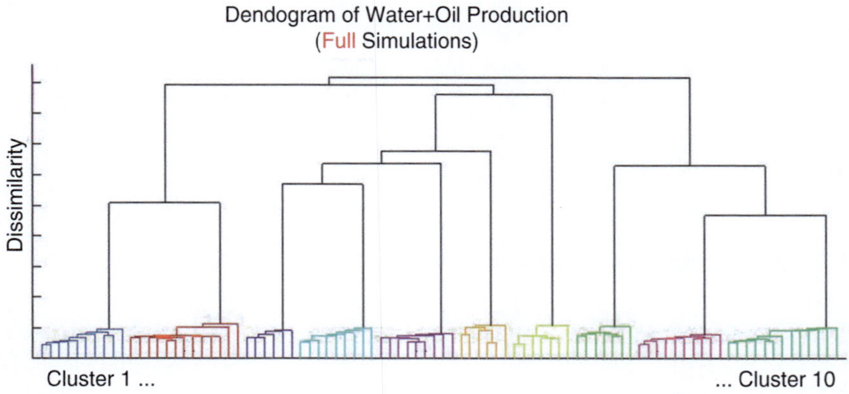

Fig. 12 A dendrogram of oil and water production measurements displays well-defined and distinct separation between ten clusters. Obtaining this chart requires complete simulation and is performed here only for assessment purposes

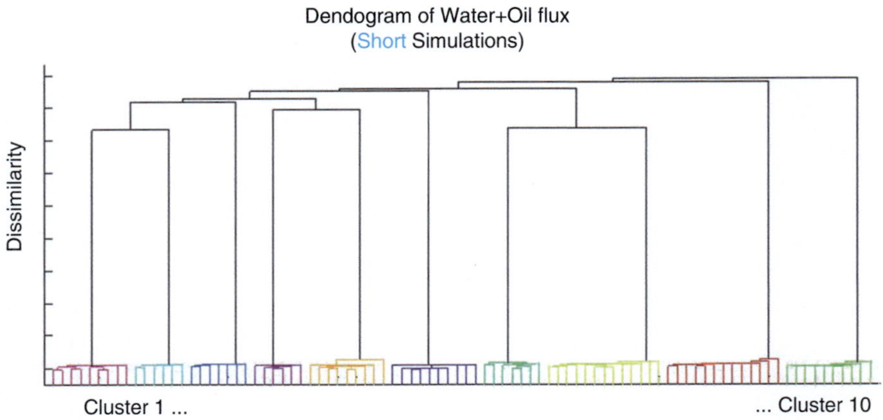

Fig. 13 Dendrogram of reduced space mass flux clustering of the small permeability variations set

dendrogram) member against its counterpart in the production dendrogram. Another approach, which is more illustrative, would be to observe the corresponding future forecasts (production data) for each realization such that it is color-coded according to the flow indicator clustering Figure 14 demonstrates that despite the use of a narrow time window for the simulation (using control modulations) we were able to consistently pick up long-term trends in the production data. Note that even in-between flow-equivalent scenarios (including when they cross each other) are clustered distinctly. This analysis can be performed on a field level, for each well individually (see Figure 15, for Injector 1), or for an aggregation of several wells. This flexibility allows us to reveal spatio-temporal correlations and consequently, enables refined operational decision capability.

The rather distinct "rainbow" stripes indicate that realizations that were identified as flow-equivalent indeed provided equivalent future forecasts. These results suggests that for operational and business decisions associated with future production forecasts, picking a single representative from each cluster for the computationally intensive history matching process suffices. In situations where other decisions are required, a similar framework can be followed, assuming appropriate indicators are devised.

6.1.3 Are the Results Consistent as the Variations Grow?

Next, we need to ensure that the results are also consistent with respect to various levels of variability within each of the prototypical models mentioned in the setup phase. In order to make this assessment we have repeated the clustering procedure for the set of models that involved both medium and high levels of variations. The dendrograms (Figure 16) show that even for a set of high variations, relatively well-defined clustering is obtained. More importantly, as expected, the links within each cluster are longer, indicating a higher level of dissimilarity

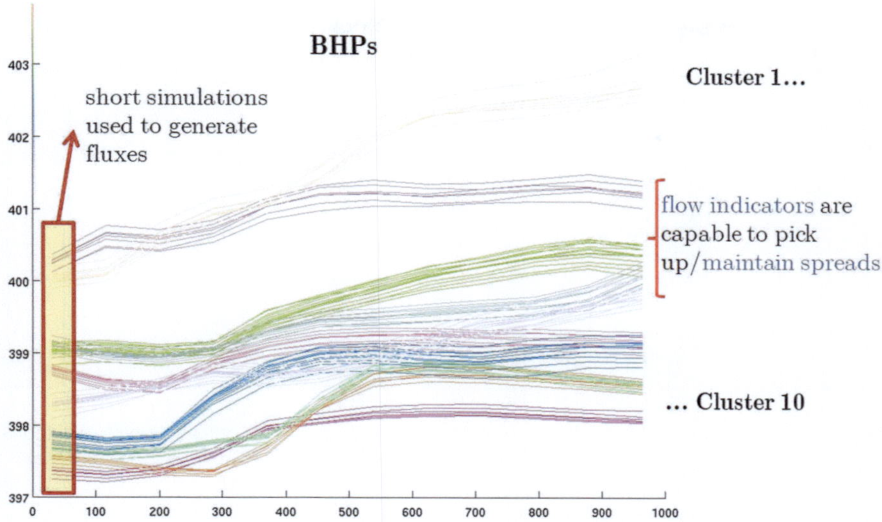

Fig. 14 BHP measurements forecast, color-coded according to the reduced space mass flux clustering. One sees a clearly distinct "rainbow" of stripes, which even manage to resolve intertwines, indicating that the realizations that were identified as flow-equivalent provide analogous future production forecasts

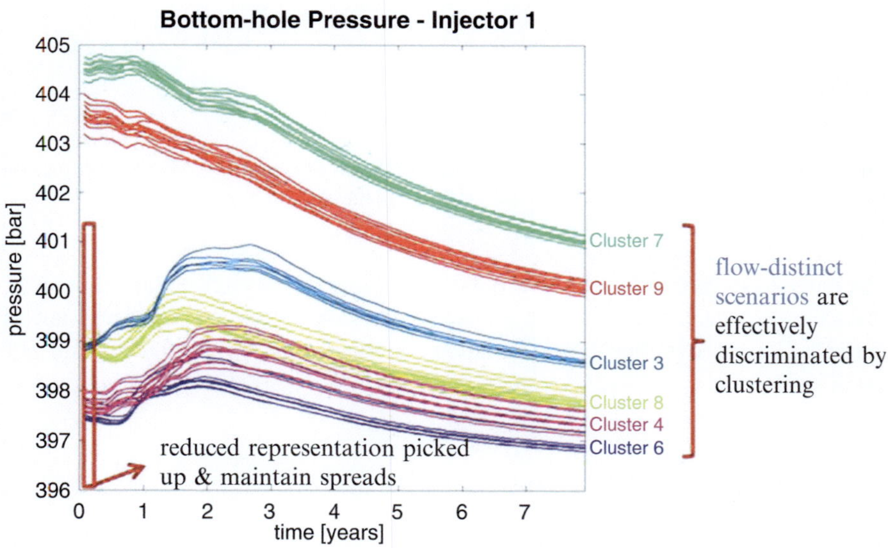

Fig. 15 BHP measurements forecast for injector # 1, color-coded according to the reduced space mass fluxes clustering. Again one sees a clearly distinct "rainbow" of stripes that is able to resolve intertwines, indicating that the realizations that were identified as flow-equivalent provide analogous future production forecasts

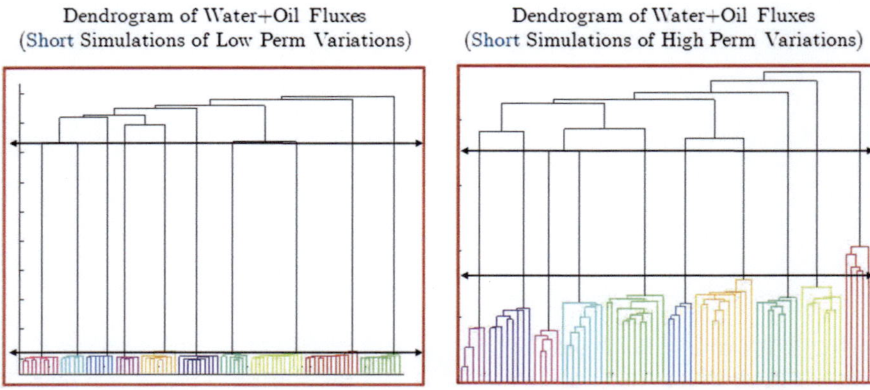

Fig. 16 Dendrograms for mass fluxes of low variation realizations (*left*) and high variations (*right*). As expected the separation between clusters is becoming less distinct as variations grow

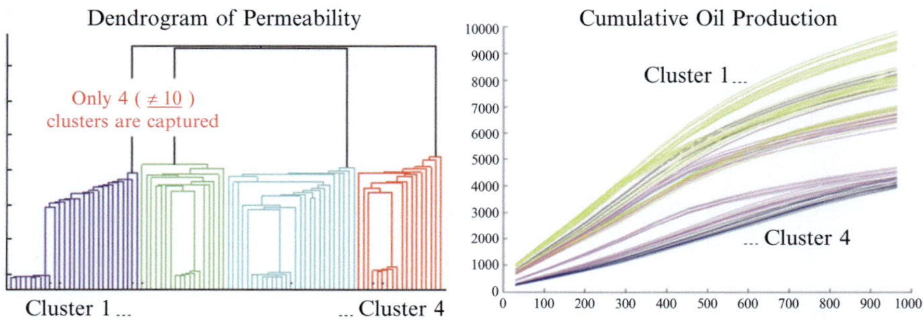

Fig. 17 *Left*: a dendrogram based upon permeability (static geology), *right*: production profiles from our full blown simulation color-coded according to the clustering given on the *left*

compared to such links in the low variations dendrogram. Also, the relatively shorter links between each cluster indicate that the separation between clusters is not as distinct as it was for the lower variations. Both results are consistent with our expectations.

6.1.4 Could Static Geology Provide Similar Results?

The answer to the question above, at least to us, is obviously negative. However, since there remain numerous groups that still attempt to do so, we decided to demonstrate how reliance upon static geology, which is agnostic to the underlying physics, is prone to failure. On the right of Figure 17, in the same manner as before, we see production profiles from our full-blown simulation, color coded according to the clustering given by the one on the left, a dendrogram, which is based upon permeability (static geology).

Evidently, at least based upon this synthetic and relatively simple example, a model agnostic approach using static data is incapable of resolving all facets of flow relevancy. For more sophisticated examples it is reasonable to postulate that the results would be at least as poor. In the above dendrogram the clustering failed to identify the intended ten flow equivalent patterns. Even worse, from observation of the color-coded production profiles, it seems that the realizations of unique dynamics are mixed up within the clusters. Clearly, prior sampling based upon these representations will almost certainly fail to reliably reflect uncertainty distributions.

6.2 *Field Scale Case*

While the above results may seem striking, the outstanding question is whether the proposed methodology would work for a more complex setup such as a real field. In order to test our framework we have considered the Draugen field model, comprising of 56 by 148 by 24 grid cells, seven injectors, 12 producers, and a simulation horizon of 14 years. For this model, we generated 3,125 model realizations spanning variations in: facies occurrences (80–92 %), with variogram major and minor direction (3,000–9,000 ft and 600–9,000 ft, respectively), variogram azimuth (8–20°), and variogram vertical range (5–25 ft), see Figure 18.

Once the models have been generated, we followed the proposed reduced space clustering on the high-dimensional vector space of all reservoir dynamic scenarios provided (Figure 19).

The rather distinct clusters obtained for this problem suggests that indeed good separation was obtained for flow equivalent model realizations. It is important to note here that for such a large number of realizations it is typically impossible to plot production measures color-coded by the devised clusters, as the number of line curves and colors becomes excessive to usefully visualize. Since often managerial decisions are based upon the spread of production profiles, we can easily determine whether such a spread was effectively captured by a small number of representatives. In the two illustrations given in Figure 20, the reservoir oil and water production for a long future horizon is provided, both using the complete set (red) of 3,125 model realizations and with a representative set (blue), chosen from each set of clusters. The order of magnitude smaller representative set, of 300 realizations, clearly captures very successfully the spread of the larger set.

In principle, more advanced statistical analysis can be performed using the dynamics eigen-vector space, along with the clusters. For instance, the spread of the realizations for each eigen-dynamic subspace can be visualized through projection. Such analytical tools can be instrumental in the exploration of the variability space for uncertainty quantification. Furthermore, they can essentially form a bridge between prior sampling analysis and sample construction through synthesis.

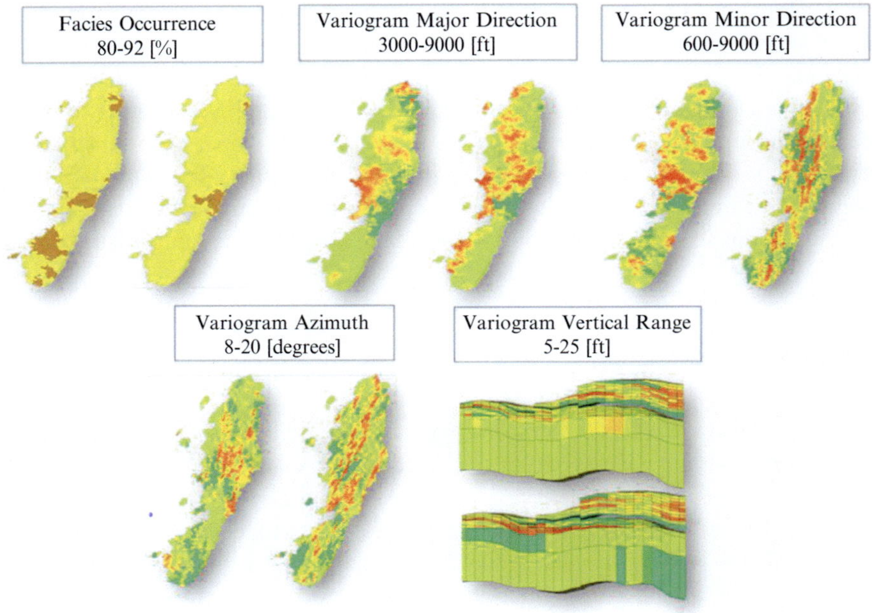

Fig. 18 Variations types of the Draugen field model

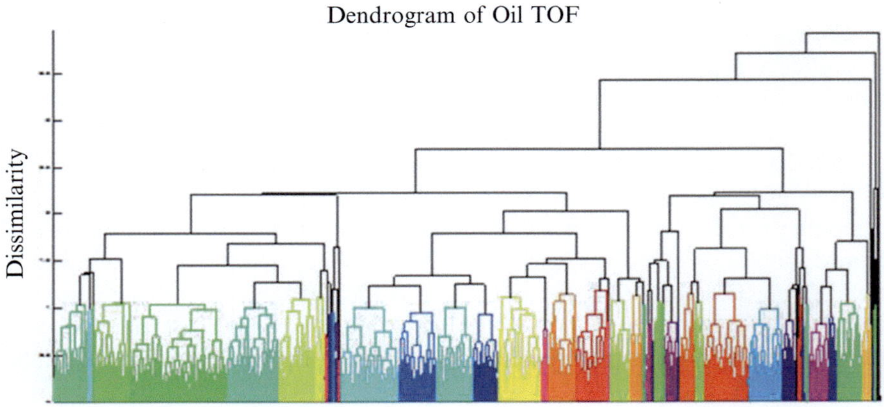

Fig. 19 Dendrogram for the Draugen field model realizations. In this example, Time of Flight is used as a flow indicator. The dendrogram indicates that even this flow measure is capable of providing good separation between distinct flow patterns

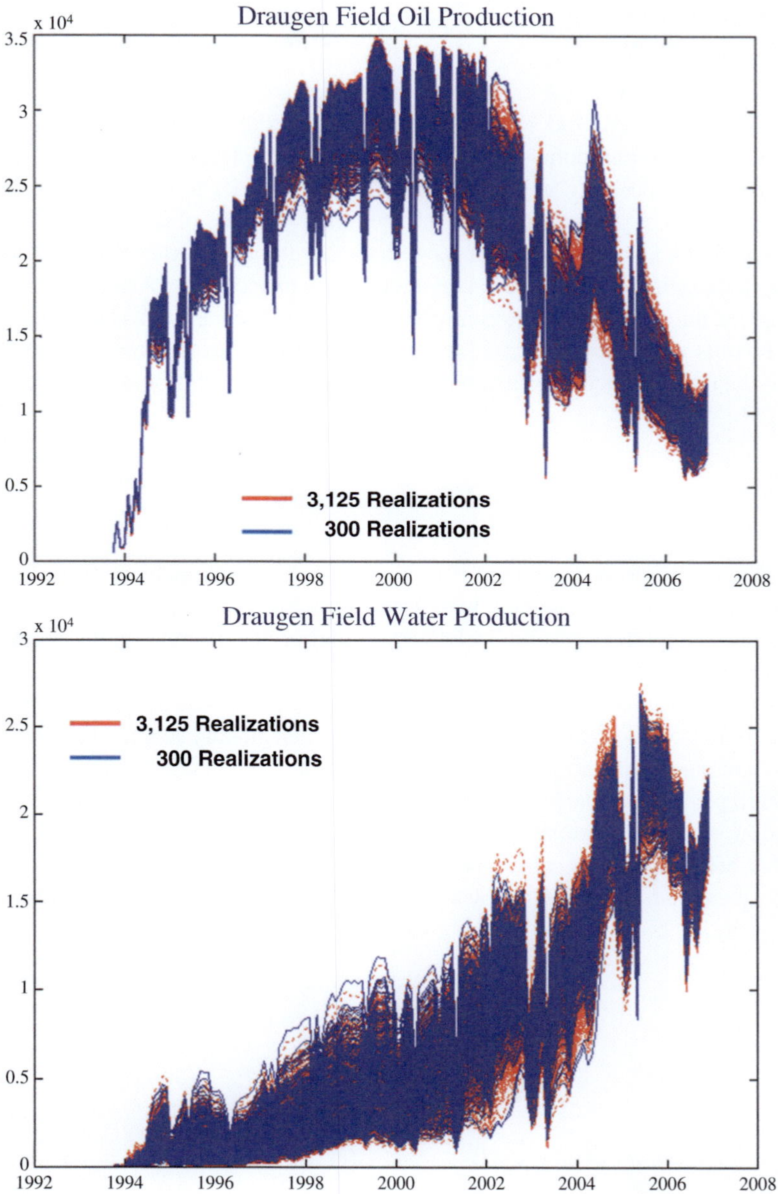

Fig. 20 Oil and water production for the Draugen field as captured by simulation of 3,125 model realizations (*red*), and by 300 selected representatives (*blue*)

7 Conclusions

In this study, we have developed a generic approach for geo-statistical prior sampling using a reduced order dynamic representation and clustering. Our framework is based upon the computation of flow indicators, which leverages our knowledge regarding the underlying physics of the problem in order to capture the principle dynamics efficiently. To ensure that the method is capable of dealing with the full complexity of realistic scale problems we have developed several computational and conceptual extensions. Rigorous assessment of the proposed prior sampling strategy was carried out with a synthetic controlled test case as well as with a large-scale field case. Results were extremely encouraging, suggesting potential functionality of the approach and its related advancements to a broader range of problems.

8 Possible Next Steps and Recommendations

The proposed framework can readily be applied to well placement analysis. It would be interesting to investigate how well it performs for such objective. The proposed method seems to identify effectively flow similarity of different model realizations and appears capable of grouping realizations that indeed correspond to similar future forecasts. Another step forward would be to attempt to identify realizations that are likely to provide similar future forecasts of their history matched models. The history matching search space normally involves multiple local minima. Intuitively, with high probability, the history matching process can be regarded as a contracting operation w.r.t. the distances between the adjacent samples, i.e., two adjacent samples are more likely to converge to the same local minimum of the active search space than disparate samples (see Figure 21 for an illustrative explanation).

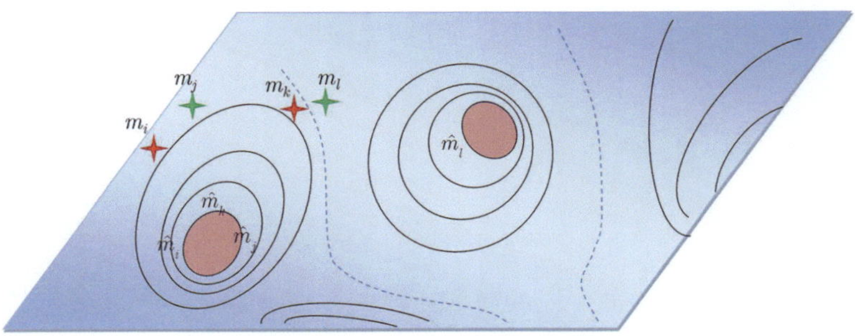

Fig. 21 An illustration of convergence of adjacent realizations through the history matching process. The prior samples m_i, m_j, and m_k converge (as $\hat{m}_i$, $\hat{m}_j$, and $\hat{m}_k$) to the same active space of a local minimum, conversely, the sample m_l, converges ($\hat{m}_l$) to the active space of another local minima. The *blue dashed lines* represent region (optimization algorithm dependent) of attraction of each local minima, aggregating realizations inside each region into the same local minima

For many problems, these regions are rather broad, implying that many prior samples that were regarded close in some sense are likely to provide similar posterior predictions. Attempting to exploit such structure rigorously is definitely challenging and worth further study.

Acknowledgements The authors wish to thank Ulisses Mello, Jorn van Doren, and Jan Dirk Jansen for their insightful comments and support throughout the evolution of the study.

References

1. Aanonsen, S.I., Nævdal, G., Oliver, D.S., Reynolds, A.C., Vallès, B.: The ensemble kalman filter in reservoir engineering—a review. SPE J. **14**(03), 393–412 (2009)
2. Adler, P.M.: Multiphase Flow in Porous Media. Springer, Berlin (1995)
3. Agarwal, B., Blunt, M.J.: Full-physics, streamline-based method for history matching performance data of a north sea field. In: SPE Reservoir Simulation Symposium, Houston, TX, 11–14 February 2001
4. Asheim, H.: Maximization of water sweep efficiency by controlling production and injection rates. In: European Petroleum Conference, London, 16–19 October 1988
5. Astrid, P.: Reduction of process simulation models: a proper orthogonal decomposition approach. Ph.D. thesis, Eindhoven University of Technology, Department of Electrical Engineering (2004)
6. Astrid, P., Weiland, S., Willcox, K., Backx, T.: Missing point estimation in models described by proper orthogonal decomposition. IEEE Trans. Autom. Control **53**(10), 2237–2251 (2003)
7. Baker, R.: Streamline technology: reservoir history matching and forecasting, its success, limitations, and future. J. Can. Pet. Technol. **40**(4), 23–27 (2001)
8. Brouwer, D.R., Naevdal, G., Jansen, J.D., Vefring, E.H., van Kruijsdiik, C.P.J.W.: Improved reservoir management through optimal control and continuous model updating. In: SPE Annual Technical Conference and Exhibition, Houston, TX, 26–29 September 2004
9. Bryson, A.E., Ho, Y.: Applied Optimal Control: Optimization, Estimation and Control. Wiley, New York (1975)
10. Daly, C., Caers, J.: Multi-point geostatistics—an introductory overview. First Break **28**(9), 39–47 (2010)
11. Datta-Gupta, A., King, M.J.: Streamline Simulation: Theory and Practice. SPE, Richardson (2007)
12. Durlofsky, L.J., Aitokhuehi, I.: Optimizing the performance of smart wells in complex reservoirs using continuously updated geological models. J. Pet. Sci. Eng. **48**(3–4), 254–264 (2005)
13. El Moselhy, T.A., Marzouk, Y.M.: Bayesian inference with optimal maps. J. Comput. Phys. **231**(23), 7815–7850 (2012)
14. Evensen, G.: The ensemble kalman filter: theoretical formulation and practical implementation. Ocean Dyn. **53**(4), 343–367 (2003)
15. Fan, X., Sivo, S.A.: Sensitivity of fit indices to model misspecification and model types. Multivar. Behav. Res. **42**(3), 509–529 (2007)
16. Guardiano, F.B., Mohan Srivastava, R.: Multivariate geostatistics: beyond bivariate moments. In: Geostatistics Troia'92, pp. 133–144. Springer, New York (1993)
17. Haber, E., Oldenburg, D.: Joint inversion: a structural approach. Inverse Prob. **13**(1), 63 (1997)
18. Haber, E., Horesh, L., Tenorio, L.: Numerical methods for experimental design of large-scale linear ill-posed inverse problems. Inverse Prob. **24**(5), 055012 (2008)
19. Haber, E., Horesh, L., Tenorio, L.: Numerical methods for the design of large-scale nonlinear discrete ill-posed inverse problems. Inverse Prob. **26**(2), 025002 (2009)

20. Haber, E., van den Doel, K., Horesh, L.: Optimal design of simultaneous source encoding. Inverse Prob. Sci. Eng. **23**(5), 780–797 (2015)
21. Hansen, P.C.: Truncated singular value decomposition solutions to discrete ill-posed problems with ill-determined numerical rank. SIAM J. Sci. Stat. Comput. **11**(3), 503–518 (1990)
22. Hansen, L.P., Sargent, T.J., Turmuhambetova, G., Williams, N.: Robust control and model misspecification. J. Econ. Theory **128**(1), 45–90 (2006)
23. Hao, N., Horesh, L., Kilmer, M.E.: Model correction using a nuclear norm constraint. In: Householder Symposium XIX, Spa, pp. 120, 8–13 June 2014
24. Hao, N., Horesh, L., Kilmer, M.: Nuclear norm optimization and its application to observation model specification. In: Carmi, A.Y., et al. (eds.) Compressed Sensing & Sparse Filtering, pp. 95–122. Springer, Berlin/Heidelberg (2014)
25. Horesh, L., Haber, E.: Sensitivity computation of the ℓ_1 minimization problem and its application to dictionary design of ill-posed problems. Inverse Prob. **25**(9), 095009 (2009)
26. Horesh, L., Haber, E., Tenorio, L.: Optimal experimental design for the large-scale nonlinear ill-posed problem of impedance imaging. In: Large-Scale Inverse Problems and Quantification of Uncertainty, pp. 273–290, Wiley, New York (2010)
27. Horesh, L., Conn, A.R., Mello, U.T., van Essen, G.M., Jimenez, E.A.: Induced control excitation for enhanced reservoir flow characterization, US Patent Disclosure, YOR8-2012-0156 (2012)
28. Horesh, L., Conn, A.R., Mello, U.T., van Essen, G.M., Jimenez, E.A.: Reduced space clustering representatives and its application to long term prediction, US Patent Disclosure, YOR8-2012-0053 (2012)
29. Horesh, L., Conn, A.R., Nahamoo, D., van Essen, G.M., Jimenez, E.A.: Method and system for augmenting singular value decomposition (svd) on large scale matrices. IP.com Prior Art Database Disclosure, IPCOM000221945D, Sep 2012
30. Horesh, L., Conn, A.R., van Essen, G.M., Jimenez, E.A.: Method for distributed computation of large scale spanning sets. IP.com Prior Art Database Disclosure, IPCOM000224837D, Jan 2013
31. Huan, X., Marzouk, Y.: Gradient-based stochastic optimization methods in Bayesian experimental design. Int. J. Uncertain. Quantif. **4**(1), 479–510 (2014)
32. Huan, X., Marzouk, Y.M.: Simulation-based optimal bayesian experimental design for nonlinear systems. J. Comput. Phys. **232**(1), 288–317 (2013)
33. Huang, X., Meister, L., Workman, R.: Reservoir characterization by integration of time-lapse seismic and production data. In: SPE Annual Technical Conference, pp. 439–447 (1997)
34. Huang, H., Haber, E., Horesh, L.: Optimal estimation of ℓ_1-regularization prior from a regularized empirical Bayesian risk standpoint. Inverse Prob. Imaging **6**(3), 447–464 (2012)
35. Idrobo, E.A., Choudhary, M.K., Datta-Gupta, A.: Swept volume calculations and ranking of geostatistical reservoir models using streamline simulation. In: SPE/AAPG Western Regional Meeting, Long Beach, CA, 19–22 June 2000
36. Jansen, J.-D., Brouwer, D.R., Naevdal, G., van Kruijsdijk, C.P.J.W.: Closed-loop reservoir management. First Break **23**(8), 43–48 (2005)
37. Jimenez, E., Sabir, K., Datta-Gupta, A., King, M.J.: Spatial error and convergence in streamline simulation. SPE **10**(3), 221–232 (2007)
38. Johansen, T.A.: On tikhonov regularization, bias and variance in nonlinear system identification. Automatica **33**(3), 441–446 (1997)
39. Liu, N., Oliver, D.S.: Ensemble kalman filter for automatic history matching of geologic facies. J. Pet. Sci. Eng. **47**(3), 147–161 (2005)
40. Liu, N., Oliver, D.S., et al.: Critical evaluation of the ensemble kalman filter on history matching of geologic facies. SPE Reserv. Eval. Eng. **8**(06), 470–477 (2005)
41. Møyner, O., Krogstad, S., Lie, K.-A.: Flow diagnostics for use in reservoir management. Technical Report, Center for Integrated Operations in the Petroleum Industry (2013)
42. Papaioannou, G., Astrid, P., Vink, J.C., Jansen, J.D.: Pressure preconditioning using proper orthogonal decomposition. In: SPE Reservoir Simulation Symposium, The Woodlands, TX, 21–23 Feb 2011

43. Pinder, G.F., Gray, W.G.: Essentials of Multiphase Flow in Porous Media. Wiley, New York (2008)
44. Raykov, T.: On sensitivity of structural equation modeling to latent relation misspecifications. Struct. Equ. Model. **7**(4), 596–607 (2000)
45. Sarma, P., Aziz, K., Durlofsky, L.J.: Implementation of adjoint solution for optimal control of smart wells. In: SPE Reservoir Simulation Symposium, Houston, TX, 31 January–2 February 2005
46. Sarma, P., Durlofsky, L.J., Aziz, K.: Efficient closed-loop production optimization under uncertainty. In: SPE Europec/EAGE Annual Conference, Madrid, 13–16 June 2005
47. Scales, J.A., Smith, M.L., Fischer, T.L.: Global optimization methods for multimodal inverse problems. J. Comput. Phys. **103**(2), 258–268 (1992)
48. Scheidt, C., Caers, J.: A new method for uncertainty quantification using distances and kernel methods: application to a deepwater turbidite reservoir. SPE **14**(4), 680–692 (2008)
49. Schmid, P.J.: Dynamic mode decomposition of numerical and experimental data. J. Fluid Mech. **656**, 5–28 (2010)
50. Sudaryanto, B., Yortsos, Y.: Optimization of displacements in porous media using rate control. In: SPE Annual Technical Conference and Exhibition, New Orleans, LA, 30 September–3 October 2001
51. Suzuki, S., Caers, J.: A distance-based prior model parameterization for constraining solutions of spatial inverse problems. Math. Geosci. **40**(4), 445–469 (2008)
52. Suzuki, S., Caumon, G., Caers, J.: Dynamic data integration into structural modeling: model screening approach using a distance-based model parameterization. Comput. Geosci. **12**, 105–119 (2008)
53. Tenorio, L., Lucero, C., Ball, V., Horesh, L.: Experimental design in the context of Tikhonov regularized inverse problems. Stat. Model. **13**(5–6), 481–507 (2013)
54. Titterington, D.M.: General structure of regularization procedures in image reconstruction. Astron. Astrophys. **144**, 381 (1985)
55. van Essen, G.M., Jimenez, E.A., Przybysz-jarnut, J.K., Horesh, L., Douma, S.G., van den Hoek, P., Conn, A.R., Mello, U.T.: Adjoint-based history-matching of production and time-lapse seismic data. In: SPE Europec/EAGE Annual Conference (2012)
56. Vozoff, K., Jupp, D.L.B.: Joint inversion of geophysical data. Geophys. J. R. Astron. Soc. **42**(3), 977–991 (1975)
57. Wang, Y., Kovscek, A.R.: A streamline approach for history matching production data. SPE **5**(4), 353–362 (2000)
58. Yeh, W.W.-G.: Review of parameter identification procedures in groundwater hydrology: the inverse problem. Water Resour. Res. **22**(2), 95–108 (1986)
59. Zakirov, I., Aanonsen, S.I., Zakirov, E.S., Palatnik, B.M.: Optimizing reservoir performance by automatic allocation of well rates. In: 5th European Conference on the Mathematics of Oil Recovery (1996)

Solving Multiscale Linear Programs Using the Simplex Method in Quadruple Precision

Ding Ma and Michael A. Saunders

Abstract Systems biologists are developing increasingly large models of metabolism and integrated models of metabolism and macromolecular expression. These Metabolic Expression (ME) models lead to sequences of multiscale linear programs for which small solution values of order 10^{-6} to 10^{-10} are meaningful. Standard LP solvers do not give sufficiently accurate solutions, and exact simplex solvers are extremely slow. We investigate whether double-precision and quadruple-precision simplex solvers can together achieve reliability at acceptable cost.

A double-precision LP solver often provides a reasonably good starting point for a Quad simplex solver. On a range of multiscale examples we find that 34-digit Quad floating-point achieves exceptionally small primal and dual infeasibilities (of order 10^{-30}) when no more than 10^{-15} is requested. On a significant ME model we also observe robustness in almost all (even small) solution values following relative perturbations of order 10^{-6} to non-integer data values.

Double and Quad Fortran 77 implementations of the linear and nonlinear optimization solver MINOS are available upon request.

Keywords Flux balance analysis • Metabolic expression model • Multiscale linear program • Simplex method • Quadruple precision • Gfortran libquadmath • MINOS

1 Introduction

We consider the solution of large, multiscale linear programs (LPs) of the form

$$\min_x \; c^T x \;\text{ s.t. }\; \ell \le \begin{pmatrix} x \\ Ax \end{pmatrix} \le u, \tag{1}$$

D. Ma • M.A. Saunders (✉)
Department of Management Science and Engineering, Stanford University, Stanford, CA, USA
e-mail: dingma@stanford.edu; saunders@stanford.edu; metsaunders@gmail.com

M. Al-Baali et al. (eds.), *Numerical Analysis and Optimization*, Springer Proceedings in Mathematics & Statistics 134, DOI 10.1007/978-3-319-17689-5_9

where A is a sparse matrix whose entries, like the variables in x, may be of widely varying magnitude. Such problems arise in systems biology in the modeling of biochemical reaction networks, notably Metabolic Expression (ME) models [20, 36]. Reliable solution methods are of such importance to systems biologists that *exact simplex solvers* have been employed [20], even though the typical solution time for an exact solver is measured in weeks for genome-scale models (compared to minutes for a conventional solver using double-precision floating-point arithmetic).

Exact solvers are based on rational arithmetic. There has been considerable work on their implementation and application to important problems [1, 2, 18, 32]. The use of quadruple-precision floating-point has also been mentioned in passing [1, 18].

Let Single, Double, and Quad denote the main floating-point options, with about 7, 15, 16, and 34 digits of precision, respectively. Single is not useful in the present context, and Double may not ensure adequate accuracy. This is the reason for our work. On today's machines, Double is implemented in hardware, while Quad (if available) is typically implemented in a software library such as libquadmath [8]. Fortunately, the GCC Fortran compiler now makes Quad available via the real(16) data type. We have therefore been able to make a Quad version of the Fortran 77 linear and nonlinear optimization solver MINOS [26, 27] using the gfortran compiler. Our aim is to explore combined use of the Double and Quad MINOS simplex solvers for the solution of large multiscale linear programs. We seek greater efficiency than is normally possible with exact simplex solvers.

Kahan [16] notes that "*carrying somewhat more precision in the arithmetic than twice the precision carried in the data and available for the result will vastly reduce embarrassment due to roundoff-induced anomalies.*" He further notes that Quad precision is unlikely to be adopted widely in the foreseeable future because of the cost in CPU time and memory (especially cache) relative to Double, but in terms of finding ways to avoid unexpected total loss of accuracy, "*default evaluation in Quad is the humane option.*"

We apply the "humane" approach to difficult LP problems by using the Double simplex solver first, saving the final solution, and warm-starting Quad simplex from that point. For a sequence of related problems, warm-starting each problem in Quad is simplest, but warm-starting in Double and then in Quad may be more efficient.

2 Motivating Applications

The Constraint-Based Reconstruction and Analysis (COBRA) approach [31, 33] has been successfully applied to biological processes such as metabolism and macromolecular synthesis, which when integrated result in inherently multiscale models. In the COBRA approach, a biochemical network is represented by a stoichiometric matrix S with m rows corresponding to metabolites and n columns representing reactions. Mathematically, S is part of the ordinary differential equation that governs the time-evolution of concentrations in the network:

$$\frac{d}{dt}x(t) = Sv(t), \tag{2}$$

where $x(t) \in \mathbf{R}^m$ is a vector of time-dependent concentrations and $v(t) \in \mathbf{R}^n$ is a vector of reaction fluxes. With $c^T v$ being a biologically motivated objective function (such as maximizing the growth rate at steady state), the constraint-based approach constructs the following LP:

$$\max_{v} \; c^T v \tag{3a}$$

$$\text{s.t.} \; Sv = 0, \tag{3b}$$

$$l \leq v \leq u, \tag{3c}$$

where growth is defined as the biosynthetic requirements of experimentally determined biomass composition, and biomass generation is a set of reaction fluxes linked in appropriate ratios [31].

The following applications have motivated our work.

Flux Balance Analysis (FBA). FBA is a mathematical and computational approach widely used for studying biochemical reaction networks [30, 31]. The biochemical networks reconstructed in FBA with a linear objective function are essentially LPs as in (3), where the fluxes in vector v may have widely varying values in the range 0–100 say, with small values such as $v_j = 10^{-10}$ being meaningful. With the increasingly large, sparse, and multiscale nature of biochemical networks, a Quad solver has become more necessary, practical, and even efficient.

ME models (FBA with Coupling Constraints). FBA has been used by Thiele et al. [36] for the first integrated stoichiometric multiscale model of metabolism and macromolecular synthesis for *Escherichia coli* K12 MG1655. The model modifies (3) by adding constraints that couple enzyme synthesis and catalysis reactions to (3b). Coupling constraints of the form

$$c_{\min} \leq \frac{v_i}{v_j} \leq c_{\max} \tag{4}$$

become linear constraints

$$c_{\min} v_j \leq v_i, \quad v_i \leq c_{\max} v_j \tag{5}$$

for various pairs of fluxes v_i, v_j. They are linear approximations of nonlinear constraints and make S in (3b) even less well-scaled because of large variations in reaction rates. Quad precision is evidently more appealing in this case.

ME Models with Nonlinear Constraints. As coupling constraints are often functions of the organism's growth rate μ, O'Brien et al. [29] consider growth-rate optimization nonlinearly with the single μ as the objective in (3a) instead of via a linear biomass objective function. Nonlinear constraints of the form

$$\frac{v_i}{v_j} \leq \mu \tag{6}$$

represented as

$$v_i \leq \mu v_j \tag{7}$$

are added to (3b), where v_i, v_j, μ are all variables. Constraints (7) are linear if μ is fixed at a specific value μ_k. O'Brien et al. [29] employ a binary search on a discrete set of values within an interval $[\mu_{\min}, \mu_{\max}]$ to find the largest $\mu_k \equiv \mu^*$ that keeps the associated linear program feasible. Thus, the procedure requires reliable solution of a sequence of related LPs.

Flux Variability Analysis (FVA). After FBA (3) returns an optimal objective value $c^T v^* = Z_0$ (3a), FVA examines how far a particular flux v_j can vary within the feasible region without changing the optimal objective significantly (if $\gamma \approx 1$):

$$\begin{aligned} \max_v \text{ or } \min_v \quad & v_j \\ \text{s.t.} \quad & Sv = 0, \\ & c^T v \geq \gamma Z_0, \\ & l \leq v \leq u, \end{aligned} \tag{8}$$

where $0 < \gamma < 1$. Potentially $2n$ LPs (8) are solved if all reactions are of interest, with warm starts being used when j increases to $j+1$ [12].

Other Challenging LPs. A set of difficult LP problems has been collected by Mészáros [22], who names them *problematic* and notes that "*modeling mistakes made these problems "crazy," but they are excellent examples to test numerical robustness of a solver*." Our procedure for handling these *problematic* problems seems appropriate for the systems biology models as well.

3 Algorithm and Implementation

The primal simplex solver in MINOS includes geometric-mean scaling of the constraint matrix, the EXPAND anti-degeneracy procedure [10, 14], and partial pricing (but no steepest-edge pricing, which would generally reduce total iterations and time). Basis LU factorizations and updates are handled by LUSOL [9, 21]. Cold starts use a Crash procedure to find a triangular initial basis. Basis files are used to preserve solutions between runs.

For Double MINOS, floating-point variables are declared double precision ($\approx$15 digits). For Quad MINOS, they are real(16) ($\approx$34 digits). The LP data A, b, c, S, ℓ, u are stored in Quad even though they are not known to that precision. This allows operations such as Ax and $A^T y$ to be carried out directly on the elements of A and the Quad vectors x, y. If A were stored in Double, such products would require each entry A_{ij} to be converted from Double to Quad at runtime.

To achieve reliability on the Mészáros problems, we developed the following three-step procedure for solving challenging examples of problems (1)–(8):

Step 1 (Cold start in Double with scaling) Apply Double MINOS with somewhat strict options. Save a final basis file.

Table 1 MINOS runtime options (defaults and those selected for Steps 1–3)

	Default Double	Step1 Double	Step2 Quad	Step3 Quad
Scale option	2	2	2	0
Feasibility tol	1e−6	1e−7	1e−15	1e−15
Optimality tol	1e−6	1e−7	1e−15	1e−15
LU factor tol	100.0	10.0	10.0	5.0
LU update tol	10.0	10.0	10.0	5.0
Expand frequency	10,000	100,000	100,000	100,000

Step 2 (Warm start in Quad with scaling) Start Quad MINOS from the saved file with stricter Feasibility and Optimality tolerances. Save a final basis file.

Step 3 (Warm start in Quad without scaling) Start Quad MINOS from the second saved file with no scaling but stricter LU tolerances.

Steps 1 and 2 are "obvious" and should usually be sufficient. In case Step 2 is interrupted, Step 3 provides some insurance and ensures that the Feasibility and Optimality tolerances are imposed upon the original problem (not the scaled problem).

Table 1 shows the default runtime options for Double MINOS and the options chosen for Steps 1–3 above. The Feasibility tolerance δ_1 is applied in absolute form. Thus, a (possibly scaled) solution v is considered feasible for problem (3) if $\ell - \delta_1 \leq v \leq u + \delta_1$. The Optimality tolerance δ_2 is applied in a relative way. If the current basic solution is of the form $Sv \equiv Bv_B + Nv_N = b$ and if $B^T y = c_B$ for the nonsingular basis matrix B, the current v is considered optimal if $z \equiv c - A^T y$ has the correct sign to within the tolerance $(1 + \|y\|_\infty)\delta_2$.

For conventional Double solvers it is reasonable to set δ_1 and δ_2 in the range 10^{-6} to 10^{-8}. For Quad MINOS we set $\delta_1 = \delta_2 = 10^{-15}$ to be sure of capturing accurately any fluxes v_j as small as $O(10^{-10})$.

4 Numerical Results

We report results from Double and Quad versions of the primal simplex solver in MINOS. All runs were on a 2.93 GHz Apple iMac with quad-core Intel i7, using the gfortran compiler with -O flag. Double MINOS uses 64-bit hardware floating-point throughout. Quad MINOS uses 128-bit software floating-point throughout via gfortran's libquadmath library.

We applied our three-step procedure to three sets of LP problems. Table 2 lists the problem dimensions and the norms of the optimal primal and dual solution vectors x^*, y^*. Table 3 summarizes the results of Steps 1–3 for each problem.

All problems were input from files in the classical MPS format of commercial mathematical programming systems [24] with 12-character fields for all data values. This was a fortuitous limitation for the ME models, as we mention below. The MPS files for these 14 LP models are downloadable from [25].

Table 2 Three pilot models from Netlib [28], eight Mészáros *problematic* LPs [22], and three ME biochemical network models [19, 35, 36]

Model	m	n	nnz(A)	$\max \lvert A_{ij} \rvert$	$\lVert x^* \rVert_\infty$	$\lVert y^* \rVert_\infty$
pilot4	411	1,000	5,145	2.8e+04	9.6e+04	2.7e+02
pilot	1,442	3,652	43,220	1.5e+02	4.1e+03	2.0e+02
pilot87	2,031	4,883	73,804	1.0e+03	2.4e+04	1.1e+01
de063155	853	1,488	5,405	8.3e+11	3.1e+13	6.2e+04
de063157	937	1,488	5,551	2.3e+18	2.3e+17	6.2e+04
de080285	937	1,488	5,471	9.7e+02	1.1e+02	2.6e+01
gen1	770	2,560	64,621	1.0e+00	3.0e+00	1.0e+00
gen2	1,122	3,264	84,095	1.0e+00	3.3e+00	1.0e+00
gen4	1,538	4,297	110,174	1.0e+00	3.0e+00	1.0e+00
l30	2,702	15,380	64,790	1.8e+00	1.0e+09	4.2e+00
iprob	3,002	3,001	12,000	9.9e+03	3.1e+02	1.1e+00
TMA_ME	18,210	17,535	336,302	2.1e+04	5.9e+00	1.1e+00
GlcAerWT	68,300	76,664	926,357	8.0e+05	6.3e+07	2.4e+07
GlcAlift	69,529	77,893	928,815	2.6e+05	6.3e+07	2.4e+07

Dimensions of $m \times n$ constraint matrices A (= S for the ME models), and size of the largest optimal primal and dual variables x^*, y^*

The Pilot Problems

These are economic models developed by Prof George Dantzig's group in the Systems Optimization Laboratory at Stanford University during the 1980s. They are available from Netlib [28]. For the middle example (pilot), MINOS required about 24 h on a DEC MicroVAX II during 1987, and did not perform reliably until the EXPAND anti-degeneracy procedure was developed.

Line 1 for pilot in Table 3 shows that Double MINOS with cold start and scaling required 16,060 primal simplex iterations and 5.7 CPU seconds. The final unscaled primal solution x satisfied the bounds ℓ and u in (1) to within $O(10^{-6})$, and the dual solution y satisfied the optimality conditions to within $O(10^{-3})$.

Line 2 for pilot shows that Quad MINOS starting from that point with scaling needed only 29 iterations and 0.7 s to obtain a very accurate solution (where Pinf $=10^{-99}$ means that the maximum primal infeasibility was 0.0).

Line 3 for pilot shows that in the "insurance" step, Quad MINOS warm-starting again but with no scaling gave a full quad-precision solution at almost no cost: maximum infeasibilities 0.0 and $O(10^{-32})$. The final Double and Quad objective values differ in the fourth significant digit, as suggested by removal of Step 1's $O(10^{-3})$ dual infeasibility.

Results for the bigger problem pilot87 are analogous.

Table 3 Iterations and runtimes in seconds for Step 1 (Double MINOS) and Steps 2 and 3 (Quad MINOS)

Model	Iterations	Times	Final objective	Pinf	Dinf
pilot4	1,571	0.1	−2.5811392602e+03	−05	−13
	6	0.0	−2.5811392589e+03	−39	−31
	0	0.0	−2.5811392589e+03	**−99**	**−30**
pilot	16,060	5.7	−5.5739887685e+02	−06	−03
	29	0.7	−5.5748972928e+02	−99	−27
	0	0.2	−5.5748972928e+02	**−99**	**−32**
pilot87	19,340	15.1	3.0171038489e+02	−09	−06
	32	2.2	3.0171034733e+02	−99	−33
	0	1.2	3.0171034733e+02	**−99**	**−33**
de063155	921	0.0	1.8968704286e+10	−13	+03
	78	0.1	9.8830944565e+09	−99	−17
	0	0.0	9.8830944565e+09	**−99**	**−24**
de063157	488	0.0	1.4561118445e+11	+20	+18
	476	0.5	2.1528501109e+07	−27	−12
	0	0.0	2.1528501109e+07	**−99**	**−12**
de080285	418	0.0	1.4495817688e+01	−09	−02
	132	0.1	1.3924732864e+01	−35	−32
	0	0.0	1.3924732864e+01	**−99**	**−32**
gen1	369,502	205.3	−1.6903658594e−08	−06	−12
	246,428	9,331.3	1.2935699163e−06	−12	−31
	2,394	81.6	1.2953925804e−06	**−45**	**−30**
gen2	44,073	60.0	3.2927907828e+00	−04	−11
	1,599	359.9	3.2927907840e+00	−99	−29
	0	10.4	3.2927907840e+00	**−99**	**−32**
gen4	45,369	212.4	1.5793970394e−07	−06	−10
	53,849	14,812.5	2.8932268196e−06	−12	−30
	37	10.4	2.8933064888e−06	**−54**	**−30**
l30	1,229,326	876.7	9.5266141574e−01	−10	−09
	275,287	7,507.1	−7.5190273434e−26	−25	−32
	0	0.2	−4.2586876849e−24	**−24**	**−33**
iprob	1,087	0.2	2.6891551285e+03	+02	−11
	0	0.0	2.6891551285e+03	+02	−31
	0	0.0	2.6891551285e+03	+02	**−28**

(continued)

Table 3 (continued)

Model	Iterations	Times	Final objective	Pinf	Dinf
TMA_ME	12,225	37.1	8.0051076669e-07	−06	−05
	685	61.5	8.7036315385e-07	−24	−30
	0	6.7	8.7036315385e-07	**−99**	**−31**
GlcAerWT	62,856	9,707.3	−2.4489880182e+04	+04	−05
	5,580	3,995.6	−7.0382449681e+05	−07	−26
	4	60.1	−7.0382449681e+05	**−19**	**−21**
GlcAlift	134,693	14,552.8	−5.1613878666e+05	−03	−01
	3,258	1,067.1	−7.0434008750e+05	−09	−26
	2	48.1	−7.0434008750e+05	**−20**	**−22**

Pinf and Dinf = final maximum primal and dual infeasibilities ($\log_{10}$ values tabulated). Problem iprob is infeasible. Bold figures show Pinf and Dinf at the end of Step 3. Pinf $= 10^{-99}$ means Pinf $= 0$. Note that Pinf$/\|x^*\|_\infty$ and Dinf$/\|y^*\|_\infty$ are all $O(10^{-30})$ or smaller, even though only $O(10^{-15})$ was requested. This is an unexpectedly favorable empirical finding

The Mészáros Problems

The *problematic* LPs were provided as MPS files by Ed Klotz [17]. The first two problems have unusually large entries in the constraint matrix A. The Step 1 Double MINOS solution has at best one digit of precision in the objective value for de063155, and is quite erroneous for de063157. Nevertheless, the Steps 2 and 3 Quad solutions are seen to be highly accurate when the solution norms are taken into account.

The gen* problems come from image reconstruction, with no large entries in A, x, y, but highly degenerate primal solutions x. (In both Steps 1 and 2 for gen1, 60 % of the iterations made no improvement to the objective, and the final solution has 30 % of the basic variables on their lower bound.) For gen1, warm-starting Quad MINOS from the Step 1 basis gave an almost feasible initial solution (266 basic variables outside their bounds by more than 10^{-15} with a sum of infeasibilities of only $O(10^{-8})$), yet nearly 250,000 iterations were needed in Step 2 to reach optimality. These examples show that Quad precision does not remove the need for a more rigorous anti-degeneracy procedure (such as Wolfe's method as advocated by Fletcher [6]), and/or steepest-edge pricing [7], to reduce significantly the total number of iterations.

Problem 130 behaved similarly (80 % degenerate iterations in Steps 1 and 2). The tiny objective value is essentially zero, so we can't expect the Steps 2 and 3 objectives to agree in their leading digits.

Problem iprob is an artificial one that was intended to be feasible with a very ill-conditioned optimal basis, but the MPS file provided to us contained low-precision data (many entries like 0.604 or 0.0422). Our Double and Quad runs agree that the problem is infeasible. This is an example of Quad removing some doubt that was inevitable with just Double.

The Systems Biology ME Problems

Like the gen* problems, the ME models showed 40–60 % degenerate iterations in Step 1, but fortunately not so many total iterations in Step 2. This is important for FVA and for ME with nonlinear constraints, where there are many warm starts.

Problem TMA_ME developed by Lerman [19] has some large matrix entries $|S_{ij}|$ and many small solution values v_j that are meaningful to systems biologists. The ME part of the model also contains small matrix entries. In Step 1, almost all iterations went on finding a feasible solution, and the objective then had the correct order of magnitude.

This was the first ME model that we used for Quad experiments (in April 2012). The data S, c, ℓ, u in (3) came as a Matlab structure with $c_j = 0$, $l_j = 0$, $u_j = 1{,}000$ for most j, except $c_{17533} = 1$ (meaning maximize flux v_{17533}), four variables had smaller positive upper bounds, the last variable had moderate positive bounds, and 64 variables were fixed at zero. We output the data to a plain text file. Most entries of S are integers (represented exactly), but about 5,000 S_{ij} values are of the form 8.037943687315e-01 or 3.488862338191e-06 with 13 significant digits. The text data was read into Double and Quad versions of a prototype Fortran 90 implementation of SQOPT [11].

For the present paper, we used the same Matlab data to generate an MPS file for input into MINOS. Since this is limited to six significant digits, the values in the preceding paragraph were rounded to 8.03794e-01 and 3.48886e-06 and in total about 5,000 S_{ij} values had $O(10^{-6})$ relative perturbations of this kind. We have been concerned that such data perturbations could alter the FBA solution greatly because the final basis matrices could have condition number as large as 10^6 or even 10^{12} (as estimated by LUSOL). In comparing Quad SQOPT on Matlab data with Quad MINOS on MPS data, we fortunately observe that the final objective values for TMA_ME agree to five digits and match the results from SoPlex [34] and the exact simplex solver QSopt_ex [32], as reported by Lerman [19]:

	Optimal objective	
SoPlex 80bit	8.703671403e-07	Matlab data
QSopt_ex	8.703646169e-07	Matlab data
Quad SQOPT	8.703646169e-07	Matlab data
Quad MINOS	8.703631539e-07	MPS data

More importantly, for the most part *even small solution values* are perturbed in only the 5th or 6th significant digit. Let v and w be the solutions obtained by the two Quad solvers on slightly different data. Some example solution values follow:

j	107	201	302	
Quad SQOPT v_j	2.336815e-06	8.703646e-07	1.454536e-11	Matlab data
Quad MINOS w_j	2.336823e-06	8.703632e-07	1.454540e-11	MPS data

Among all j for which $\max(v_j, w_j) > 10^{-15}$ (the feasibility tolerance), the largest relative difference $|v_j - w_j| / \max(v_j, w_j)$ was less than 10^{-5} for all but 31 variables. For 22 of these pairs, either v_j or w_j was primal or dual degenerate (meaning one of them was zero and there are alternative solutions with the same objective value). The remaining nine variables had these values:

j	v_j	w_j	Relative difference
16383	6.0731e-07	2.0374e-06	0.70
16459	1.7090e-06	2.1778e-06	0.22
16483	2.4675e-06	5.9936e-07	0.76
16730	1.4432e-06	7.8685e-07	0.46
17461	1.7090e-06	2.1778e-06	0.22
17462	2.4675e-06	5.9936e-07	0.76
17478	6.0731e-07	2.0374e-06	0.70
17507	1.4432e-06	7.8685e-07	0.46
17517	8.7036e-07	2.9740e-06	0.71

We see that the v_j, w_j values are quite small (the same magnitude as the data perturbation), and for each of the nine pairs there is about one digit of agreement. In general we could expect thousands of small solution pairs to differ this much, yet for almost *all* 17,535 pairs, there are at least five digits of agreement.

These observations about two forms of problem TMA_ME are welcome empirical evidence of the robustness of this particular multiscale model. Quad solvers can help evaluate the robustness of future (increasingly large) models of metabolic networks by enabling similar comparison of high-accuracy solutions for slightly different problems.

Problem GlcAerWT is an ME model from the detailed study by Thiele et al. [36]. Difficulties with solving TMA_ME and GlcAerWT led to the lifting technique of Sun et al. [35] (and to problem GlcAlift).

After 33,000 iterations on GlcAerWT, Double MINOS began to report singularities every 50–100 iterations following updates to the basis LU factors. After another 30,000 iterations, MINOS terminated Step 1 with maximum infeasibility $O(10^4)$. Step 2 required significant work to achieve a reasonably accurate solution. Step 3 quickly confirmed the final objective value with high accuracy considering the $O(10^7)$ primal and dual solutions norms.

Problem GlcAlift is a reformulated version of GlcAerWT in which some large matrix entries $c_{\max}$ in (5) have been reduced via the lifting technique [35]. In Step 1, Double MINOS again reported frequent singularities and required twice as many iterations as GlcAerWT, but a near-optimal solution was found (allowing for the primal and dual solution norms of $O(10^7)$), and Steps 2–3 were more efficient and accurate. The objective function for both GlcA models is to maximize variable v_{60069}. The fact that the Step 1 objective values have no correct digits illustrates the challenge these models present and emphasizes the benefits that Quad precision offers. Theoretically the optimal objectives for GlcAerWt and GlcAlift should agree. We assume that the limited data precision in the MPS files is responsible for only three-digit agreement. Fortunately the Tomlab interface used by Thiele et al. [36] allows full double-precision data [37]. We can do the same for MINOS, as we did for SQOPT.

5 Discussion

While today's advanced LP solvers such as CPLEX, Gurobi, Mosek, and Xpress [3, 5, 13, 23] are effective on a wide range of large and challenging linear optimization models, the study by Thiele et al. [36] emphasizes the need for improved reliability in solving FBA and ME models in systems biology. Fortunately, reformulation [35] and careful use of the commercial solvers CPLEX and Gurobi permitted successful analysis in Thiele et al. [36] of the GlcAerWT and GlcAlift models discussed here, and we encourage this approach for Step 1 of our proposed three-step procedure (Section 3). The bulk of the work in solving multiscale LP problems can still be performed there by conventional Double solvers (possibly including Barrier solvers with simplex "cross-over" to provide a basis).

Our aim has been to demonstrate that the Step 2 and 3 warm starts with Quad solvers will be acceptably efficient, and that the accuracy achieved exceeds requirements by a very safe margin. The "humane" approach of Kahan [16]—use of Quad LP solvers—is certainly more efficient than applying exact simplex solvers, even though the latter have proved their value in several applications [1, 2, 18, 20].

An intriguing question remains concerning the bold figures in Table 3. The primal and dual solutions obtained with Quad precision are *substantially more accurate than the* 10^{-15} *requested.* The same has been true for all of the classic set of Netlib problems [28] that we have run. Kahan [16] explains that "perturbations get amplified by singularities near the data." He describes a "pejorative surface" of data points where singularity exists, and expects loss of accuracy as data approaches the surface. The volume surrounding the pejorative surface is the danger zone, but: "*Arithmetic precision is usually extravagant enough if it is somewhat more than twice as [great] as the data's and the desired result's. Often that shrunken volume contains no data.*" We can surmise that Kahan has anticipated our observed situation, wherein LP problems defined with double-precision data appear unlikely to be too ill-conditioned for a Quad solver.

We believe that quadruple-precision solutions are now practical for multiscale LP applications such as FBA and FVA models in systems biology [12, 30, 31, 36], and that they justify increased confidence as systems biologists build ever-larger models to explore new hypotheses about metabolism and macromolecular synthesis. Our three-step procedure of Section 3 allows combined use of Double and Quad solvers and should lead to solutions of exceptional accuracy in other areas of computational science involving multiscale optimization problems. For example, Dattorro [4] has derived an approach to analog filter design that requires a Quad LP or nonlinear solver to deal with a wide range of frequencies (which must be raised to the fourth power). We look forward to implementing this approach, as well as treating the nonlinear constraints (7) directly to take advantage of the nonlinear algorithms in Quad MINOS.

Funding

This work was supported by the National Institute of General Medical Sciences of the National Institutes of Health [award U01GM102098]. The content is solely the responsibility of the authors and does not necessarily represent the official views of the funding agencies.

Acknowledgements We thank Ronan Fleming, Philip Gill, Ed Klotz, Joshua Lerman, Edward O'Brien, Yuekai Sun, Ines Thiele, and Elizabeth Wong for much help during the course of this work and for valuable comments on the manuscript. Joshua Lerman at UC San Diego provided the model named TMA_ME here (originally model_final_build_unscaled.mat) and advised us of the final objective values obtained by SoPlex and QSopt_exact. Ines Thiele at the University of Luxembourg provided the GlcAerWT model and insight into the ME coupling constraints. Yuekai Sun at Stanford University created the reformulated version named GlcAlift here. Ed Klotz of IBM in Carson City NV provided MPS files for the Mészáros *problematic* LPs and lengthy discussions of their properties. A referee also provided valuable feedback.

References

1. Applegate, D.L., Cook, W., Dash, S., Espinoza, D.G.: Exact solutions to linear programming problems. Oper. Res. Lett. **35**, 693–699 (2007)
2. Cook, W., Dash, S., Espinoza, D.G.: Exact solutions to linear programming problems. Oper. Res. Lett. **35**, 693–699 (2007)
3. CPLEX: IBM ILOG CPLEX optimizer. http://www.ibm.com/software/commerce/optimization/cplex-optimizer/ (2014)
4. Dattorro, J.: Private Communication. Stanford University, Stanford (2014)
5. FICO Xpress Optimization Suite: http://www.fico.com/en/products/fico-xpress-optimization-suite/ (2015)
6. Fletcher, R.: On Wolfe's method for resolving degeneracy in linearly constrained optimization. SIAM J. Optim. **24**(3), 1122–1137 (2014)
7. Forrest, J.J., Goldfarb, D.: Steepest-edge simplex algorithms for linear programming. Math. Program. **57**, 341–374 (1992)
8. GCC libquadmath: The GCC Quad-precision math library application programming interface (API). http://gcc.gnu.org/onlinedocs/libquadmath/ (2014)
9. Gill, P.E., Murray, W., Saunders, M.A., Wright, M.H.: Maintaining LU factors of a general sparse matrix. Linear Algebra Appl. **88/89**, 239–270 (1987)
10. Gill, P.E., Murray, W., Saunders, M.A., Wright, M.H.: A practical anti-cycling procedure for linear and nonlinear programming. Math. Program. **45**, 437–474 (1989)
11. Gill, P.E., Murray, W., Saunders, M.A.: SNOPT: an SQP algorithm for large-scale constrained optimization. SIAM Rev. **47**(1), 99–131 (2005). SIGEST article

12. Gudmundsson, S., Thiele, I.: Computationally efficient flux variability analysis. BMC Bioinf. **11**(489), 3 pp. (2010)
13. Gurobi: Gurobi optimization system for linear and integer programming. http://www.gurobi.com (2014)
14. Hall, J.A.J., McKinnon, K.I.M.: The simplest examples where the simplex method cycles and conditions where EXPAND fails to prevent cycling. Math. Progam. Ser. B **100**, 133–150 (2004)
15. IEEE standard for floating-point arithmetic: IEEE Std 754-2008. IEEE Computer Society (2008)
16. Kahan, W.: Desperately needed remedies for the undebuggability of large floating-point computations in science and engineering. In: IFIP/SIAM/NIST Working Conference on Uncertainty Quantification in Scientific Computing, Boulder (2011). http://www.eecs.berkeley.edu/~wkahan/Boulder.pdf
17. Klotz, E.: Private Communication. IBM, Carson City (2014)
18. Koch, T.: The final NETLIB-LP results. Oper. Res. Lett. **32**, 138–142 (2004)
19. Lerman, J.A.: Private Communication. University of California, San Diego (2012)
20. Lerman, J.A., Hyduke, D.R., Latif, H., Portnoy, V.A., Lewis, N.E., Orth, J.D., Schrimpe-Rutledge, A.C., Smith, R.D., Adkins, J.N., Zengler, K., Palsson, B.Ø.: In silico method for modelling metabolism and gene product expression at genome scale. Nat. Commun. **3**(929), 10 pp. (2012)
21. LUSOL: Sparse LU factorization package. http://web.stanford.edu/group/SOL/software/lusol (2013)
22. Mészáros, C.: A collection of challenging LP problems. http://www.sztaki.hu/~meszaros/public_ftp/lptestset/problematic (2004)
23. MOSEK: MOSEK Optimization Software. http://www.mosek.com/ (2014)
24. MPS: Input format for LP data. http://lpsolve.sourceforge.net/5.5/mps-format.htm (1960)
25. MPS files: Data files in MPS format for models in Table 2. http://web.stanford.edu/group/SOL/multiscale/models.html (2014)
26. Murtagh, B.A., Saunders, M.A.: Large-scale linearly constrained optimization. Math. Program. **14**, 41–72 (1978)
27. Murtagh, B.A., Saunders, M.A.: A projected Lagrangian algorithm and its implementation for sparse nonlinear constraints. Math. Program. Stud. **16**, 84–117 (1982)
28. Netlib: Netlib collection of LP problems in MPS format. http://www.netlib.org/lp/data (1988)
29. O'Brien, E.J., Lerman, J.A., Chang, R.L., Hyduke, D.R., Palsson, B.Ø.: Genome-scale models of metabolism and gene expression extend and refine growth phenotype prediction. Mol. Syst. Biol. **9**(693), 13 pp. (2013)
30. Orth, J.D., Thiele, I., Palsson, B.Ø.: What is flux balance analysis? Nat. Biotechnol. **28**(3), 245–248 (2010)
31. Palsson, B.Ø.: Systems Biology: Properties of Reconstructed Networks. Cambridge University Press, New York (2006)
32. QSopt_ex: Qsopt_ex: a simplex solver for computing exact rational solutions to LP problems. http://www.math.uwaterloo.ca/~bico/qsopt/index.html (2008)
33. Schellenberger, J., Que, R., Fleming, R.M.T., Thiele, I., Orth, J.D., Feist, A.M., Zielinski, D.C., Bordbar, A., Lewis, N.E., Rahmanian, S., et al.: Quantitative prediction of cellular metabolism with constraint-based models: the COBRA Toolbox v2.0. Nat. Protoc. **6**(9), 1290–1307 (2011)
34. SoPlex: Soplex: The sequential object-oriented simplex solver. http://soplex.zib.de (1996)
35. Sun, Y., Fleming, R.M.T., Thiele, I., Saunders, M.A.: Robust flux balance analysis of multiscale biochemical reaction networks. BMC Bioinf. **14**, 240 (2013)
36. Thiele, I., Fleming, R.M.T., Que, R., Bordbar, A., Diep, D., Palsson, B.Ø.: Multiscale modeling of metabolism and macromolecular synthesis in *E. coli* and its application to the evolution of codon usage. PLoS One **7**(9), 18 pp. (2012)
37. TOMLAB: Optimization environment for MATLAB. http://tomopt.com (2014)

Real and Integer Extended Rank Reduction Formulas and Matrix Decompositions: A Review

Nezam Mahdavi-Amiri and Effat Golpar-Raboky

Abstract We have recently developed an extended rank reducing process for rank reduction of a matrix leading to various matrix decompositions containing the Abaffy-Broyden-Spedicato (ABS) and Wedderburn processes. Notably, the extended process contains both the Wedderburn biconjugation process and the scaled extended ABS class of algorithms. The process provides a general finite iterative approach for constructing factorizations of a matrix and its transpose under a common framework of a general decomposition having various useful structures such as triangular, orthogonal, diagonal, banded and Hessenberg and many others. One main new result is the derivation of an extended rank reducing process for an integer matrix leading to the so-called Smith normal form. For this process, to solve the arising quadratic Diophantine equations, we have proposed two algorithms. Here, we report some numerical results on randomly generated test problems showing a better performance of one algorithm, based on a recent ABS algorithm, in controlling the size of the solution. We also report results obtained by our algorithm on the Smith normal form having a more balanced distribution of the intermediate values as compared to the ones obtained by Maple.

Keywords Linear systems • Matrix decomposition • Wedderburn rank reduction • ABS Algorithms • Smith normal form • Quadratic Diophantine equation

1 Introduction

Wedderburn was the first to use a rank reducing algorithm in reducing quadratic forms [30]. Later, independently of him, Egervary also developed a rank reduction procedure [9]. Egervary proved the basic theorem of rank reduction, which really is

N. Mahdavi-Amiri (✉)
Faculty of Mathematical Sciences, Sharif University of Technology, Tehran, Iran
e-mail: nezamm@sharif.edu

E. Golpar-Raboky
Department of Mathematics, University of Qom, Qom, Iran
e-mail: g.raboky@qom.ac.ir

M. Al-Baali et al. (eds.), *Numerical Analysis and Optimization*, Springer Proceedings in Mathematics & Statistics 134, DOI 10.1007/978-3-319-17689-5_10

the only constructive characterization of the matter. The sufficient part of this result was given in Wedderburn's book [30], which was later extended to block matrices by Guttman [18, 19]. Galantai [1, 13–15] established some results concerned with the rank reduction algorithm developed by Egervary.

Recently, Golpar-Raboky and Mahdavi-Amiri [23] gave an extended rank reduction formula transforming rows, columns or both rows and columns simultaneously of a given matrix. The formula makes use of null space transformations on rows and columns leading to the establishment of necessary and sufficient conditions for rank reduction transformations. By repeatedly applying the formula to reduce the rank, a general extended rank reducing process (GERRP) was derived. The process provides a finite iterative approach for constructing a general decomposition $V^TAP = \Omega$, with V, P and Ω having various useful structures such as triangular, orthogonal, diagonal, banded, Hessenberg and others.

The main results established for the extended rank reduction formula and GERRP are outlined below:

- The Wedderburn rank reduction formula is a special case of the extended rank reduction formula. The formula provides a new formulation of the Wedderburn rank reduction formula; see Section 2.
- GERRP computes all possible A-conjugate pairs as well as A-biconjugate pairs (P,V), resulting in various matrix factorizations.
- GERRP provides new formulations for the biconjugation process associated with the Wedderburn rank reduction and the scaled Abaffy-Broyden-Spedicato (ABS) class of algorithms, to be explained later. Using the new formulations, we establish new properties for the biconjugation process and the scaled extended ABS algorithms.
- The new formulations assure that both the biconjugation process associated with the Wedderburn rank reducing process and the scaled extended ABS class of algorithms belong to our proposed class of algorithms.
- The scaled ABS algorithms contain the Wedderburn rank reducing process and the biconjugation process.
- The biconjugation process produces all possible A-biconjugate pairs (P,V) resulting in computation of a variety of matrix factorizations.
- A main result is the derivation of the extended rank reduction formula for integer matrices. Similar properties of the extended rank reduction and GERRP are also preserved for the integer case; see Section 3.
- An integer Wedderburn rank reduction formula and its associated integer biconjugation process are developed. All properties we state for the extended rank reduction process, the scaled ABS algorithms and the biconjugation process are preserved for the integer case. Both the integer biconjugation process and the scaled extended integer ABS class of algorithms are shown to be special cases of the integer rank reducing process.
- The Smith normal form of an arbitrary integer matrix is computed by GERRP, as well as the scaled extended integer ABS algorithm and the integer biconjugation process. For the Smith normal form, having the need to solve a quadratic Diophantine equation, we have proposed two algorithms for solving such

equations. The first algorithm makes use of a special integer basis for the row space of the matrix, and the second algorithm, with the intention of controlling the growth of intermediate results and making use of our given conjecture, is based on a recently proposed integer ABS algorithm; see Section 4. Some numerical results are reported on randomly generated test problems showing a better performance of the second algorithm in controlling the size of the solution. We also report the results obtained by our proposed algorithm for the Smith normal form and compare them with the ones obtained by using Maple, observing a more balanced distribution of the values of the results obtained by our algorithm; see Section 5.

2 Extended Rank Reduction Formulas

Definition 1. A transformation f on $M_{m,n}$, the vector space of $m \times n$ matrices, is a rank reduction formula, if for $A \in M_{m,n}$, $rank(f(A)) < rank(A)$.

The extended rank reduction process (ERRP) introduced by Golpar-Raboky and Mahdavi-Amiri [23] makes use of the null space transformations on rows and/or columns of a given matrix. The formulas provide the necessary and sufficient conditions for rank reduction transformations. Other existing rank reduction formulas can now be derived as special cases of the extended rank reduction formulas.

Theorem 1 (Block Extended Rank k Reduction Formula). *Let $A \in R^{m\times n}$ have rank r and $r \geq k$, where k is a positive integer. Assume that $\bar{G} \in R^{m\times m}$ and $G \in R^{n\times n}$. We have*

$$rank(\bar{G}AG^T) = rank(A) - k \tag{1}$$

if and only if one of the following conditions holds:

(i) G is nonsingular, $dim(N(\bar{G})) = k$ and there is $X \in R^{n\times k}$ so that X^TA^T has rank k and $\bar{G}^T$ generates the null space of X^TA^T.
(ii) $\bar{G}$ is nonsingular, $dim(N(G)) = k$ and there is $Y \in R^{m\times k}$ so that Y^TA has rank k and G^T generates the null space of Y^TA.
(iii) $Dim(N(G)) = dim(N(\bar{G})) = k$, and there are $X \in R^{n\times k}$ and $Y \in R^{m\times k}$ so that Y^TAX is a nonsingular matrix, G^T generates the null space of Y^TA, and $\bar{G}^T$ generates the null space of X^TA^T.

For $k = 1$, a rank one reduction formula is obtained as follows.

Corollary 1 (Extended Rank One Reduction Formula). *Let $A \in R^{m\times n}$ be a nonzero matrix, $\bar{G} \in R^{m\times m}$ and $G \in R^{n\times n}$. We have*

$$rank(\bar{G}AG^T) = rank(A) - 1 \tag{2}$$

if and only if one of the following conditions holds:

(i) (left rank reduction) G is nonsingular, $dim(N(\bar{G})) = 1$ and there is a vector $x \in R^n$ so that $\bar{G}^T$ generates the null space of $x^TA^T \neq 0$.

(ii) (right rank reduction) $\bar{G}$ *is nonsingular,* $dim(N(G)) = 1$ *and there is a vector* $y \in R^m$ *so that* G^T *generates the null space of* $y^T A \neq 0$.

(iii) (left-right rank reduction) $Dim(N(\bar{G})) = dim(N(G)) = 1$, *and there are vectors* $x \in R^n$ *and* $y \in R^m$ *so that* $y^T Ax \neq 0$, G^T *generates the null space of* $y^T A$, *and* $\bar{G}^T$ *generates the null space of* $x^T A^T$.

By repeatedly applying the extended rank reduction formula, we next present a rank reducing processes so that the left, right and left-right reductions lead to upper triangular, lower triangular and diagonal decompositions, respectively, naming it to be the generalized extended rank reducing process (GERRP). In this process, in addition to reducing rank, we produce the independent vectors $v_i \in R^m$ and the independent vectors $p_i \in R^n$ to construct the general rank preserving transformation

$$V^T AP = \Omega, \tag{3}$$

where $rank(A) = r$, $V = (v_1, \cdots, v_r)$, $P = (p_1, \cdots, p_r)$, with Ω accordingly intended to have various useful structures.

Algorithm 1. **General extended rank one reducing process (GERRP).**
Let $A \in R^{m\times n}$ have rank r. Start with nonsingular matrices $R_1 \in R^{m\times m}$ and $H_1 \in R^{n\times n}$ (basis for the null space of the null matrices). Let $i = 1$ and $\bar{A}_1 = R_1 AH_1^T$. Choose one of the cases $(a)-(c)$ and execute the case in all iterations.
While $\bar{A}_i \neq 0$ do
execute the chosen case:

(a) (***left rank reducing process***) Choose a nonsingular matrix G_i, a vector $x_i \in R^n$ so that $\bar{A}_i x_i \neq 0$ and a matrix $\bar{G}_i$, with $dim(N(\bar{G}_i)) = 1$, so that $\bar{G}_i^T$ generates the null space of $x_i^T \bar{A}_i^T$. Choose $y_i \in R^m$ such that $y_i^T \bar{A}_i x_i \neq 0$.

(b) (***right rank reducing process***) Choose a nonsingular matrix $\bar{G}_i$, a vector $y_i \in R^m$ so that $y_i^T \bar{A}_i \neq 0$ and a matrix G_i, with $dim(N(G_i)) = 1$, so that G_i^T generates the null space of $y_i^T \bar{A}_i$. Choose $x_i \in R^n$ such that $y_i^T \bar{A}_i x_i \neq 0$.

(c) (***left-right rank reducing process***) Choose the vectors $x_i \in R^n$, $y_i \in R^m$ so that $y_i^T \bar{A}_i x_i \neq 0$, and the matrices G_i and $\bar{G}_i$, with $dim(N(\bar{G}_i)) = dim(N(G_i)) = 1$, so that G_i^T generates the null space of $y_i^T \bar{A}_i$ and $\bar{G}_i^T$ generates the null space of $x_i^T \bar{A}_i^T$.

Let $v_i = R_i^T y_i$, $p_i = H_i^T x_i$, $H_{i+1} = G_i H_i$, $R_{i+1} = \bar{G}_i R_i$, $\bar{A}_{i+1} = R_{i+1} AH_{i+1}^T$ and $i = i+1$.
End While.
Stop.

Note: The nonsingular matrices $\bar{G}_i$ or G_i can be used as permutation matrices when we need to apply row or column interchanges, or as scaling matrices, to produce various structured factorizations.

Assume that $rank(A) = r$, $V = (v_1, \cdots, v_r)$ and $P = (p_1, \cdots, p_r)$. Then, the matrices V and P have rank r and $V^T AP$ is an $r \times r$ nonsingular matrix. Here, we recall some properties of GERRP; see [23].

Theorem 2. *For GERRP, the followings hold.*

(i) In the left reduction process, let the columns of R^T generate a basis for the null space of P^TA^T. Then, V^TAP is an upper triangular matrix.
(ii) In the right reduction process, let the columns of H^T generate a basis for the null space of V^TA. Then, V^TAP is a lower triangular matrix.
(iii) In the left-right reduction process, let the columns of R^T generate a basis for the null space of P^TA^T and the columns of H^T generate a basis for the null space of V^TA. Then, V^TAP is a diagonal matrix.

For square matrices, the proposed extended formula can readily lead to conjugation and biconjugation and the corresponding decompositions.

Definition 2. Let $A \in R^{n\times n}, P \in R^{n\times n}$ and $V \in R^{n\times n}$. The pair (P,V) is said to be A-conjugate if the matrix $L = V^TAP$ is lower triangular, and the pair (P,V) is said to be A-biconjugate if $D = V^TAP$ is nonsingular diagonal.

Theorem 3. *Let $A \in R^{n\times n}$. GERRP (a) computes all possible A^T-conjugate pairs, GERRP (b) computes all possible A-conjugate pairs and finally GERRP (c) computes all possible A-biconjugate pairs.*

Remark 1. A more general reduction scheme for GERRP (c) is obtained by making use of two extra parameters z_i and $\bar{z}_i$ to be explained next, even though we were not able to prove that $rank(A) = rank(V^TAP)$ in the general case, in spite of the fact that all our numerical experiments confirmed it to be true. Thus, in [23], we gave the following conjecture.

• **Conjecture.** If $z_i \in R^m$ and $\bar{z}_i \in R^n$ are so that $y_i^T\bar{A}_i\bar{z}_i \neq 0$ and $z_i^T\bar{A}_ix_i \neq 0$, with $p_i = H_i^T\bar{z}_i$ and $v_i = R_i^Tz_i$, then $rank(V^TAP) = rank(A)$.

We should point out that the proposed conjecture allows for the development of some new effective algorithms for a variety of several factorizations such as banded and Hessenberg as defined below.

Definition 3. $A \in R^{m\times n}$ is a banded matrix if and only if the nonzero elements of A are located in a band around the main diagonal. If A is a banded matrix such that $a_{i,j} = 0$, for $i-j > k$ or $j-i > l$, then A has lower bandwidth k and upper bandwidth l. If $k = n-1$ and $l = 1$, then A is a lower Hessenberg, and if $l = n-1$ and $k = 1$, then A is an upper Hessenberg matrix.

Now, we show how to choose the parameters for computing the banded and Hessenberg factorizations. Let $A \in R^{n\times n}$ be strongly nonsingular (that is, the determinant of every leading left principal submatrix of A is nonzero). For GERRP (c), let $H_1 = I$, $R_1 = I$, $x_i = e_i$, $y_i = e_i$,

$$\bar{z}_i = \begin{cases} e_{i+k}, \; i+k \leq n \\ e_n, \quad otherwise \end{cases}, \; z_i = \begin{cases} e_{i+l}, \; i+l \leq n \\ e_n, \quad otherwise. \end{cases} \tag{4}$$

Update H_i and R_i by

$$H_{i+1} = H_i - \frac{H_i A^T R_i^T y_i x_i^T H_i}{y_i^T R_i A H_i^T x_i}, \; R_{i+1} = R_i - \frac{R_i A H_i^T x_i y_i^T R_i}{y_i^T R_i A H_i^T x_i}. \tag{5}$$

Compute $p_i = H_i^T \bar{z}_i$ and $v_i = R_i^T z_i$. Then, $V^T AP = \Omega$ is a banded matrix with lower bandwidth k and upper bandwidth l, where $V = [v_1, \cdots, v_n]$ and $P = [p_1, \cdots, p_n]$. For $k = n-1$ and $l = 1$, the matrix Ω is lower Hessenberg, and for $k = 1$ and $l = n-1$, the matrix Ω is upper Hessenberg.

In the following two subsections, we show that both the ABS algorithms and the Wedderburn rank reducing process are special cases of our extended rank reduction process. We end this section by giving a number of well-known matrix factorizations obtained as special cases of our rank reducing process.

2.1 Relations to the ABS Algorithms

Classes of scaled ABS algorithms have been introduced for solving linear systems of equations based on the basic ABS algorithms [2, 3, 26–29]. A major result of the scaled ABS algorithms has been the derivation of scaled ABS class of algorithms for linear Diophantine equations [10, 11, 26, 27]. An ABS method provides the general solution of the system by computing a particular solution and a matrix with rows generating the null space of the coefficient matrix.

Consider the following linear system,

$$Ay = b, \qquad y \in R^n, \;\; A \in R^{m \times n}, \;\; b \in R^m, \tag{6}$$

where $A = [a_1, \cdots, a_m]^T$, with $a_i \in R^n$, $1 \le i \le m$, and $rank(A)$ is arbitrary.

The ABS methods compute the null space and a matrix factorion of A implicitly.

As a key component of the basic ABS method, an arbitrary and nonsingular matrix $H_1 \in R^{n \times n}$, Spedicato's parameter, originally is set as the basis for null space of no equations. Given H_i, a matrix with rows generating the null space of the first $i-1$ rows of the coefficient matrix, a basic ABS algorithm computes H_{i+1}, with rows generating the null space of the first i rows of the coefficient matrix, by performing the following step:

- Update the Abaffian matrix H_i by

  $$H_{i+1} = H_i - \frac{H_i a_i w_i^T H_i}{w_i^T H_i a_i}$$

  with $w_i \in R^n$ (Abaffy's parameter) satisfying $w_i^T H_i a_i \neq 0$.
- Determine x_i (Broyden's parameters) such that $x_i^T H_i a_i \neq 0$ and set $p_i = H_i^T x_i$.

- Compute $s_i = H_i a_i$, where $a_i \in R^n$ is the ith row of A (note that $s_i \neq 0$ if and only if a_i is linearly independent of $a_1, \cdots, a_{i-1}$) [2].

Obviously, the original system (6) is equivalent to the following scaled system,

$$V^T A y = V^T b, \tag{7}$$

where V, the scaled matrix, is an arbitrary nonsingular m by m matrix. It is obvious that by replacing a_i with $A^T v_i$ in the above procedure, a particular solution y_{m+1} is obtained, as applied to (7), and the rows of the resulting H_{m+1} span the null space of $V^T A$ or equivalently that of A.

Now, a tailored basic ABS algorithm as applied to A can be described as follows, r_i gives the rank of the first $i-1$ rows of $V^T A$.

Algorithm 2. **The scaled ABS (SABS) algorithm for generation of the** H_i.

(1) Let $H_1 \in R^{n\times n}$ be an arbitrary nonsingular matrix. Let i=1 and $r_i = 0$.
(2) Choose v_i linearly independent of $v_1, \cdots, v_{i-1}$ and set $s_i = H_i A^T v_i$.
(3) **If** $s_i = 0$ (the ith row of $V^T A$ is dependent on its first i-1 rows) **then** let $H_{i+1} = H_i$, $r_{i+1} = r_i$ and **go to** (6).
(4) ($s_i \neq 0$ and hence the ith row of $V^T A$ is independent of its first i-1 rows) Compute the search vector p_i by

$$p_i = H_i^T x_i, \tag{8}$$

where $x_i \in R^n$ is so that $v_i^T A H_i^T x_i \neq 0$.
(5) (Updating the null space generator) Update H_i by

$$H_{i+1} = H_i - \frac{H_i A^T v_i w_i^T H_i}{w_i^T H_i A^T v_i}, \tag{9}$$

where $w_i \in R^n$ is so that $s_i^T w_i \neq 0$, and let $r_{i+1} = r_i + 1$.
(6) **If** $i = m$ **then Stop** (H_{m+1}^T generates the null space of A and r_{m+1} is its rank) **else** let $i = i+1$ and **go to** (2).

Matrices H_i, which are generalizations of projection matrices, have been named as Abaffians (due to Abaffy). Using Algorithm 2, we have an implicit matrix factorization $V^T A P = L$, where $P = [p_1, \cdots, p_r]$ and L is a lower triangular matrix. Choices of the parameters H_1, v_i, x_i and w_i determine particular methods within the class so that various matrix factorizations are derived [2–4, 16, 26].

Chen et al. [6] introduced a generalization of the ABS algorithms, called extended ABS (EABS) class of algorithms for the real case, which differs from the ABS algorithms only in updating the Abaffian matrices H_i. In the EABS algorithms, the Abaffian matrices H_i are updated as follows:

- $H_{i+1} = G_i H_i$, where $G_i \in R^{j_{i+1} \times j_i}$ is such that we have $G_i x = 0$ if and only if $x = \lambda H_i a_i$, for some $\lambda \in R$.

We next show that the scaled extended ABS class of algorithms can readily be produced by GERRP.

Consider a scaled extended ABS algorithm on A with H_1, $n \times n$, nonsingular as given in the beginning of the algorithm, the scale vectors v_i, the H_i being updated as $H_{i+1} = G_i H_i$, and let $p_i = H_i^T x_i$. Now, for GERRP, case (b), consider H_1, v_i and G_i from the scaled extended ABS algorithm and let

$$R_1 = I,\ \bar{G}_i = I,\ y_i = v_i, H_{i+1} = G_i H_i,\ p_i = H_i^T x_i. \tag{10}$$

Then, $R_{i+1} = \bar{G}_i R_i = V^T$, where V is the scale matrix in the ABS algorithm. Therefore, GERRP (b) produces the same result as the corresponding scaled extended ABS algorithm applied to A. Thus, we have the following result.

Theorem 4. *The scaled extended ABS class of algorithms is produced by GERRP.*

Remark 2. We can also compute the scale vectors v_i, step by step, using case (c). Let $A \in R^{m \times n}$, with rank m, and $H_1 \in R^{n \times n}$ and $R_1 \in R^{m \times m}$ be nonsingular and arbitrary. Starting with $i = 1$, inductively choose the vectors $x_i \in R^n$, $y_i \in R^m$ so that $y_i^T R_i A H_i^T x_i \neq 0$, and the matrices G_i and $\bar{G}_i$, with $dim(N(\bar{G}_i)) = dim(N(G_i)) = 1$, so that G_i^T generates the null space of $y_i^T R_i A H_i^T$ and $\bar{G}_i^T$ generates the null space of $x_i^T H_i A^T R_i^T$. Compute

$$H_{i+1} = G_i H_i, R_{i+1} = \bar{G}_i R_i,\ p_i = H_i^T x_i,\ v_i = R_i^T y_i. \tag{11}$$

Then, let $v_i = R_i^T y_i$. It means, we apply the ABS algorithm on A and A^T simultaneously for computing H_i as the right Abaffian, p_i as the right search vector, R_i as the left Abaffian and v_i as the left search vector. Later, we show how to compute the new factorizations such as *banded* and Hessenberg by the ABS algorithms.

2.2 Relations to the Wedderburn Rank Reducing Process

Wedderburn showed that subtracting rank one matrices of the form $\beta^{-1} A x y^T A$ from a matrix A resulted in a matrix with rank one less than that of A if $\beta = y^T A x \neq 0$ [20, 30]. The converse is also true [20]. For a comprehensive investigative treatment of the Wedderburn approach, see the work by Chu, Funderlic and Golub [7].

The following theorem gives a characterization of what we refer to as the Wedderburn rank one reduction formula.

Theorem 5 (Wedderburn Rank One Reduction Formula). *Let $A \in R^{m\times n}, u \in R^n$ and $v \in R^m$. The rank of the matrix*

$$B = A - \beta^{-1} uv^T, \tag{12}$$

is one less than that of A if and only if there are vectors $x \in R^n$ and $y \in R^m$ such that $u = Ax,\ v = A^T y$ and $\beta = y^T Ax \neq 0$.

Simultaneous multiple rank reduction is also possible. Cline and Funderlic [8] proved a block version of the rank reduction formula in the following sense.

Theorem 6. *Let $A \in R^{m\times n}$ have rank r and $r \geq k$. Suppose $U \in R^{m\times k},\ R \in R^{k\times k}$ and $V \in R^{n\times k}$. We have*

$$rank(A - UR^{-1}V^T) = rank(A) - rank(UR^{-1}V^T) = rank(A) - k \tag{13}$$

if and only if there exist $X \in R^{n\times k}$ and $Y \in R^{m\times k}$ so that

$$U = AX,\ V = A^T Y,$$

with $R = Y^T AX$ being a nonsingular matrix.

The first constructive characterization of rank reduction is due to Wedderburn (see [30], pp. 68–69). Independently of him, Egervary [9] developed a rank reduction procedure by proposing a general finitely terminating scheme unifying a variety of processes occurring in the solution of linear equation systems. In this scheme, the rank reduction formula (12) is repeatedly applied as follows. Let $A_1 = A$. While $A_k \neq 0$, apply (12) repeatedly to generate a sequence of matrices $\{A_k\}$ by using

$$A_{k+1} = A_k - \beta_k^{-1} A_k x_k y_k^T A_k = (I - \beta_k^{-1} A_k x_k y_k^T) A_k = A_k (I - \beta_k^{-1} x_k y_k^T A_k), \tag{14}$$

for any vectors $x_k \in R^n$ and $y_k \in R^m$ for which $\beta_k = y_k^T A_k x_k \neq 0$. The sequence determined by (14) must terminate in $r = rank(A)$ iterations, since $rank(A_k)$ decreases by exactly one at every iteration k. The process is called a *rank reducing process* and the A_k are called the *Wedderburn matrices*.

Any matrix pair (X,Y) of a rank reducing process can be transformed to a biconjugate pair, (N,M), such that

$$\Omega = M^T AN \tag{15}$$

is nonsingular and diagonal, by the biconjugation process:

$$n_k = x_k - \sum_{i=1}^{k-1} \frac{< x_k, m_i >}{< n_i\ ,\ m_i >} n_i, \tag{16}$$

$$m_k = y_k - \sum_{i=1}^{k-1} \frac{< n_i, y_k >}{< n_i \ , \ m_i >} m_i, \tag{17}$$

where $< x, y > \equiv y^T Ax$.

As observed in [7], many of the fundamental processes of numerical linear algebra and almost all matrix factorizations can be derived from the biconjugation process.

Note that the Wedderburn rank one reduction formula (12) is a special case of the extended rank one reduction formula (2). Choose G and $\bar{G}$ according to any one of the following three cases:

1. $G = I, \ \bar{G} = (I - \frac{Axy^T}{y^T Ax})$,
2. $G = (I - \frac{A^T yx^T}{y^T Ax}), \ \bar{G} = I$,
3. $G = (I - \frac{A^T yx^T}{y^T Ax}), \ \bar{G} = (I - \frac{Axy^T}{y^T Ax})$,

using x and y as in the Wedderburn formula. We then have $\bar{G}AG^T = A - \frac{Axy^T A}{y^T Ax}$, giving the Wedderburn rank one reduction formula (12).

Now, let $\{x_1, \cdots, x_r\}$ and $\{y_1, \cdots, y_r\}$ be vectors associated with the rank reducing process, $H_1 = I$, $R_1 = I$,

$$\bar{G}_i = I - \frac{R_i A H_i^T x_i y_i^T}{y_i^T R_i A H_i^T x_i}, \ R_{i+1} = \bar{G}_i R_i = R_i - \frac{R_i A H_i^T x_i y_i^T R_i}{y_i^T R_i A H_i^T x_i}, \tag{18}$$

and

$$G_i = I - \frac{H_i A^T R_i^T y_i x_i^T}{y_i^T R_i A H_i^T x_i}, \ H_{i+1} = G_i H_i = H_i - \frac{H_i A^T R_i^T y_i x_i^T H_i}{y_i^T R_i A H_i^T x_i}. \tag{19}$$

We can show by induction that

$$A_{k+1} = A_k - \beta_k^{-1} A_k x_k y_k^T A_k = R_{k+1} A H_{k+1}^T, \tag{20}$$

where $\beta_k = y_k^T A_k x_k$, $m_i = R_i^T y_i$ and $n_i = H_i^T x_i$ [17].

Then, we have the following results.

Theorem 7. *The biconjugation process associated with the rank reducing process is a special case of GERRP.*

Theorem 8. *The Wedderburn rank reducing process and the biconjugation process are obtained by a scaled ABS algorithm applied to A or A^T.*

It was shown in [7] that the basic ABS method can be derived from the Wedderburn rank one reducing process. However, we have recently shown that more general ABS algorithms indeed include the Wedderburn rank reducing process [23] as a special case.

Now, exchange the roles of x_i and y_i and apply the above argument to A^T. Then, we have a scaled ABS algorithm resulting in $N^T A^T M = \Omega^T$, a biconjugation process on A^T.

According to Theorem 3, we have the following result at hand.

Corollary 2. *Let $A \in R^{n\times n}$. The biconjugation process produces all possible A-biconjugate pairs (P,V).*

We observe that (18) is a special case of (11). Thus, the biconjugation process is a special case of the scaled extended ABS algorithm on A.

Next, we show that by various choices of the parameters in the extended rank reducing process, several well-known matrix decompositions are obtained.

2.3 Matrix Decompositions

According to Theorem 7, a biconjugation process is a special case of the case (c) in GERRP. In [7], it was shown how to obtain the various factorizations such as *LU*, Cholesky and *QR* by the biconjugation process. Here, we show the parameter settings of GERRP for these factorizations as established by Mahdavi-Amiri and Golpar-Raboky [23].

Theorem 9.

(i) ***(LU factorization)*** *Let $A \in R^{n\times n}$ be strongly nonsingular. The choices $X = Y = I, H_1 = I$ and $R_1 = I$ are well defined and $V^T AP$ gives an LU factorization of A.*
(ii) ***(Cholesky factorization)*** *Let $A \in R^{n\times n}$ be symmetric and positive definite. The Cholesky factorization of A is obtained by letting $X = Y = I$, $H_1 = I$ and $R_1 = I$.*
(iii) ***(QR factorization)*** *Let A have full column rank. The choices $X = H_1 = I_{n,n}$, $R_1 = I_{m,m}$ and $Y = A$ are well defined and we have $V = Q\Psi$, $P = R_1^{-1}$, where Ψ is a diagonal matrix and $P = R_1^{-1} = R^{-1}\Psi$ is an upper triangular matrix. Furthermore, $V^T A = \Psi R$, $AU = V$ and $V^T V = \Psi^2 = \Omega$.*

Proof.

(i) Use Theorem 7 here and Theorem 3.1 in [7].
(ii) Use Theorem 7 here and Theorem 3.3 in [7].
(iii) Use Theorem 7 here and Theorem 3.8 in [7]. □

For computing the singular value decomposition (SVD) of a general matrix, however, the computing process is more complicated, because of the inherent nonlinearity of the problem. The next result provides a characterization of the problem in our context.

Table 1 Matrix factorizations associated with the extended rank reduction formula

A	X	Y	H_1	R_1	Algorithm
$n \times n$ *sn*	I_n	I_n	I_n	I_n	LDM^T
$n \times n$	Arbitrary	X	*ns*	*ns*	Congruence
$n \times n$ *pd-sy*	I_n	I_n	I_n	I_n	Cholesky
$n \times n$	I_n	A	I_n	I_n	QR

Let $A \in R^{m\times n}$ have rank r. Suppose that the vectors $\{x_1,\cdots,x_r\}$ and $\{y_1,\cdots,y_r\}$ are right singular vectors and left singular vectors of A, respectively. In GERRP, let $H_1 = I$, $R_1 = I$ and starting with $i = 1$, inductively let R_{i+1} and H_{i+1} as defined by (18) and (19). Then, GERRP, case(c), produces the SVD factorization of A [7]. There is no simple way to choose the x_i and y_i in computing the SVD factorization. For numerical stability, one can choose the x_i and y_i by solving the following problem:

$$\begin{aligned} &Maximize && y_i^T \bar{A}_i x_i \\ &Subject\ to && x_i^T x_i = 1,\ y_i^T y_i = 1, \end{aligned}$$

where $\bar{A}_i = R_i A H_i^T$ [7].

Summary. We have demonstrated that the proposed extended rank one reducing process provides a general framework for producing various well-known factorizations through the selection of appropriate parameters X and Y. Table 1 gives a summary of some cases. We have assumed that the matrices X and Y satisfy $y_i^T \bar{A}_i x_i \neq 0$ and shown the choices of H_1, R_1, x_i, y_i, z_i and $\bar{z}_i$ for the corresponding factorizations of A. We used the abbreviation *ns* for nonsingular, *pd* for positive definite, *sy* for symmetric, and *sn* for strongly nonsingular.

3 Extended Integer Rank Reduction Formulas

A main result of extended rank reduction process has been the derivation of extended rank reduction formulas for integer matrices; see Golpar-Raboky and Mahdavi-Amiri [17]. The results given for the real case given in Section 2 appropriately hold for the integer case as well.

Theorem 10 (Extended Integer Rank One Reduction Formula). *Let $A \in Z^{m\times n}$ be a nonzero matrix, $\bar{G} \in Z^{m\times m}$ and $G \in Z^{n\times n}$. We have*

$$rank(\bar{G}AG^T) = rank(A) - 1 \tag{21}$$

if and only if one of the following conditions holds:

(i) (left rank reduction) G is unimodular, $dim(N(\bar{G})) = 1$ and there is a vector $x \in Z^n$ so that $\bar{G}^T$ generates the integer null space of $x^T A^T \neq 0$.
(ii) (right rank reduction) $\bar{G}$ is unimodular, $dim(N(G)) = 1$ and there is a vector $y \in Z^m$ so that G^T generates the integer null space of $y^T A \neq 0$.
(iii) (left-right rank reduction) $Dim(N(\bar{G})) = dim(N(G)) = 1$, there are vectors $x \in Z^n$ and $y \in Z^m$ so that $y^T Ax \neq 0$, G^T generates the integer null space of $y^T A$, and $\bar{G}^T$ generates the integer null space of $x^T A^T$.

The following algorithm contains the three cases $(a)-(c)$ associated with the left, right and left-right integer rank reductions. To preserve unimodularity of V and P, only one of the three cases must be used in every iteration. Furthermore, in the left reduction process we let $G_i = I$ and in right reduction process we let $\bar{G}_i = I$.

***Algorithm 3.* General extended integer rank one reducing process (GEIRRP).**

(1) Let $A \in Z^{m\times n}$ have rank r. Start with nonsingular matrices $R_1 \in Z^{m\times m}$ and $H_1 \in Z^{n\times n}$ (bases for the integer null space of the null matrices). Let $i = 1$ and $\bar{A}_1 = R_1 A H_1^T$. Below, choose one of the cases $(a)-(c)$ and execute the case in all iterations.
(2) While $\bar{A}_i \neq 0$ do
execute the chosen case, (a), (b) or (c), in step (1):

   (a) Let $G_i = I$, choose $x_i \in Z^n$ so that $x_1,\cdots,x_i$ is a part of a unimodular matrix and $\bar{A}_i x_i \neq 0$. Choose a matrix $\bar{G}_i$, with $dim(N(\bar{G}_i)) = 1$, so that $\bar{G}_i^T$ generates the integer null space of $x_i^T \bar{A}_i^T$. Choose $y_i \in Z^m$ such that $y_i^T \bar{A}_i x_i = gcd(\bar{A}_i x_i)$.
   (b) Let $\bar{G}_i = I$, choose $y_i \in Z^m$ so that $y_1,\cdots,y_i$ is a part of a unimodular matrix and $y_i^T \bar{A}_i \neq 0$. Choose a matrix G_i, with $dim(N(G_i)) = 1$, so that G_i^T generates the integer null space of $y_i^T \bar{A}_i$. Choose $x_i \in Z^n$ such that $y_i^T \bar{A}_i x_i = gcd(y_i^T \bar{A}_i)$.
   (c) Choose vectors $x_i \in Z^n$ and $y_i \in Z^m$ so that $y_i^T \bar{A}_i x_i = gcd(\bar{A}_i)$, and choose matrices G_i and $\bar{G}_i$, with $dim(N(\bar{G}_i)) = dim(N(G_i)) = 1$, so that G_i^T generates the integer null space of $y_i^T \bar{A}_i$ and $\bar{G}_i^T$ generates the integer null space of $x_i^T \bar{A}_i^T$.

   Let $H_{i+1} = G_i H_i$, $R_{i+1} = \bar{G}_i R_i$, $\bar{A}_{i+1} = R_{i+1} A H_{i+1}^T, v_i = R_i^T y_i, p_i = H_i^T x_i$ and $i = i+1$.
   End While.
(3) Stop.

The properties of the extended integer rank reduction formula and GEIRRP are similar to the ones for the real case presented in Section 2.

We next show that GEIRRP produces the so-called Smith normal form, making use of simultaneous rank reduction transformations on A and A^T.

Smith [25] proved that any integer matrix A with rank r can be transformed by elementary row and column operations into the Smith normal form. Every matrix $A \in Z^{m\times n}$ of rank r is equivalent to a diagonal matrix Λ, given by

$$\Lambda = \begin{pmatrix} \Sigma & 0 \\ 0 & 0 \end{pmatrix}, \ \Sigma = diag(\lambda_{1,1}, \cdots, \lambda_{r,r}),$$

where $\lambda_{i,i} \neq 0$, $i = 1, \cdots, r$, and $\lambda_{1,1} \mid \cdots \mid \lambda_{r,r}$, ($\lambda \mid \beta$ means λ divides β). Moreover, the $\lambda_{i,i}$ are unique, apart from their signs. The $\lambda_{i,i}$ are known as *invariant factors* of A.

Definition 4. $A \in Z^{n\times n}$ is a unimodular matrix if and only if $|det(A)| = 1$.

Note that if A is unimodular, then A^{-1} is also unimodular.

Theorem 11. *Let $A \in Z^{m\times n}$ have rank r. There exist unimodular matrices $U \in Z^{n\times n}$ and $V \in Z^{m\times m}$ such that $\Lambda(A) = (\lambda_{i,j}) = V^T AU$ is a diagonal matrix in the* ***Smith normal form (elementary divisor normal form)****; that is, $\lambda_{i,j} = 0$, for $i \neq j$ and for $i = j > r$, $\lambda_{i,i} > 0$, for $1 \leq i \leq r$, and $\lambda_{1,1} \mid \lambda_{2,2} \mid \cdots \mid \lambda_{r,r}$. The diagonal elements (****elementary divisors****) of $\Lambda(A)$ are uniquely determined.*

Now, we are ready to present an algorithm for computing the Smith normal form using GEIRRP.

Algorithm 4. **The Smith normal form using GEIRRP (SNF-GEIRRP).**

(1) Let $A \in Z^{m\times n}$ have rank r. Start with nonsingular matrices $R_1 \in Z^{m\times m}$ and $H_1 \in Z^{n\times n}$ (bases for the integer null spaces of the null matrices). Let $i = 1$ and $\bar{A}_1 = R_1 A H_1^T$.
(2) While $\bar{A}_i \neq 0$ do
Choose the vectors $x_i \in Z^n$ and $y_i \in Z^m$ so that $y_i^T \bar{A}_i x_i = gcd(\bar{A}_i)$. Choose the matrices G_i and $\bar{G}_i$, with $dim(N(\bar{G}_i)) = dim(N(G_i)) = 1$, so that G_i^T generates the integer null space of $y_i^T \bar{A}_i$ and $\bar{G}_i^T$ generates the integer null space of $x_i^T \bar{A}_i^T$. Let

$$H_{i+1} = G_i H_i, \ R_{i+1} = \bar{G}_i R_i, \ \bar{A}_{i+1} = R_{i+1} A H_{i+1}^T,$$
$$v_i = R_i^T y_i, p_i = H_i^T x_i, \ \lambda_{i,i} = v_i^T A p_i, \tag{22}$$

and $i = i + 1$.
End While.
(3) Let $r = i - 1$, $V = (v_1, \cdots, v_r)$, $P = (p_1, \cdots, p_r)$. Configure the Smith normal form of A as:

$$V^T AP = \Omega = (\lambda_{1,1}, \cdots, \lambda_{r,r}).$$

(4) Stop.

We note that for the Smith normal form, in step (2) above we need to solve a quadratic Diophantine equation of the form $x^T Ay = gcd(A)$. In Section 4, we will give two algorithms for solving such equations. In the remainder of this section, we show that GEIRRP includes both the integer ABS and integer Wedderburn processes as special cases.

3.1 Producing the Integer ABS Algorithms

The ABS algorithms have been appropriated to derive the ABS class of algorithms for linear Diophantine equations by Esmaeili, Mahdavi-Amiri and Spedicato [11] and extended to the scaled integer ABS algorithms [28]. Each integer ABS algorithm decides if the Diophantine system has an integer solution, and, if so, obtains a particular solution along with an integer matrix with possibly dependent rows generating the integer null space of the equations. In a recent work, Khorramizadeh and Mahdavi-Amiri [22] have also presented a new class of extended integer ABS algorithms for solving linear Diophantine systems by computing an integer basis for the null space to control the growth of intermediate results. They also showed that Rosser's algorithm [24] and its generalization belong to the integer ABS class of algorithms [21].

Consider the following linear system,

$$Ay = b, \qquad y \in Z^n,\ A \in Z^{m\times n},\ b \in Z^m,\ m \le n, \tag{23}$$

which is equivalent to the following scaled system,

$$V^T Ay = V^T b, \tag{24}$$

where V is an arbitrary m by m unimodular matrix.

An integer ABS algorithm starts with a unimodular matrix H_1, with its rows generating the integer null space of the null matrix. If H_i is an integer matrix and $w_i \in Z^n$ is so that $w_i^T H_i a_i$ divides all components of $H_i a_i$, then it is clear that H_{i+1} defined by (9) is an integer matrix. Conditions for the existence of an integer solution and determination of all integer solutions of a linear Diophantine system were characterized, using the integer ABS algorithms, in [10, 11].

We next present the integer ABS class of algorithms (note that below, r_i gives the rank of the first $i-1$ rows of A).

Algorithm 5. **The integer ABS algorithm (IABS).**

(1) Choose $H_1 \in Z^{n\times n}$, arbitrary and unimodular. Let i=1, and $r_i = 0$.
(2) Compute $s_i = H_i a_i$.
(3) **If** ($s_i = 0$) **then** let $H_{i+1} = H_i$, $r_{i+1} = r_i$ and **go to** (6) (the ith equation is redundant).
(4) $\{s_i \neq 0\}$ Compute $\delta_i = gcd(s_i)$ and $p_i = H_i^T x_i$, where $x_i \in Z^n$ is an arbitrary integer vector satisfying $s_i^T x_i = \delta_i$.
(5) Update H_i by

$$H_{i+1} = H_i - \frac{H_i a_i w_i^T H_i}{w_i^T H_i a_i},$$

where $w_i \in Z^n$ is an arbitrary integer vector satisfying $s_i^T w_i = \delta_i$, and let $r_{i+1} = r_i + 1$.

(6) **If** $i = m$ **then Stop** (r_{m+1} is the rank of A) **else** let $i = i+1$ and **go to** (2).

In [16], we presented a new extended integer ABS (EIABS) class of algorithms computing an integer basis for the integer null space of an integer matrix. The method updates the so-called Abaffian matrices H_i as follows:

- $H_{i+1} = G_i H_i$, where, $G_i \in Z^{j_{i+1} \times j_i}$ generates the integer null space of the vector $s_i = H_i a_i$.

Similar to the real case discussed in Section 2, we have the following results.

Theorem 12. *The scaled extended integer ABS (SEIABS) class of algorithms on A and A^T can both be derived by GEIRRP.*

In [16], we showed how to choose the parameters of a scaled integer ABS algorithm for computing the Smith normal form.

Theorem 13. *The parameters of the SEIABS algorithm can be chosen so that the algorithm generates the Smith normal form of an integer matrix.*

Proof. See [16].

3.2 Producing the Integer Wedderburn Rank Reducing Process

Here, we discuss a special case of the Wedderburn rank reduction formula for integer matrices, as established by Golpar-Raboky and Mahdavi-Amiri [17] .

Theorem 14. *Let $A \in Z^{m\times n}, x \in Z^n$ and $y \in Z^m$ so that $0 \neq \beta = y^T Ax$ and $\beta \mid gcd(Ax)gcd(y^T A)$. The matrix*

$$B = A - \beta^{-1} Axy^T A \tag{25}$$

is an integer matrix and has rank exactly one less than that of A.

Remark 3. There are some cases in which x and y can be found easily. Let $y^T Ax = gcd(Ax)$ or $y^T Ax = gcd(y^T A)$. Then, x and y satisfy the divisibility condition $y^T Ax \mid gcd(Ax)gcd(y^T A)$. In [16], we presented two algorithms for solving the quadratic Diophantine equation $y^T Ax = gcd(A)$.

Now, by repeatedly applying the rank reduction formula (25), an integer rank reducing process is obtained as follows. Let $A \in Z^{m\times n}$ have rank r. Let $A_1 = A$. While $A_k \neq 0$, apply (25) repeatedly to generate a sequence of matrices $\{A_k\}$ by using

$$A_{k+1} = A_k - \beta_k^{-1} A_k x_k y_k^T A_k = (I - \beta_k^{-1} A_k x_k y_k^T) A_k = A_k (I - \beta_k^{-1} x_k y_k^T A_k), \tag{26}$$

for any vectors $x_k \in Z^n$ and $y_k \in Z^m$, for which, $\beta_k = y_k^T A_k x_k \mid gcd(A_k x_k) gcd(y_k^T A_k)$.

Integer Biconjugation Process: Let $\{x_1, \cdots, x_\gamma\}$ and $\{y_1, \cdots, y_r\}$ be integer vectors so that $y_k^T A_k x_k = gcd(A_k)$, for $k = 1, \cdots, r$. Then,

$$n_k = x_k - \sum_{i=1}^{k-1} \frac{< x_k, m_i >}{< n_i \ , \ m_i >} n_i, \tag{27}$$

$$m_k = y_k - \sum_{i=1}^{k-1} \frac{< n_i, y_k >}{< n_i \ , \ m_i >} m_i, \tag{28}$$

are well-defined integer vectors, for $k = 1, \ldots, r$. Furthermore, any matrix pair (X, Y) of an integer rank reducing process can be transformed to a biconjugate pair (N, M) with

$$\Omega = M^T AN, \tag{29}$$

as an integer diagonal matrix.

A main problem arising from the integer biconjugation process is finding integer vectors x_k and y_k as a solution set of the quadratic Diophantine equation $y_k^T A_k x_k = gcd(A_k)$. We will present two algorithms for solving such equations in Section 4.

The next two results can readily be established (see [17]).

Theorem 15. *The integer biconjugation process associated with the rank reducing process defined by (27) and (28) is a special case of GEIRRP.*

Corollary 3. *The integer biconjugation process is obtained by the scaled integer ABS class of algorithms applied to A or A^T.*

Let us now choose the parameters of the biconjugation process defined by (18) based on Algorithm 4. Then, the integer biconjugation process computes the Smith normal form specially without the need to compute the matrices G_i, $\bar{G}_i$, R_i and H_i as given in Algorithm 6 below.

Algorithm 6. **Smith normal form using the integer biconjugation process (SNF-IBP).**

(1) Let $i = 1$ and $A_1 = A$.
(2) While $A_i \neq 0$ do
Choose the vectors $x_i \in Z^n$ and $y_i \in Z^m$ so that $\beta_i = y_i^T A_i x_i = gcd(A_i)$, and compute

$$n_i = x_i - \sum_{k=1}^{i-1} \frac{< x_i, m_k >}{< n_k \ , \ m_k >} n_k, \ m_i = y_i - \sum_{k=1}^{i-1} \frac{< n_k, y_i >}{< n_k \ , \ m_k >} m_k.$$

Let $A_{i+1} = A_i - \beta^{-1} A_i x_i y_i^T A_i$ and $i = i + 1$.
End While.

(3) Let $r = i - 1$, $M = (m_1, \cdots, m_r)$, $N = (n_1, \cdots, n_r)$. Configure the Smith normal form of A as:

$$M^T AN = \Omega = (\beta_1, \cdots, \beta_r).$$

(4) Stop.

4 Solving Quadratic Diophantine Equations

Let $A \in Z^{m\times n}$, $b \in Z$, and consider the quadratic equation,

$$x^T Ay = \sum_{i=1}^{m} \sum_{j=1}^{n} x_i a_{i,j} y_j = b,\ x \in Z^m,\ y \in Z^n,\ b \in Z, \tag{30}$$

where $x \in Z^m$ and $y \in Z^n$ are solution vectors to be found. In [16], we showed that (30) has integer solutions if and only if $gcd(A) \mid b$.

The quadratic equation (30) is equivalent to a linear Diophantine system and a single equation as follows:

$$\begin{cases} Ay = d,\ gcd(A) = gcd(d) & (a) \\ d^T x = b & (b). \end{cases}$$

Next, we present two algorithms for solving (30). These algorithms would first compute an integer vector y so that $gcd(Ay) = gcd(A)$, and set $Ay = d$. Then, x is computed as a solution of the single linear Diophantine equation, $x^T Ay = x^T d = b$ (this computation may be done by Rosser's approach [21, 24]).

4.1 *Divisibility Sequence Approach*

Let $A \in Z^{m\times n}$ have row rank r. One can always obtain a set of vectors $b_i \in Z^n$, $1 \le i \le n$, as an integer basis for Z^n, a divisible sequence $m_1, \cdots, m_r$ so that $m_i \mid m_{i+1}$, $1 \le i \le r-1$ and $c_i = m_i b_i$, $1 \le i \le r$, as a basis for the integer row space of A (integer range of A^T). The c_i form a divisibility sequence. A constructive proof for this claim exists that provides the m_i, b_i and hence c_i (see [5]); the m_i and the c_i can be obtained by elementary operations. We can now present the algorithm using the divisibility sequence.

Algorithm 7. Solution of the quadratic equation by the divisibility sequence (QEDS) approach.
Input: $A \in Z^{m\times n}$ with row rank r, c_1 (the first element of a divisibility sequence c_i, $1 \le i \le r$, as the basis for the integer range of A^T) and $b \in Z^m$.

(1) **If** $gcd(A) \not| b$ **then** declare that the quadratic equations lacks integer solution and **stop**.
(2) Compute y so that $c_1^T y = gcd(c_1)$.
(3) Compute $k = Ay$.
(4) Solve the single Diophantine equation $k^T x = b$ (x and y are the solution vectors for $x^T Ay = b$) **else** declare that the quadratic equations lacks integer solution.
(5) **Stop**.

4.2 *The Integer ABS Approach Based on a Conjecture*

The quadratic equation is solved by an integer ABS algorithm, named QEIABS, with the intention of controlling the growth of intermediate results (see [22]). Using our given conjecture based on the Dirichlet's Theorem [12], a method is developed based on recently proposed integer ABS algorithms [16].

Algorithm 8. Solution of the quadratic equation by the integer ABS (QEIABS) algorithms.

(1) **If** $gcd(A) \not| b$ **then** declare that the quadratic equations lacks integer solution and **stop**.
(2) Choose $y_1 \in Z^n$, arbitrary, $H_1 \in Z^{n\times n}$, arbitrary and unimodular. Let $i = 1$.
(3) Compute $s_i = H_i a_i$.
(4) **If** $s_i = 0$ **then** let $y_{i+1} = y_i$, $H_{i+1} = H_i$, $d_i = a_i^T y_i$ and **go to** (9).
(5) $\{s_i \neq 0\}$ Compute $\delta_i = gcd(s_i)$ and $p_i = H_i^T z_i$, with $z_i \in Z^n$ any arbitrary integer vector satisfying $s_i^T z_i = \delta_i$. Compute $w_i \in Z^n$ an integer vector satisfying $s_i^T w_i = \delta_i$.
(6) **If** i=1 **then** set $d_1 = a_1^T y_1$ **else** compute the smallest integer number k_i among $k_i = 0, 1, \ldots, |d_{l_i}| - 1$, such that
$$gcd(k_i.\delta_i + a_i^T y_i, d_{l_i}) = gcd(\delta_i, a_i^T y_i, d_{l_i}),$$
where $1 \le l_i < i$ is the largest index such that $d_{l_i} \neq 0$ and set $d_i = k_i\delta_i + a_i^T y_i$.
(7) Set $\alpha_i = k_i$. Compute all pairs of integer numbers λ_i and θ_i so that $\alpha_i = \lambda_i\theta_i$ and compute $t_i = y_i + \theta_i H_i^T z_i$, where $s_i^T z_i = \lambda_i\delta_i$. Choose t_i with a minimal value of $\|t_i\|_2$, and let $y_{i+1} = t_i$.
(8) Update H_i by
$$H_{i+1} = H_i - H_i a_i w_i^T H_i / \delta_i.$$
(9) **If** $i < m$ **then** set $i = i + 1$ and **go to** (2) **else** y_{m+1} is a solution.
(10) Compute an integer vectors $x \in Z^n$ such that $d^T x = b$.
(11) **Stop**.

5 Numerical Experiments

Here, we generate some random matrices of various size for testing the QEDS and QEIABS algorithms. We also compute the Smith normal form of the generated matrices and compare the results with the ones obtained by Maple. We will see that the QEIABS algorithm outperforms the QEDS algorithm in controlling the size of the obtained solutions, and the components of U and V in the Smith normal form by our algorithm are more balanced than the ones obtained by Maple.

The random integer matrices are generated by Maple's

$$RandomMatrix(m,n,generator = -2^{BL}..2^{BL}).$$

Table 2 shows the numerical results after finding the Smith normal form of A for the randomly generated problems of increasing size. In this table, *TN* refers to test number, m, n and BL are as defined above for the random integer matrices of Maple, BL-QEIABS and BL-QEDS give the maximum bit lengths of the vectors x and y computed by the QEIABS and QEDS algorithms, respectively.

We implemented the QEIABS algorithm, starting with y_1 as the zero vector and H_1 as the identity matrix. The algorithms were implemented using Maple 9.5 and the programs were executed on a Pentium 4 having 2.4 GHz processor and 5.12 MB storage. The results in Table 2 show that in most of the test problems, the QEIABS algorithm outperforms the QEDS algorithm significantly by having smaller mean values of the bit lengths of the obtained results. In the few instances, on the contrary, however, the outperformance of QEDS algorithm over QEIABS algorithm is not significant.

The results for the Smith normal forms obtained for the same problems 1–7 of Table 2 using Algorithm 8 and the procedure *ismith* from Maple are given in Table 3. The headings $\|V_{SNF}\|_2$ and $\|U_{SNF}\|_2$ are Euclidean norms of V and U, respectively, obtained by Algorithm QEIABS, and $\|V_{Maple}\|_2$ and $\|U_{Maple}\|_2$ are Euclidean norms of V and U, respectively, obtained by *ismith*. The numerical results show that Algorithm QEIABS generates a more balanced U and V as compared to the ones obtained by *ismith*.

Table 2 Comparative results for the QEIABS and QEDS algorithms

TN	*m*	*n*	*BL*	BL-QEIABS	BL-QEDS
1	5	5	13	7	13
2	7	8	9	9	11
3	8	16	4	2	1
4	9	9	11	7	12
5	10	12	11	3	2
6	12	17	3	4	3
7	19	17	4	7	9

Table 3 Euclidean norms of unimodular matrices V and U produced by Algorithm 8 and the *ismith* procedure of Maple

TN	m	n	$\|V_{SNF}\|_2$	$\|V_{Maple}\|_2$	$\|U_{SNF}\|_2$	$\|U_{Maple}\|_2$
1	5	5	2.3604E+006	0.2912E+014	3.1361E+006	0.3138E+009
2	7	8	3.0321E+006	0.6310E+07	2.0673E+004	0.8901E+011
3	8	16	899.8250	67.4070	135.4543	4,799.9348
4	9	9	3.3391E+004	0.1823E+008	2.0802E+005	0.6753E+008
5	10	12	119.8925	130.596	66.4436	1,365.5846
6	12	17	1.1157E+005	967.5534	179.9770	28,673.7232
7	19	17	1.8384E+006	0.3090E+008	1.5038E+007	28,262.9158

6 Conclusions

We presented extended rank reduction formulas transforming row and columns of a matrix A. Using these formulas, we proposed a general extended rank reducing process and developed a general finite iterative approach giving a unified treatment of various iterative procedures occurring in matrix factorizations for A and A^T. The Wedderburn rank reduction process and the scaled extended ABS class of algorithms were shown to be special cases of the general approach here. We also showed a new general result that the biconjugate decomposition associated with the Wedderburn rank reducing process belonged to the scaled ABS class of algorithms. We presented extended integer rank reduction formulas and the associated general extended integer rank reducing process (GEIRRP). Moreover, we presented the integer Wedderburn rank reduction formula and its integer biconjugation process. Then, we showed that the integer biconjugate process associated with the integer Wedderburn rank reducing process belonged to the scaled integer ABS class of algorithms, and hence was a special case of GEIRRP. We computed the Smith normal form using GEIRRP as well as the scaled integer extended ABS algorithms and the integer biconjugation process. For the Smith normal form, having the need to solve a quadratic Diophantine equation, we presented two algorithms for solving such equations. The first algorithm makes use of a special integer basis for the row space of the matrix, and the second one, with the intention of controlling the growth of intermediate results and making use of our given conjecture, is based on a recently proposed integer ABS algorithm.

References

1. Abaffy, J., Galantai, A.: Conjugate direction methods for linear and nonlinear systems of algebraic equations. Colloq. Math. Soc. Janos Bolyai **50**, 481–502 (1986)
2. Abaffy, J., Spedicato, E.: ABS Projection Algorithms, Mathematical Techniques for Linear and Nonlinear Equations. Halsted Press, Chichester (1989)

3. Abaffy, J., Broyden, C.G., Spedicato, E.: A class of direct methods for linear systems. Numer. Math. **45**, 361–376 (1984)
4. Adib, M., Mahdavi-Amiri, N., Spedicato, E.: Broyden method as an ABS algorithm. Publ. Univ. Miskolc Ser. D Nat. Sci., Math. **40**, 3–13 (1999)
5. Cassels, J.W.S.: Rational Quadratic Forms. Academic, New York (1979)
6. Chen, Z., Deng, N.Y., Xue, Y.: A general algorithm for underdetermined linear systems. In: The Procedings of the First International Conference on ABS Algorithms, pp. 1–13 (1992)
7. Chu, M.T., Funderlic, R.E., Golub, G.H.: A rank one reduction formula and its applications to matrix factorizations. SIAM Rev. **37**(4), 512–530 (1995)
8. Cline, R.E., Funderlic, R.E.: The rank of a difference of matrices and associated generalized inverse. Linear Algebra Appl. **24**, 185–215 (1979)
9. Egervary, E.: On rank-diminshing operators and their applications to the solution of linear equations. Z. Angew. Math. Phys. **11**, 376–386 (1960)
10. Esmaeili, H., Mahdavi-Amiri, N., Spedicato, E.: Generating the integer null space and conditions for determination of an integer basis using the ABS algorithms. Bull. Iran. Math. Soc. **27**(1), 1–18 (2001)
11. Esmaeili, H., Mahdavi-Amiri, N., Spedicato, E.: A class of ABS algorithms for linear Diophantine systems. Numer. Math. **90**, 101–115 (2001)
12. Frumkin, M.A.: An application of modular arithmetic to the constuction of algorithms for solving systems of linear equations. Soviet Math. Dok. **17**, 1165–1168 (1976)
13. Galantai, A.: Rank reduction, factorization and conjugation. Linear Multilinear Algebra **49**, 195–207 (2001)
14. Galantai, A.: Rank reduction and borderd inversion. Univ. Miskolc Math. Notes **2**(2), 117–126 (2001)
15. Galantai, A.: The rank reduction procedure of Egervary. CEJOR **18**(1), 5–24 (2010)
16. Golpar-Raboky, E., Mahdavi-Amiri, N.: Diophantine quadratic equation and Smith normal form using scaled extended integer ABS algorithms. J. Optim. Theory Appl. **152**(1), 75–96 (2012)
17. Golpar-Raboky, E., Mahdavi-Amiri, N.: Extended integer rank reduction formulas containing Wedderburn and Abaffy-Broyden-Spedicato rank reducing processes. Linear Multilinear Algebra **61**(12), 1641–1659 (2013)
18. Guttman, L.: General theory and methods for matric factoring. Psychometrika **9**, 1–16 (1944)
19. Guttman, L.: A necessary and sufficient formula for matrix factoring. Psychometrika **22**, 79–91 (1957)
20. Householder, A.S.: The Theory of Matrices in Numerical Analysis. Blasidell/Dover, New York/New York (1964/1975)
21. Khorramizadeh, M., Mahdavi-Amiri, N.: On solving linear Diophantine systems using generalized Rosser's algorithm. Bull. Iran. Math. Soc. **34**(2), 1–25 (2008)
22. Khorramizadeh, M., Mahdavi-Amiri, N.: Integer extended ABS algorithms and possible control of intermediate results for linear Diophantine systems. 4OR **7**, 145–167 (2009)
23. Mahdavi-Amiri, N., Golpar-Raboky, E.: Extended rank reduction formulas containing Wedderburn and Abaffy-Broyden-Spedicato rank reducing processes. Linear Algebra Appl. **439**, 3318–3331 (2013)
24. Rosser, J.B.: A note on the linear Diophantine equation. Am. Math. Mon. **48**, 662–666 (1941)
25. Smith, H.J.S.: On systems of linear indeterminate equations and congruences. Phil. Trans. R. Soc. Lond. **151**, 293–326 (1861)
26. Spedicato, E., Xia, Z. Zhang, L.: The implicit LX method of the ABS class. Optim. Methods Softw. **8**, 99–110 (1997)
27. Spedicato, E., Bodon, E., Del Popolo, A., Xia, Z.: ABS algorithms for linear systems and optimization: a review and a bibliography. Ric. Oper. **29**, 39–88 (2000)
28. Spedicato, E., Bodon, E., Del Popolo, A. Mahdavi-Amiri, N.: ABS methods and ABSPACK for linear systems and optimization: a review. 4OR **1**, 51–66 (2003)

29. Spedicato, E., Bodon, E., Zunquan, X., Mahdavi-Amiri, N.: ABS methods for continous and integer linear equations and optimization. CEJOR **18**, 73–95 (2010)
30. Wedderburn, J.H., Lectures on Matrices, Colloquium Publications, vol. XVII. American Mathematical Society/Dover, New York/New York (1934/1964)

Distributed Block Coordinate Descent for Minimizing Partially Separable Functions

Jakub Mareček, Peter Richtárik, and Martin Takáč

Abstract A distributed randomized block coordinate descent method for minimizing a convex function of a huge number of variables is proposed. The complexity of the method is analyzed under the assumption that the smooth part of the objective function is partially block separable. The number of iterations required is bounded by a function of the error and the degree of separability, which extends the results in Richtárik and Takác (Parallel Coordinate Descent Methods for Big Data Optimization, Mathematical Programming, DOI:10.1007/s10107-015-0901-6) to a distributed environment. Several approaches to the distribution and synchronization of the computation across a cluster of multi-core computer are described and promising computational results are provided.

Keywords Distributed coordinate descent • Empirical risk minimization • Support vector machine • Big data optimization • Partial separability • Huge-scale optimization • Iteration complexity • Expected separable over-approximation • Composite objective • Convex optimization • Communication complexity

1 Introduction

With the ever increasing availability of data comes the need to solve ever larger instances of problems in data science and machine learning, many of which turn out to be convex optimization problems of enormous dimensions. A single machine is often unable to store the complete data in its main memory. This suggests the

J. Mareček
IBM Research – Ireland, Dublin, Ireland
e-mail: jakub.marecek@ie.ibm.com

P. Richtárik (✉)
School of Mathematics, University of Edinburgh, Edinburgh, UK
e-mail: peter.richtarik@ed.ac.uk

M. Takáč
Department of Industrial & Systems Engineering, Lehigh University, Bethlehem, PA, USA
e-mail: takac.mt@gmail.com

M. Al-Baali et al. (eds.), *Numerical Analysis and Optimization*, Springer Proceedings in Mathematics & Statistics 134, DOI 10.1007/978-3-319-17689-5_11

need for efficient algorithms, which can benefit from distributing the data and computations across many computers.

In this paper, we study optimization problems of the form:

$$\min_{x\in\mathbf{R}^N} \left[F(x) := f(x) + \Omega(x)\right], \tag{1}$$

where f is a smooth, convex and partially block separable function, and Ω is a possibly non-smooth, convex, block separable, and "simple" extended real valued function. The technical definitions of these terms are given in Section 2.

1.1 Contributions

We propose and study the performance of a *distributed block coordinate descent method* applied to problem (1).

In our method, the blocks of coordinates are first partitioned among C computers of a cluster. Likewise, data associated with these blocks are partitioned accordingly and stored in a distributed way. In each of the subsequent iterations, each computer chooses τ blocks out of those stored locally, uniformly at random. Then, each computer computes and applies an update to the selected blocks, in parallel, out of information available to it locally. An update, which happens to be the residual in data-fitting problems, is then transmitted to other computers, which receive it either by the beginning of the next iteration or at some later time. In the former case, we denote the methods "sychronous" and we analyse them in detail. In the latter case, we denote the methods "asynchronous" and we include them for the sake of comparison in Section 7.

The main contributions of this paper are, in no particular order:

1. *Partial separability.* This is the first time such a distributed block-coordinate descent method is analyzed under the assumption that f is partially separable.
2. *New step-length.* Our method and analyse is based on an expected separable overapproximation (ESO) inequality for partially separable functions and distributed samplings in Theorem 4 in Section 4. The length of the step we take in each iteration is given by the optimum of this ESO.
3. *Iteration complexity.* We show that the iteration complexity of the method depends on the degree of block separability of f: the more separable the instance, the fewer iterations the method requires. The complexity results are stated in two theorems in Section 5 and are of the order of $O(\log(1/\varepsilon))$ for strongly convex F and $O(1/\varepsilon)$ for general convex F. At the same time, the separability also reduces the run-time per iteration.
4. *Efficient implementation.* When we replace the natural synchronous communications between computers, as analysed in Section 5, with asynchronous communication, we obtain a major speed-up in the computational performance. An efficient open-source implementation of both synchronous and asynchronous methods is available as part of the package http://www.code.google.com/p/ac-dc/.

Our method and results are valid not only for a cluster setting, where there really are C computers which do not share any memory, and hence have to communicate by sending messages to each other, but also for computers using the Non-Uniform Memory Access (NUMA) architecture, where the memory-access time depends on the memory location relative to a processor, and accessing local memory is much faster than accessing memory elsewhere. NUMA architectures are increasingly more common in multi-processor machines.

1.2 Related Work

Before we proceed, we give a brief overview of some existing literature on coordinate descent methods. For further references, we refer the reader to [3, 5, 26].

Block-Coordinate Descent. Block-coordinate descent is a simple iterative optimization strategy, where two subsequent iterates differ only in a single block of coordinates. In a very common special case, each block consists of a single coordinate. The choice of the block can be deterministic, e.g., cyclic [30], greedy [25], or randomized. Recent theoretical guarantees for randomized coordinate-descent algorithms can be found in [6, 12, 16, 19, 21, 29]. Coordinate descent algorithms are also closely related to coordinate relaxation, linear and non-linear Gauss-Seidel methods, subspace correction, and domain decomposition (see [2] for references). For classical references on non-randomized variants, we refer to the work of Tseng [17, 40–42].

Parallel Block-Coordinate Descent. Clearly, one can parallelize coordinate descent by updating several blocks in parallel. The related complexity issues were studied by a number of authors. Richtárik and Takáč studied a broad class of parallel methods for the same problem we study in this paper, and introduced the concept of ESO [26]. The complexity was improved by Tappenden et al. [39]. An efficient accelerated version was introduced by Fercoq and Richtárik [5] and an inexact version was studied in [37]. An asynchronous variant was studied by Liu et al. [15]. A non-uniform sampling and a method for dealing with non-smooth functions were described in [28] and [6], respectively. Further related work can be found in [22, 32, 38, 43].

Distributed Block-Coordinate Descent. Distributed coordinate descent was first proposed by Bertsekas and Tsitsiklis [3]. The literature on this topic was rather sparse, c.f. [9], until the research presented in this paper raised the interest, which lead to the analyses of Richtárik and Takáč [27] and Fercoq et al. [7]. These papers do not consider blocks, and specialize our results to convex functions admitting a quadratic upper bound.

In the machine-learning community, distributed algorithms have been studied for particular problems, e.g., training of support vector machines (SVM) [31]. Google [4] developed a library called PSVM, where parallel row-based incomplete Cholesky factorization is employed in an interior-point method. A MapReduce-based distributed algorithm for SVM was found to be effective in automatic image annotation [1]. Nevertheless, none of these papers use coordinate descent.

2 Notation and Assumptions

In this section, we introduce the notation used in the rest of the paper and state our assumptions formally. We aim to keep our notation consistent with that of Nesterov [21] and Richtárik and Takáč [26].

Block Structure. We decompose $\mathbf{R}^N$ into n subspaces as follows. Let $U \in \mathbf{R}^{N\times N}$ be the $N \times N$ identity matrix and further let $U = [U_1, U_2, \dots, U_n]$ be a column decomposition of U into n submatrices, with U_i being of size $N \times N_i$, where $\sum_i N_i = N$. It is easy to observe that any vector $x \in \mathbf{R}^N$ can be written uniquely as $x = \sum_{i=1}^n U_i x^{(i)}$, where $x^{(i)} \in \mathbf{R}^{N_i}$. Moreover, $x^{(i)} = U_i^T x$. In view of the above, from now on we write $x^{(i)} := U_i^T x \in \mathbf{R}^{N_i}$, and call $x^{(i)}$ the *block* i of x.

Projection onto a Set of Blocks. Let us denote $\{1,2,\dots,n\}$ by $[n]$, a set of blocks $S \subseteq [n]$, $x \in \mathbf{R}^N$, and let $x_{[S]}$ be the vector in $\mathbf{R}^N$ whose blocks $i \in S$ are identical to those of x, but whose other blocks are zeroed out. Block-by-block, we thus have $(x_{[S]})^{(i)} = x^{(i)}$ for $i \in S$ and $(x_{[S]})^{(i)} = 0 \in \mathbf{R}^{N_i}$, otherwise. It will be more useful to us however to write

$$x_{[S]} := \sum_{i\in S} U_i x^{(i)}, \tag{2}$$

where we adopt the convention that if $S = \emptyset$, the sum is equal $0 \in \mathbf{R}^N$.

Norms. Spaces $\mathbf{R}^{N_i}$, $i \in [n]$, are equipped with a pair of conjugate norms: $\|t\|_{(i)}$ and $\|t\|_{(i)}^* := \max_{\|s\|_{(i)}\le 1} \langle s,t\rangle$, $t \in \mathbf{R}^{N_i}$. For $w \in \mathbf{R}^n_{>0}$, where $\mathbf{R}_{>0}$ is a set of positive real numbers, define a pair of conjugate norms in $\mathbf{R}^N$ by

$$\|x\|_w = \left[\sum_{i=1}^n w_i \|x^{(i)}\|_{(i)}^2\right]^{1/2}, \quad \|y\|_w^* := \max_{\|x\|_w\le 1} \langle y,x\rangle = \left[\sum_{i=1}^n w_i^{-1}(\|y^{(i)}\|_{(i)}^*)^2\right]^{1/2}. \tag{3}$$

We shall assume throughout the paper that f has the following properties.

Assumption 1 (Properties of f). *Function $f : \mathbf{R}^N \to \mathbf{R}$ satisfies:*

1. ***Partial separability.*** *Function f is of the form*

$$f(x) = \sum_{J\in\mathscr{J}} f_J(x), \tag{4}$$

where $\mathscr{J}$ is a collection of subsets of $[n]$ and function f_J depends on x through blocks $x^{(i)}$ for $i \in J$ only. The quantity $\omega := \max_{J\in\mathscr{J}} |J|$ is the degree of separability of f.

2. ***Convexity.*** *Functions f_J, $J \in \mathscr{J}$ in (4) are convex.*

3. ***Smoothness.*** *The gradient of f is block Lipschitz, uniformly in x, with positive constants $L_1,\dots,L_n$. That is, for all $x \in \mathbf{R}^N$, $i \in [n]$ and $t \in \mathbf{R}^{N_i}$,*

$$\|\nabla_i f(x+U_i t) - \nabla_i f(x)\|_{(i)}^* \leq L_i \|t\|_{(i)}, \tag{5}$$

where $\nabla_i f(x) := (\nabla f(x))^{(i)} = U_i^T \nabla f(x) \in \mathbf{R}^{N_i}$.

A few remarks are in order:

1. Note that every function f is trivially of the form (4): we can always assume that $\mathscr{J}$ contains just the single set $J=[n]$ and let $f_J = f$. In this case we would have $\omega = n$. However, many functions appearing in applications can naturally be decomposed as a sum of a number of functions each of which depends on a small number of blocks of x only. That is, many functions have degree of separability ω that is much smaller than n.
2. Note that since f_J are convex, so is f. While it is possible to remove this assumption and provide an analysis in the non-convex case, this is beyond the scope of this paper.
3. An important consequence of (5) is the following standard inequality [20]:

$$f(x+U_i t) \leq f(x) + \langle \nabla_i f(x), t\rangle + \tfrac{L_i}{2}\|t\|_{(i)}^2. \tag{6}$$

Assumption 2 (Properties of Ω). *We assume that $\Omega : \mathbf{R}^N \to \mathbf{R} \cup \{+\infty\}$ is (block) separable, i.e., that it can be decomposed as follows:*

$$\Omega(x) = \sum_{i=1}^{n} \Omega_i(x^{(i)}), \tag{7}$$

where the functions $\Omega_i : \mathbf{R}^{N_i} \to \mathbf{R} \cup \{+\infty\}$ are convex and closed.

3 Distributed Block Coordinate Descent Method

In this section we describe our distributed block coordinate descent method (Algorithm 1). It is designed to solve convex optimization problems of the form (1), where the data describing the instance are so large that it is impossible to store these in memory of a single computer.

Pre-processing. Before the method is run, the set of blocks is partitioned into C sets $P^{(c)}$, $c = 1,2,\dots,C$. Each computer "owns" one partition and will only store and update blocks of x it owns. That is, the blocks $i \in P^{(c)}$ of x are stored on and updated by computer c only. Likewise, "all data" relevant to these blocks are stored on computer c. We deal with the issues of data distribution and communication only in Section 6.

Algorithm Schema 1: Distributed block coordinate descent

1 choose $x_0 \in \mathbf{R}^N$
2 $k \longleftarrow 0$
3 **while termination criteria are not satisfied**
4 $\quad x_{k+1} \longleftarrow x_k$
5 $\quad$ **for each computer** $c \in \{1, \dots, C\}$ **in parallel do**
6 $\quad\quad$ sample a set of coordinates $Z_k^{(c)} \subseteq P^{(c)}$ of size τ, uniformly at random
7 $\quad\quad$ **for each thread** $i \in Z_k^{(c)}$ **in parallel do**
8 $\quad\quad\quad$ compute an update $h^{(i)}(x_k)$
9 $\quad\quad\quad$ $x_{k+1} \longleftarrow x_{k+1} + U_i h^{(i)}(x_k)$
10 $\quad k \longleftarrow k+1$

Distributed Sampling of Blocks. In Step 6 of Algorithm 1, each computer c chooses a random subset $Z_k^{(c)}$ of blocks from its partition $P^{(c)}$. We assume that $|Z_k^{(c)}| = \tau$, and that it is chosen uniformly at random from all subsets of $P^{(c)}$ of cardinality τ. Moreover, we assume the choice is done independently from all history and from what the other computers do in the same iteration. Formally, we say that the set of blocks chosen by all computers in iteration k, i.e., $Z_k = \cup_{c=1}^{C} Z_k^{(c)}$, is a (C, τ)-*distributed sampling*.

For easier reference in the rest of the paper, we formalize the setup described above as Assumption 3 at the end of this section (where we drop the subscript k, since the samplings are independent of k).

Computing and Applying Block Updates. In Steps 7–9, each computer c first computes and then applies updates to blocks $i \in Z_k^{(c)}$ to x_k. This is done on each computer in parallel. Hence, we have two levels of parallelism: across the nodes/computers and within each computer. The update to block i is denoted by $h^{(i)}(x_k)$ and arises as a solution of an optimization problem in the lower dimensional space R^{N_i}:

$$h^{(i)}(x_k) \leftarrow \arg\min_{t \in \mathbf{R}^{N_i}} \langle \nabla_i f(x_k), t \rangle + \frac{\beta w_i}{2} \|t\|_{(i)}^2 + \Omega_i(x_k^{(i)} + t). \tag{8}$$

Our method is most effective when this optimization problem has a closed form solution, which is the case in many applications. Note that *nearly all* information that describes problem (8) for $i \in P^{(c)}$ is available at node c. In particular, $x_k^{(i)}$ is stored on c. Moreover, we can store the description of Ω_i, norm $\|\cdot\|_{(i)}$ and the pair (β, w_i), for $i \in P^{(c)}$, on node c and only there.

Note that we did not specify yet the values of the parameters β and $w = (w_1, \dots, w_n)$. These depend on the properties of f and sampling $\hat{Z}$. We shall give theoretically justified formulas for these parameters in Section 4.

Communication. Finally, note that in order to find $h^{(i)}(x_k)$, each computer needs to be able to compute $\nabla_i f(x_k)$ for blocks $i \in Z_k^{(c)} \subseteq P^{(c)}$. This is the only information that an individual computer can *not* obtain from the data stored locally. We shall describe an efficient communication protocol that allows each node to compute $\nabla_i f(x_k)$ in Section 6.

Assumption 3 (Distributed sampling). *We make the following assumptions:*

1. ***Balanced partitioning.*** *The set of blocks is partitioned into C groups $P^{(1)},\dots,P^{(C)}$, each of size $s := n/C$. That is,*

   a. $\{1,2,\dots,n\} = \cup_{c=1}^{C} P^{(c)}$,
   b. $P^{(c')} \cap P^{(c'')} = \emptyset$ *for* $c' \neq c''$,
   c. $|P^{(c)}| =: s$ *for all* c.

2. ***Sampling.*** *For each $c \in \{1,\dots,C\}$, the set $\hat{Z}^{(c)}$ is a random subset of $P^{(c)}$ of size $\tau \in \{1,2,\dots,s\}$, where each subset of size τ is chosen with equal probability.*

We refer call the random set-valued mapping $\hat{Z} := \cup_{c=1}^{C} \hat{Z}^{(c)}$ by the name (C,τ)-distributed sampling.

4 Expected Separable Overapproximation

The following concept was first defined in [26]. It plays a key role in the complexity analysis of randomized coordinate descent methods.

Definition 1 (ESO). Let $\hat{Z}$ be any uniform sampling, i.e., a random sampling of blocks for which $\mathbf{Prob}(i \in \hat{Z}) = \mathbf{Prob}(j \in \hat{Z})$ for all $i,j \in [n]$. We say that function f admits an ESO with respect to sampling $\hat{Z}$, with parameters $\beta > 0$ and $w \in \mathbf{R}^n_{>0}$, if the following inequality holds for all $x,h \in \mathbf{R}^N$:

$$\mathbf{E}[f(x + h_{[\hat{Z}]})] \le f(x) + \tfrac{\mathbf{E}[|\hat{Z}|]}{n}\left(\langle \nabla f(x), h\rangle + \tfrac{\beta}{2}\|h\|_w^2\right). \tag{9}$$

For simplicity, we will sometimes write $(f,\hat{Z}) \sim ESO(\beta,w)$.

In the rest of this section we derive an ESO inequality for f satisfying Assumption 1 (smooth, convex, partially separable) and for sampling $\hat{Z}$ satisfying Assumption 3 ((C,τ)-distributed sampling). This has not been done before in the literature. In particular, we give simple closed-form formulas for parameters β and w, which we shall use in Section 5 to shed light on the performance of the method.

We first need to establish an auxiliary result. We use $[n]$ to denote $\{1,2,\dots,n\}$.

Lemma 1. *Let $\hat{Z} = \cup_{c=1}^{C} \hat{Z}^{(c)}$ be a (C,τ)-distributed sampling. Pick $J \subseteq [n]$ and assume that $|P^{(c)} \cap J| = \xi$ for some $\xi \ge 1$ and all c. Let $\kappa = \kappa(|\hat{Z} \cap J|, i)$ be any function that depends on $|\hat{Z} \cap J|$ and $i \in [n]$ only. Then*

$$\mathbf{E}\left[\sum_{i \in \hat{Z} \cap J} \kappa(|\hat{Z} \cap J|, i)\right] = \mathbf{E}\left[\frac{|\hat{Z} \cap J|}{C\xi} \sum_{i \in J} \kappa(|\hat{Z} \cap J|, i)\right]. \tag{10}$$

Proof. Let us denote by $J^{(c)} = J \cap P^{(c)}$, $\zeta = |\hat{Z} \cap J|$ and $\zeta^{(c)} = |\hat{Z} \cap J^{(c)}|$. Then

$$\begin{aligned}
\mathbf{E}\left[\sum_{i\in\hat{Z}\cap J} \kappa(\zeta,i)\right] &= \mathbf{E}\left[\mathbf{E}\left[\sum_{i\in\hat{Z}\cap J} \kappa(\zeta,i) \mid \zeta\right]\right] \\
&= \mathbf{E}\left[\mathbf{E}\left[\mathbf{E}\left[\sum_{i\in\hat{Z}\cap J} \kappa(\sum_{c=1}^{C} \zeta^{(c)},i) \mid \zeta^{(1)},\dots,\zeta^{(C)}, \sum_{c=1}^{C} \zeta^{(c)} = \zeta\right] \mid \zeta\right]\right] \\
&= \mathbf{E}\left[\mathbf{E}\left[\mathbf{E}\left[\sum_{c=1}^{C} \sum_{i\in\hat{Z}^{(c)}\cap J^{(c)}} \kappa(\zeta,i) \mid \zeta^{(1)},\dots,\zeta^{(C)}\right] \mid \sum_{c=1}^{C} \zeta^{(c)} = \zeta\right]\right] \\
&= \mathbf{E}\left[\mathbf{E}\left[\sum_{c=1}^{C} \frac{\zeta^{(c)}}{\xi} \sum_{i\in\cap J^{(c)}} \kappa(\zeta,i) \mid \sum_{c=1}^{C} \zeta^{(c)} = \zeta\right]\right] \\
&= \mathbf{E}\left[\sum_{c=1}^{C} \frac{\zeta}{\xi C} \sum_{i\in\cap J^{(c)}} \kappa(\zeta,i)\right] = \mathbf{E}\left[\frac{\zeta}{\xi C}\sum_{i\in J} \kappa(\zeta,i)\right].
\end{aligned}$$

□

The main technical result of this paper follows. This is a generalization of a result from [26] for partially separable f and τ-nice sampling to the distributed ($c > 1$) case. Notice that for $C = 1$ we have $\xi = \omega$.

Theorem 4 (ESO). *Let f satisfy Assumption 1 and $\hat{Z}$ satisfy Assumption 3. Let*[1] $\xi := \max\{|P^{(c)} \cap J| \;:\; c \in \{1,\dots,C\},\; J \in \mathcal{J}\}$. *Then $(f,\hat{Z})$ admits ESO with parameters β and w given by*

$$\beta = 1 + \frac{(\xi-1)(\tau-1)}{\max\{1, s-1\}} + (C-1)\frac{\xi\tau}{s}, \tag{11}$$

and $w_i = L_i$, $i = 1,2,\dots,n$.

Proof. For fixed $x \in \mathbf{R}^N$, define $\phi(h) := f(x+h) - f(x) - \langle \nabla f(x), h\rangle$. Likewise, for all $J \in \mathcal{J}$ we define $\phi_J(h) := f_J(x+h) - f_J(x) - \langle \nabla f_J(x), h\rangle$. Note that

$$\phi(h) = \sum_{J\in\mathcal{J}} \phi_J(h). \tag{12}$$

Also note that the functions ϕ_J and ϕ are convex and minimized at $h = 0$, where they attain the value of 0. For any uniform sampling, and hence for $\hat{Z}$ in particular, and any $a \in \mathbf{R}^N$, one has $\mathbf{E}[\langle a, h_{[\hat{Z}]}\rangle] = \frac{\mathbf{E}[|\hat{Z}|]}{n}\langle a, h\rangle$, and therefore

$$\mathbf{E}[\phi(h_{[\hat{Z}]})] = \mathbf{E}[f(x + h_{[\hat{Z}]})] - f(x) - \frac{\mathbf{E}[|\hat{Z}|]}{n}\langle \nabla f(x), h\rangle. \tag{13}$$

[1] Note that $\xi \in \{\lceil \frac{\omega}{C} \rceil, \dots, \omega\}$.

Because of this, and in view of (9) and the fact that as $\mathbf{E}[|\hat{Z}|] = C\tau$,[2] we only need to show that

$$\mathbf{E}[\phi(h_{[\hat{Z}]})] \leq \tfrac{C\tau}{n}\tfrac{\beta}{2}\|h\|_w^2. \tag{14}$$

Our starting point in establishing (14) will be the observation that from (6) used with $t = h^{(i)}$ we get

$$\phi(U_i h^{(i)}) \leq \tfrac{L_i}{2}\|h^{(i)}\|_{(i)}^2, \quad i \in [n]. \tag{15}$$

To simplify the proof, we shall without loss of generality assume that $|P^{(c)} \cap J| = \xi$ for all $c \in \{1,2,\ldots,C\}$ and $J \in \mathscr{J}$ for some constant $\xi > 1$. This can be achieved by extending the sets $J \in \mathscr{J}$ by introducing dummy dependencies (note that the assumptions of the theorem are still satisfied after this change). For brevity, let us write $\theta_{J,\hat{Z}} := |J \cap \hat{Z}|$ and $h_{[i]} := U_i h^{(i)}$. Fixing $J \in \mathscr{J}$ and $h \in \mathbf{R}^N$, we can estimate

$$\begin{aligned}
\mathbf{E}[\phi_J(h_{[\hat{Z}]})] &\overset{(2)}{=} \mathbf{E}\left[\phi_J\left(\sum_{i\in\hat{Z}} h_{[i]}\right)\right] = \mathbf{E}\left[\phi_J\left(\sum_{i\in\hat{Z}\cap J} h_{[i]}\right)\right] \\
&= \mathbf{E}\left[\phi_J\left(\tfrac{1}{\theta_{J,\hat{Z}}}\sum_{i\in\hat{Z}\cap J} \theta_{J,\hat{Z}} h_{[i]}\right)\right] \leq \mathbf{E}\left[\tfrac{1}{\theta_{J,\hat{Z}}}\sum_{i\in\hat{Z}\cap J} \phi_J\left(\theta_{J,\hat{Z}} h_{[i]}\right)\right] \\
&\overset{(10)}{=} \mathbf{E}\left[\tfrac{1}{\theta_{J,\hat{Z}}}\left(\tfrac{\theta_{J,\hat{Z}}}{C\xi}\sum_{i\in J}\phi_J\left(\theta_{J,\hat{Z}} h_{[i]}\right)\right)\right] = \frac{1}{C\xi}\mathbf{E}\left[\sum_{i\in J}\phi_J\left(\theta_{J,\hat{Z}} h_{[i]}\right)\right] \\
&= \frac{1}{C\xi}\mathbf{E}\left[\sum_{i\in[n]}\phi_J\left(\theta_{J,\hat{Z}} h_{[i]}\right)\right].
\end{aligned} \tag{16}$$

In the second equation above we have used the assumption that ϕ_J depends on blocks $i \in J$ only. The only inequality above follows from convexity of ϕ_J. Note that this step can only be performed if the sum is over a nonempty index set, which happens precisely when $\theta_{J,\hat{Z}} \geq 1$. This technicality can be handled at the expense of introducing a heavier notation (which we shall not do here), and (16) still holds. Finally, in one of the last steps we have used (10) with $\kappa(|\hat{Z} \cap J|, i) \leftarrow \phi_J(\theta_{J,\hat{Z}} h_{[i]})$.

[2] In fact, $|\hat{Z}| = C\tau$ with probability 1.

By summing up inequalities (16) for $J \in \mathscr{J}$, we get

$$\mathbf{E}\left[\phi(h_{[\hat{Z}]})\right] \overset{(12)}{=} \sum_{J\in\mathscr{J}} \mathbf{E}\left[\phi_J(h_{[\hat{Z}]})\right] \overset{(16)}{\leq} \frac{1}{C\xi} \sum_{J\in\mathscr{J}} \mathbf{E}\left[\sum_{i\in[n]} \phi_J\left(\theta_{J,\hat{Z}} h_{[i]}\right)\right]$$

$$\overset{(12)}{=} \frac{1}{C\xi}\mathbf{E}\left[\sum_{i\in[n]} \phi\left(\theta_{J,\hat{Z}} h_{[i]}\right)\right] \overset{(15)}{\leq} \frac{1}{C\xi}\mathbf{E}\left[\sum_{i\in[n]} \tfrac{L_i}{2}\|\theta_{J,\hat{Z}} h^{(i)}\|_{(i)}^2\right]$$

$$= \frac{1}{2C\xi}\mathbf{E}\left[\theta_{J,\hat{Z}}^2 \sum_{i\in[n]} L_i\|h^{(i)}\|_{(i)}^2\right] \overset{(3)}{=} \frac{1}{2C\xi}\|h\|_w^2 \mathbf{E}\left[\theta_{J,\hat{Z}}^2\right]. \tag{17}$$

We now need to compute $\mathbf{E}[\theta_{J,\hat{Z}}^2]$. Note that the random variable $\theta_{J,\hat{Z}}$ is the sum of C independent random variables $\theta_{J,\hat{Z}} = \sum_{c=1}^{C} \theta_{J,\hat{Z}^{(c)}}$, where $\theta_{J,\hat{Z}^{(c)}}$ has the simple law

$$\mathbf{Prob}(\theta_{J,\hat{Z}^{(c)}} = k) = \binom{\xi}{k}\binom{s-\xi}{\tau-k} \Big/ \binom{s}{\tau}.$$

We therefore get

$$\mathbf{E}[\theta_{J,\hat{Z}}^2] = \mathbf{E}[(\sum_{c=1}^{C} \theta_{J,\hat{Z}^{(c)}})^2] = C\mathbf{E}[(\theta_{J,\hat{Z}^{(c)}})^2] + C(C-1)(\mathbf{E}[\theta_{J,\hat{Z}^{(c)}}])^2$$

$$= C\tfrac{\xi\tau}{s}\left(1 + \tfrac{(\xi-1)(\tau-1)}{\max\{1,s-1\}}\right) + C(C-1)\left(\tfrac{\xi}{s}\tau\right)^2. \tag{18}$$

It only remains to combine (17) and (18) to get (14). □

Note that ESO inequalities have recently been used in the analysis of distributed coordinate descent methods by Richtárik and Takáč [27] and Fercoq et al. [7]. However, their assumptions on f and derivation of ESO are very different and hence our results apply to a different class of functions.

5 Iteration Complexity

In this section, we state two iteration complexity results for Algorithm 1. Theorem 5 deals with a non-strongly convex objective and shows that the algorithm achieves sub-linear rate of convergence $\mathscr{O}(\frac{1}{\varepsilon})$. Theorem 6 shows Algorithm 1 achieves linear convergence rate $\mathscr{O}(\log\frac{1}{\varepsilon})$ for a strongly convex objective.

However, we wish to stress that in high dimensional settings, and especially in applications where low- or medium-accuracy solutions are acceptable, the dependence of the method on ε is somewhat less important than its dependence on data size through quantities such as the dimension N and the number of blocks n, and on quantities such as the number of computers C and number of parallel updates per computer τ, which is related to the number of cores.

Notice that once the ESO is established by Theorem 4, the complexity results, Theorems 5 and 6, follow from the generic complexity results in [26, 39], respectively.

5.1 Convex Functions

Theorem 5 (Based on [39]). *Let f satisfy Assumption 1 and sampling $\hat{Z}$ satisfy Assumption 3. Let x_k be the iterates of Algorithm 1 applied to problem* (1)*, where parameters β and w are chosen as in Theorem 4 and the random sets Z_k are iid, following the law of $\hat{Z}$. Then for all $k \geq 1$,*

$$\mathbf{E}\left[F(x_k) - F^*\right] \leq \frac{n}{n + C\tau k}\left(\frac{\beta}{2}\|x_0 - x^*\|_w^2 + F(x_0) - F^*\right). \tag{19}$$

Note that the leading term in the bound decreases as the number of blocks updated in a single (parallel) iteration, $C\tau$, increases. However, notice that the parameter β also depends on C and τ. We shall investigate this phenomenon in Section 5.3 and show that the level of speed-up one gets by increasing C and/or τ (where by speed-up we mean the decrease of the upper bound established by the theorem) depends on the degree of separability ω of f. The smaller ω is, the more speed-up one obtains.

5.2 Strongly Convex Functions

If we assume that F is strongly convex with respect to the norm $\|\cdot\|_w$ then the following theorem shows that $F(x_k)$ converges to F^* linearly, with high probability.

Definition 2 (Strong Convexity). Function $\phi : \mathbf{R}^N \to \mathbf{R} \cup \{+\infty\}$ is strongly convex with respect to the norm $\|\cdot\|_w$ with convexity parameter $\mu_\phi(w) \geq 0$ if

$$\phi(y) \geq \phi(x) + \langle \phi'(x), y - x\rangle + \tfrac{\mu_\phi(w)}{2}\|y - x\|_w^2, \quad \forall x, y \in \operatorname{dom}\phi, \tag{20}$$

where $\phi'(x)$ is any subgradient of ϕ at x.

Notice that by setting $\mu_\phi(w) = 0$, one obtains the usual notion of convexity. Strong convexity of F may come from f or Ω (or both); we write $\mu_f(w)$ (resp. $\mu_\Omega(w)$) for the (strong) convexity parameter of f (resp. Ω). It follows from (20) that if f and Ω are strongly convex, then F is strongly convex with, e.g., $\mu_F(w) \geq \mu_f(w) + \mu_\Omega(w)$.

Theorem 6 (Based on [26]). *Let us adopt the same assumptions as in Theorem 5. Moreover, assume that F is strongly convex with $\mu_f(w)+\mu_\Omega(w)>0$. Choose initial point $x_0 \in \mathbf{R}^N$, target confidence level $0<\rho<1$, target accuracy level $0<\varepsilon<F(x_0)-F^*$ and*

$$K \geq \frac{n}{C\tau}\frac{\beta+\mu_\Omega(w)}{\mu_f(w)+\mu_\Omega(w)}\log\left(\frac{F(x_0)-F^*}{\varepsilon\rho}\right). \tag{21}$$

If $\{x_k\}$ are the random points generated by Algorithm 1, then $\mathbf{Prob}(F(x_K)-F^* \leq \varepsilon) \geq 1-\rho$.

Notice that now both ε and ρ appear inside a logarithm. Hence, it is easy to obtain accurate solutions with high probability.

5.3 *Parallelization Speed-Up Is Governed by Sparsity*

If we assume that $\|x_0-x^*\|_w^2 \gg F(x_0)-F^*$, then in view of Theorem 5, the number of iterations required by our method to get an ε solution in expectation is $O(\frac{\beta}{C\tau\varepsilon})$. Hence, the smaller $\frac{\beta}{C\tau\varepsilon}$ is, the fewer are the iterations required. If β were a constant independent of C and τ, one would achieve linear speed-up by increasing workload (i.e., by increasing $C\tau$). However, this is the case for $C=1$ and $\omega=1$ only (see Theorem 4). Let us look at the general case. If we write $\eta := \frac{\xi}{s}$ (this a measure of sparsity of the partitioned data), then

$$\frac{\beta}{C\tau} \overset{(11)}{=} \frac{1+\frac{(\xi-1)(\tau-1)}{\max\{1,s-1\}}+(C-1)\frac{\xi\tau}{s}}{C\tau} \leq \frac{1+\frac{\xi(\tau-1)}{s}+(C-1)\frac{\xi\tau}{s}}{C\tau}$$
$$= \frac{1+\eta(\tau-1)+(C-1)\eta\tau}{C\tau} = \frac{1+\eta(C\tau-1)}{C\tau} = \frac{1}{C\tau}+\eta\left(1-\frac{1}{C\tau}\right).$$

As expected, the first term represents linear speed-up. The second term represents a penalty for the lack of sparsity (correlations) in the data. As $C\tau$ increases, the second term becomes increasingly dominant, and hence slows the speed-up from almost linear to none. Notice that for fixed η, the ratio $\frac{\beta}{C\tau}$ as a function of $C\tau$ is decreasing and hence we always get *some* speed-up by increasing $C\tau$.

Figure 1 (left) shows the speed-up factor ($\frac{C\tau}{\beta}$; high values are good) as a function of $C\tau$ for different sparsity levels η. One can observe that sparse problems achieve almost linear speed-up even for bigger value of $C\tau$, whereas for, e.g., $\eta=0.2$, almost linear speed-up is possible only up to $C\tau=10$. For sparser data with $\eta=0.01$, linear speed-up can be achieved up to $C\tau=100$. For $\eta=0.001$, we can use $C\tau=10^3$. The right part of Figure 1 shows how sparsity affects speed-up for a fixed number of updates $C\tau$. Again, the break-point of almost linear speed-up is visibly present.

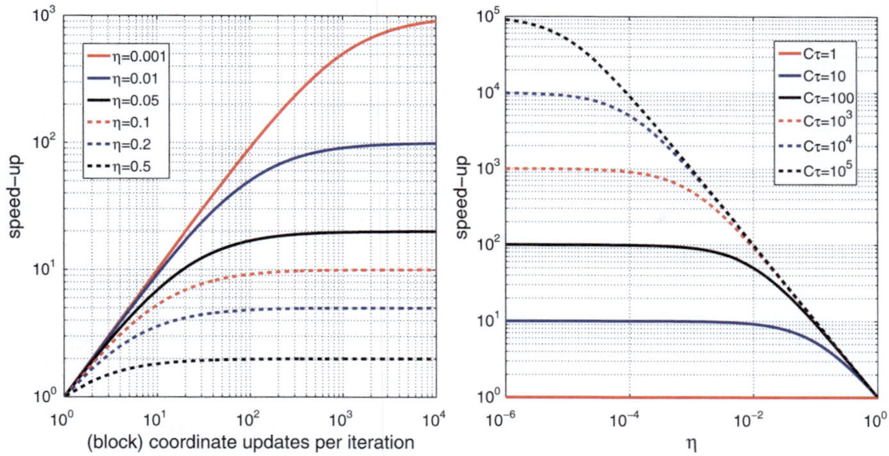

Fig. 1 Speed-up gained from updating more blocks per iteration is almost linear initially and depending on sparsity level η, may become significantly sublinear

Similar observations in the non-distributed setting were reported in [26]. The phenomenon is not merely a by-product of our theoretical analysis; it also appears in practice.

5.4 *The Cost of Distribution*

Notice that in a certain intuitive sense, variants of Algorithm 1 are comparable, as long as each iteration updates the same number $C\tau$ of blocks. This allows us to vary C and τ, while keeping the product constant. In particular, let us consider two scenarios:

1. Consider C computers, each updating τ blocks in parallel, and
2. Consider 1 computer updating $C\tau$ blocks in each iteration in parallel.

For the sake of comparison, we assume that the underlying problem is small enough so that it can be stored on and solved by a single computer. Further, we assume that F is strongly convex, $\mu(\Omega) = 0$ and $s = \frac{n}{C} \geq 2$. Similar comparisons can be made in other settings as well, but given the page restrictions, we restrict ourselves to this case only.

In the iteration-complexity bound (21), we notice that the only difference is in the value of β. Let β_1 be the β parameter in the first situation with C computers, and β_2 be the β parameter in the second situation with one computer. The ratio of the complexity bounds (21) is hence equal to the ratio

$$\frac{\beta_1}{\beta_2} = \frac{\left(1 + \frac{(\xi-1)(\tau-1)}{s-1} + (C-1)\frac{\xi\tau}{s}\right)}{1 + \frac{(\omega-1)(C\tau-1)}{Cs-1}}.$$

Table 1 Lower and upper bounds on β_1/β_2 for a selection parameters n, ω, C and τ

n	ω	C	τ	β_2	LB	UB
10^6	10^2	10	50	1.049	1.0000086	1.4279673
10^7	10^2	10	50	1.005	1.0000009	1.0446901
10^8	10^2	100	100	1.009	1.0000010	1.9801990

Notice that $\frac{\omega}{C} \leq \xi \leq \omega$. The ratio β_1/β_2 is increasing in ξ. We thus obtain the following bounds:

$$\text{LB} := \frac{1+\frac{(\omega-C)(\tau-1)}{n-C}+(C-1)\frac{\omega\tau}{n}}{1+\frac{(\omega-1)(C\tau-1)}{n-1}} \leq \frac{\beta_1}{\beta_2} \leq \frac{1+\frac{(\omega-1)(C\tau-C)}{n-C}+(C-1)\frac{\omega C\tau}{n}}{1+\frac{(\omega-1)(C\tau-1)}{n-1}} =: \text{UB}.$$

Table 1 presents the values of LB and UB for various parameter choices and problem sizes. We observe that the value of β_2 is around 1. The value of β_1 depends on a particular partition, but we are sure that $\beta_1 \in [\beta_2 \cdot \text{LU}, \beta_2 \cdot \text{UB}]$. In Table 1, UB is less than 2, which means that by distributing the computation, the method will at most double the number of iterations. However, larger values of UB, albeit UB $\precsim C$, are possible for different settings of the parameters. For a different class of functions f, an upper bound of 2 was proven in [27] and improved in [7] to the factor $1+1/(\tau-1)$ whenever $\tau > 1$.

Of course, if the problem size exceeds the memory available at a single computer, the option of not distributing the data and computation may not be available. It is reassuring, though, to know that the price we pay for distributing the data and computation, in terms of the number of iterations, is bounded. Having said that, a major complication associated with any distributed method is the communication, which we discuss in the two following sections.

6 Two Implementations

Although our algorithm and results apply to a rather broad class of functions, we focus on two important problems in statistics and machine learning in describing our computational experience, so as to highlight the finer details of the implementations.

6.1 *An Implementation for Sparse Least Squares*

In many statistical analyses, e.g., linear regression, one hopes to find a solution x with only a few non-zero elements, which improves interpretability. It has been recognized, however, that the inclusion of the number of non-zero elements, $\|x\|_0$, in the objective function raises the complexity of many efficiently solvable problems to NP-Hard [8, 18]. Recently, a number of randomized coordinate descent methods try to handle the ℓ_0-norm directly [24], but only local convergence can be guaranteed. Fortunately, the inclusion of the sum of absolute values, $\|x\|_1$, provides a provably good proxy, which is also known as ℓ_1 regularization. There is a large and growing body of work on both practical solvers for non-smooth convex problems, obtained by such a regularization, and their convergence properties, when one restricts oneself to a single computer storing the complete input. Such solvers are, however, most useful in high-dimensional applications, where the size of the data sets often exceeds the capacity of random-access memory of any single computer available today.

Hence, the first implementation we present is a distributed coordinate-descent algorithm for ℓ_1-regularized ("sparse") least squares. The key components needed by Algorithm 1 are the computation of L_i, $\nabla_i f(x_k)$, and solving of a block-wise minimization problem. Note that $\nabla_i f(x) = \sum_{j=1}^m -A_{j,i}(y^{(j)} - A_{j:}x)$, where $A_{j:}$ denotes j-th row of matrix A, and $L_i = \|A_{:i}\|_2^2$. The only difficulty is that given the data partition $\{P^{(c)}\}_{c=1}^C$, no single computer c is able to compute $\nabla_i f(x)$ for any $i \in P^{(c)}$. The reasoning follows from a simple observation: If we wanted to compute $\nabla_i f(x_k)$ for a given x_k from scratch, we would have to access all coordinates of x_k, vector y, and all non-zero elements of the input matrix A. This could be avoided by introducing an auxiliary vector $g_k := g(x_k)$ defined as

$$g_k := Ax_k - y. \tag{22}$$

Once the value of $g_k = g(x_k)$ is available, a new iterate is

$$x_{k+1} = x_k + \sum_{c=1}^{C} \sum_{i \in Z_k^{(c)}} U_i h^{(i)}(x_k). \tag{23}$$

and $g_{k+1} = g(x_{k+1})$ can be easily expressed as

$$g_{k+1} = g_k + \sum_{c=1}^{C} \underbrace{\sum_{i \in Z_k^{(c)}} A_{:i} h^{(i)}(x_k)}_{\delta g^{(c)}}. \tag{24}$$

Note that the value $\delta g^{(c)}$ can be computed on computer c as all required data are available on computer c. Subsequently, g_{k+1} can be obtained by summation and the formula for $\nabla_i f(x)$ will take the form $\nabla_i f(x) = A_{:i}^T g = \sum_{j=1}^m A_{j,i} g^{(j)}$. Once we know how to compute $\nabla_i f(x)$ and L_i, all that remains to be done is to solve the problem

$$\min_{t \in \mathbf{R}} a + bt + \frac{c}{2}t^2 + \lambda|d+t|, \tag{25}$$

where $a, b, d \in \mathbf{R}$ and $c, \lambda \in \mathbf{R}_{>0}$, which is given by a *soft-thresholding* formula $t^* = \operatorname{sgn}(\zeta)(|\zeta| - \frac{\lambda}{c})_+ - d$, where $\zeta = d - \frac{b}{c}$.

6.2 An Implementation for Training SVM

Let us present another example implementation. The key problem in supervised machine learning is the training of classifiers. Given a matrix $A \in \mathbf{R}^{m\times N}$, a compatible vector $y \in \mathbf{R}^m$, and constant $\gamma > 0$, the goal is to find a vector $x \in \mathbf{R}^N$ which solves the following optimization problem:

$$\min_{x\in\mathbf{R}^N} F(x) := \underbrace{\gamma\|x\|_1}_{\Omega(x)} + \underbrace{\sum_{j=1}^{m} \mathscr{L}(x, A_{j:}, y^{(j)})}_{f(x)}, \tag{26}$$

where $A_{j:}$ again denotes j-th row of matrix A and $\mathscr{L}$ is a loss function, such as

$$\mathscr{L}_{SL}(x, A_{j:}, y^{(j)}) := \frac{1}{2}(y^{(j)} - A_{j:}x)^2, \qquad \text{square loss,} \tag{SL}$$

$$\mathscr{L}_{LL}(x, A_{j:}, y^{(j)}) := \log(1 + e^{-y^{(j)}A_{j:}x}), \qquad \text{logistic loss,} \tag{LL}$$

$$\mathscr{L}_{HL}(x, A_{j:}, y^{(j)}) := \frac{1}{2}\max\{0, 1 - y^{(j)}A_{j:}x\}^2, \qquad \text{hinge square loss.} \tag{HL}$$

The input (A, y) is often referred to as the training data. Rows of matrix A represent observations of N features each and y are the corresponding classifications to train the classifier on.

Square hinge loss is a popular choice of $\mathscr{L}$, but is not smooth. It is well known that the dual has the form [9, 33, 36]:

$$\min_{x\in\mathbf{R}^m} F(x) := \underbrace{\frac{1}{2\lambda m^2}x^T Q x - \frac{1}{m}x^T \mathbf{1}}_{f(x)} + \underbrace{\sum_{i=1}^{m} \Phi_{[0,1]}(x^{(i)})}_{\Omega(x)}, \tag{SVM-DUAL}$$

where $\Phi_{[0,1]}$ is the characteristic (or "indicator") function of the interval $[0,1]$ and $Q \in \mathbf{R}^{m\times m}$ is the Gram matrix of the data, i.e., $Q_{i,j} = y^{(i)}y^{(j)}A_{i:}A_{j:}^T$. If x^* is an optimal solution of (SVM-DUAL), then $w^* = w^*(x^*) = \frac{1}{\lambda m}\sum_{i=1}^m y^{(i)}(x^*)^{(i)}A_{i:}^T$ is an optimal solution of the primal problem

$$\min_{w\in\mathbf{R}^N} P(w) := \frac{1}{N}\sum_{i=1}^{N} \mathscr{L}(w, A_{i:}, y^{(i)}) + \frac{\lambda}{2}\|w\|^2, \tag{27}$$

where $\mathscr{L}(w, A_{i:}, y^{(i)}) = \max\{0, 1 - y^{(i)}A_{i:}w\}$.

Our second example implementation is a distributed coordinate-descent algorithm for SVM in the (SVM-DUAL) formulation. In this case, we define

$$g_k := \frac{1}{\lambda m} \sum_{i=1}^{m} x_k^{(i)} y^{(i)} A_{i:}^T. \tag{28}$$

Then

$$\nabla_i f(x) = \frac{y^{(i)} A_{i:} g_k - 1}{m}, \qquad L_i = \frac{\|A_{i:}\|^2}{\lambda m^2}. \tag{29}$$

The optimal step length is then solution of a one-dimensional problem:

$$h^{(i)}(x_k) = \arg\min_{t \in \mathbf{R}} \nabla_i f(\alpha) t + \frac{\beta}{2} L_i t^2 + \Phi_{[0,1]}(\alpha^{(i)} + t) \tag{30}$$

$$= \mathrm{clip}_{[-\alpha^{(i)}, 1-\alpha^{(i)}]} \left(\frac{\lambda m (1 - y^{(i)} A_{i:} g_k)}{\beta \|A_{i:}\|^2} \right), \tag{31}$$

where for $a < b$

$$\mathrm{clip}_{[a,b]}(\zeta) = \begin{cases} a, & \text{if } \zeta < a, \\ b, & \text{if } \zeta > b, \\ \zeta, & \text{otherwise.} \end{cases}$$

The new value of the auxiliary vector $g_{k+1} = g(x_{k+1})$ is given by

$$g_{k+1} = g_k + \sum_{c=1}^{C} \underbrace{\sum_{i \in Z_k^{(c)}} \frac{1}{\lambda m} h^{(i)}(x_k) y^{(i)} A_{i:}^T}_{\delta g^{(c)}} \tag{32}$$

and the duality gap $G(x_k) = P(g_k) + F(x_k)$ can be easily obtained [11, 33, 36] as

$$G(x_k) = \frac{1}{m} \sum_{i=1}^{m} (\mathscr{L}(g_k, A_{i:}, y^{(i)}) - x_k^{(i)}) + \lambda \|g_k\|^2. \tag{33}$$

7 Per-iteration Complexity

Using the auxiliary vector g_k, which was introduced in the previous section, Algorithm 1 has two alternating and time-consuming sub-procedures, namely:

1. computation of an update $\sum_{i \in Z_k^{(c)}} U_i h^{(i)}(x_k)$ and the accumulation of g_k: $\delta g^{(c)}$,
2. updating g_k to g_{k+1}.

Let us denote the run-time of the first sub-procedure by $\mathscr{T}_1(\tau)$, considering this depends on τ, and the run-time of a second one by $\mathscr{T}_2$. We will neglect the rest of the run-time cost, such as managing a loop, evaluation of termination criteria, measuring a computation time, etc. The total run-time cost $\mathcal{T}_T$ is hence given by

$$\mathcal{T}_T = \mathcal{O}\left(\frac{\beta}{C\tau}(\mathcal{T}_1(\tau)+\mathcal{T}_2)\right) \tag{34}$$

where we consider the case when $\mu_\Omega(w) \equiv 0$ in (21). Let us now for simplicity assume that the first sub-procedure is linear in τ, i.e., $\mathcal{T}_1(\tau) = \tau\mathcal{T}_1(1) =: \tau\mathcal{T}_1$. Then

$$\mathcal{T}_T = \mathcal{O}\left(\frac{\beta}{C\tau}(\tau\mathcal{T}_1+\mathcal{T}_2)\right). \tag{35}$$

Numerical values of $\mathcal{T}_1$ and $\mathcal{T}_2$ could be estimated, given problem sparsity and underlying hardware, or can be measured during the run.

Optimal Choice of Sampling Parameter τ. In the previous paragraph, we gave an estimate of the complexity of a single iteration. In this paragraph, we answer the question of how to choose a τ given times $\mathcal{T}_1, \mathcal{T}_2$. For variable β, we have more options, but we stick to the most general one given in (11). Given that $s \geq 2$, we have

$$\mathcal{T}_T = \mathcal{O}\left(\frac{1+\frac{(\xi-1)(\tau-1)}{s-1}+(C-1)\frac{\xi\tau}{s}}{C}\left(r_{1,2}+\frac{1}{\tau}\right)\mathcal{T}_2\right) = \mathcal{O}\left(\left(\frac{s}{\xi C}+\tau\right)\left(r_{1,2}+\frac{1}{\tau}\right)\right), \tag{36}$$

where $r_{1,2} = \frac{\mathcal{T}_1}{\mathcal{T}_2}$ is a work to communication ratio. The optimal parameter τ^* can be obtained by minimizing (36) and is given by

$$\tau^* = \sqrt{\frac{s}{r_{1,2}\,\xi\,C}}. \tag{37}$$

Therefore, smaller values of $r_{1,2}$ imply that we should do more work in each iteration, and hence bigger values of τ should be chosen. This is quite natural, as one should tune the parameters in such a way that time spent in communication should be in comparable with that of effective computation (Figure 2).

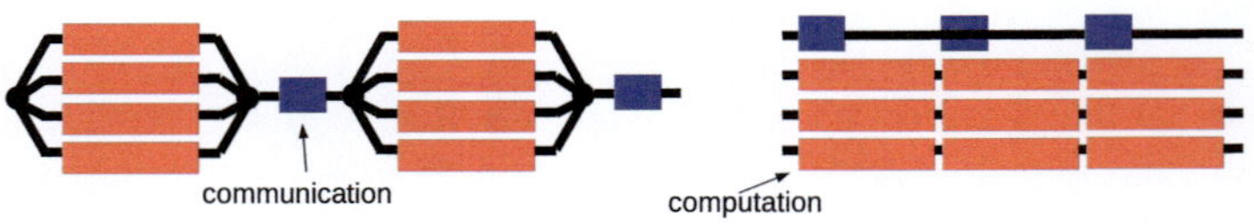

Fig. 2 An illustration of a naïve (PS) approach (*left*), which alternates between parallel regions, where computations take place, and serial regions dedicated to MPI communications with other computers. An alternative (FP) approach (*right*) dedicates the communication task to one thread and uses other threads for computation

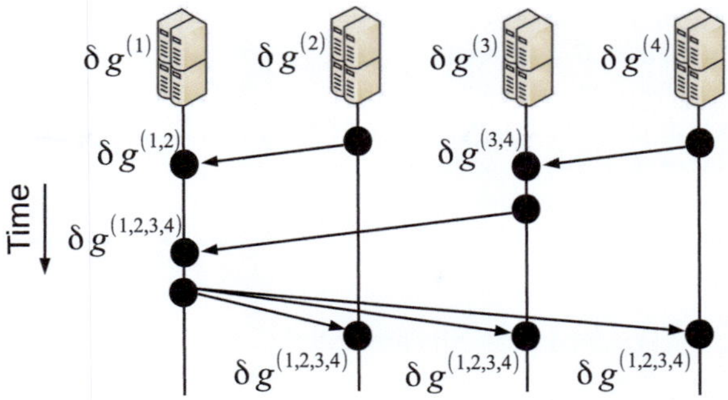

Fig. 3 Schematic diagram of a standard reduce all implementation. The goal is to compute $\sum_{c=1}^{C} \delta g^{(c)}$. The *arrows* show data flow between computers

Message Passing Interface (MPI). In order to discuss finer details of the implementations, we need to introduce the architecture we use. We use OpenMP [23] for dealing with concurrency within a single computer and MPI [35] as the abstraction layer for network communication. In MPI, one passes data from one MPI process to another MPI process, which may run on another computer. (We disregard the concept of groups for brevity.) Communication can involve any subset of computers, which run MPI processes. Communication can be either blocking ("synchronous") or non-blocking ("asynchronous"). A *collective* operation involves the communication among two or more MPI processes. An example of a collective operation is a *barrier*, where computers wait until all of them reach the same point in the algorithm. Another common collective operation is *reduce all*, which is parametrized by an arbitrary operation that takes a set of elements and produces a single element of the same type. This "reduce" operation is applied to all elements of the particular type stored across all MPI processes and the result is returned to all MPI processes. For example, let us assume that each computer stores a vector $\delta g^{(c)} \in \mathbf{R}^m$ and the goal is to sum it up, i.e., to compute $\delta g^{(1,\dots,C)} = \sum_{c=1}^{C} \delta g^{(c)}$ and to make this result available on each computer. Figure 3 shows a standard approach, which leads to the desired result. From the performance point of view, however, the use of *reduce all* should be minimized, as it involves an implicit synchronization and leaves most of the computers idle throughout the collective operation.

This suggests the following range of progressively better-performing variants:

Alternating Parallel and Serial Regions (PS). The naïve implementation alternates two sub-procedures. One, which is computationally heavy and is done in parallel, but with no MPI communication, and another one, which is purely communicational. As an easy fix, one can dedicate one thread to the communication and other threads within the same computer to computation. We call this approach **Fully Parallel** (FP). Figure 2 compares the naïve strategy (left) with the FP (right).

Reduce All (RA). As mentioned above, the use of *reduce all* operations significantly decreases the performance of many distributed algorithms. It is, however, the preferred form of communication between computers close to each other in the computer network, such as computers directly connected by a network cable. The use of asynchronous methods is also preferred over synchronous methods.

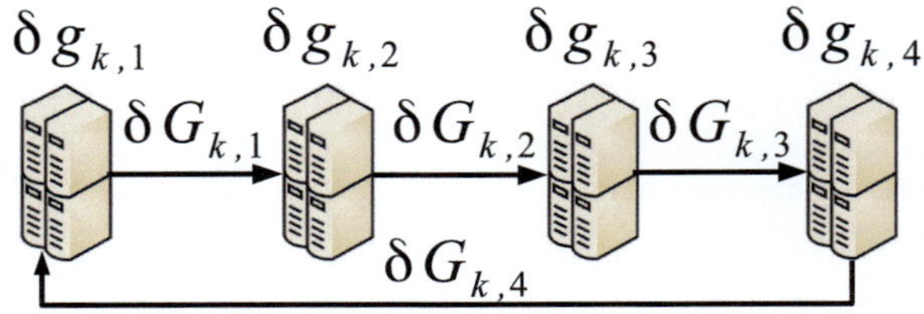

Fig. 4 Illustration of ASL method for $C = 4$. During k-th iteration, computer c obtains its contribution $\delta g_k^{(c)}$ but asynchronically sends an accumulated update $\delta G_k^{(c)}$ to its successor

Asynchronous StreamLined (ASL). We propose another pattern of communication, where each computer in one iteration sends only one message to the closest computer, asynchronously, and receives only one message from another computer close-by, asynchronously. The communication hence takes place in a ring. This tweak, however, requires a significant change in the algorithm. Figure 4 illustrates the data flow of messages at the end of iteration k for $C = 4$. We fix an order of computers in a ring, denoting $\text{pred}_R(c)$ and $\text{succ}_R(c)$ the two computers neighbouring computer c along the two directions on the ring. Computer c always receives data only from computer $\text{pred}_R(c)$ and sends data only to computer $\text{succ}_R(c)$. Let us denote by $\delta G_k^{(c)}$ the data, which computer c sends to computer $\text{succ}_R(c)$ at the end of iteration k. When computer c starts iteration k, it has already received $\delta G_{k-1}^{(\text{pred}_R(c))}$.[3] Hence the data, which will be sent at the end of iteration k by computer c are:

$$\delta G_k^{(c)} = \delta G_{k-1}^{(\text{pred}_R(c))} - \delta g_{k-C}^{(c)} + \delta g_k^{(c)}. \tag{38}$$

It should be noticed that at the end of each iteration in the ASL procedure, each computer has a different vector g_k, which we denote $g_k^{(c)}$. The update rule is

$$g_{k+1}^{(c)} = g_k^{(c)} + \delta g_k^{(c)} + \delta G_k^{(\text{pred}_R(c))} - \delta g_{k-C+1}^{(c)}. \tag{39}$$

The clear advantage of the ASL method is a decrease in communication time. On the other hand, it comes with a cost of slower propagation of information. Indeed, it takes $C-1$ iterations to propagate information to all computers. It also comes with bigger storage requirements, as at iteration k, we have to have all vectors $\delta g_l^{(c)}$ for $k - C \le l \le k$ stored on computer c.

Asynchronous Torus (AST). There is a compromise solution, though, which inherits many desirable features of both RA and ASL. This employs a toroidal networking topology, which is common in high-performance computing (HPC) in general, and HPC using InfiniBand networks [10], in particular. Let us assume that C is a multiple of $r \in N$, where r represents the width of a torus, i.e., C computers are partitioned into subsets R_i each with size r. Each group R_i has a root computer. These root computers aggregate updates from their respective groups, e.g., using a local

[3]For the start of the algorithm we define $\delta g_l^{(c)} = \delta G_l^{(c)} = \mathbf{0}$ for all $l < 0$.

Table 2 Summary of additional memory and computation requirements for strategies RA, SLA, AST

Strategy	Memory for g's	Communication	Extra computation
RA	$2m$	$\mathscr{T}_{ra}$	0
SLA	$(2+C)m$	$\mathscr{T}_{p2p}$	$4m$ additions
AST	$(2+C/r)m$	$\mathscr{T}_{p2p}+\mathscr{T}_{ra}/r$	$8m$ additions

reduce all operation, in each iteration and exchange those update in an asynchronous ring with two other adjacent root computers. Thus the communication between the root nodes follows the ASL communication pattern. The AST approach decreases the propagation time from C to $\frac{C}{r}$, additional storage is also decreased by factor r, and the overall communication complexity remains low.

The Comparison. Changing from the FP approach to the PS approach does not require much computational or storage overhead, but can reduce the idle time of processors. However, changing from RA to SLA or AST brings significant storage requirements, while it reduces both communication and idle time significantly. Table 2 summarizes maximum memory requirements on each single node of the cluster, time spent in communication, and amount of data transferred over the network. Once the time spent in communication is measured or estimated, one can pick the most appropriate strategy. Notice that the wall-clock time required for the *reduce all* operation, $\mathscr{T}_{ra}$, is typically of the order $\mathscr{O}(\log C)\cdot\mathscr{T}_{p2p}$, where $\mathscr{T}_{p2p}$ is the time required by the point-to-point transmission.

8 Numerical Experiments

In this section we present numerical evidence of the efficiency of the distributed (block) coordinate-descent method.

The Code. The code of the distributed (block) coordinate-descent solver is part of our AC-DC library, available at http://code.google.com/p/ac-dc/. The library is written in C++ using OpenMP. The extensive use of template classes, Boost::MPI, and Boost.Serialization makes it easy to change the composite function and the precision of the computation. Both wall-clock and CPU-time were measured using Boost::Timers, which achieve nano-second accuracy on recent processors running recent versions of Linux.

The Facility. Our empirical tests were conducted in UK's HPC facility, HECToR, equipped with multi-core computers connected using Infiniband [10]. In particular, in Phase 3 of the facility, which is a Cray XE6 cluster, we have used up to 128 nodes, equipped with two AMD Opteron Interlagos 16-core processors and 32 GB of memory each. This gave us 4,096 cores in total, interconnected using Cray Gemini routers in a 3D torus. Each Gemini router was connected to processors and random-access memory of two nodes via HyperTransport links. Each router is then connected to ten other routers. In practice, the latency is about 1–1.5 μs and the capacity of each link is 8 GBs^{-1}. The facility ran a Cray Linux Environment, based on SuSE Linux.

SVM. One of the goals of this paper is to train huge sparse support-vector machines (SVM) that do not fit into the memory of a single computer. In the machine learning literature, one often performs experiments on instances of moderate size, e.g., 100 MB [9, 34, 36]. Well-known instances of this scale include, e.g., CCAT variant of RCV1 [13], Astro-ph [34], and COV [34]. In this section, we focus on a larger dataset, known as WebSpam [14]. This dataset consists of 350,000 observations (rows) and 16,609,143 features (columns). The size of the instance is 25 GB. Figure 5 shows the execution time and duality gap for WebSpam dataset, using $C = 16$ MPI processes, with each process using 8 threads. τ is the number of coordinates updated by one MPI process during one iteration. As expected, the main run-time cost it not computing the updates, but updating g. Let us remark that ε is usually not particularly small in the machine-learning community. In experimenting with small ε, we just wanted to demonstrate that our algorithm is able to close the duality gap within the limits of machine precision. The truly important measures of the performance of the classifier, e.g., 0–1 loss or prediction error, are actually within 10 % after the first minute, which is the first time we compute it. In practice, a duality gap of 0.1 or 0.01 can be sufficient for machine learning problems.

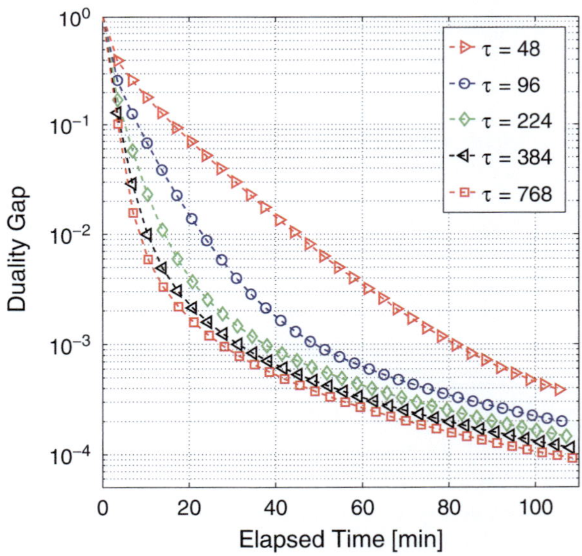

Fig. 5 Evolution of duality gap for the WebSpam dataset for various choices of τ

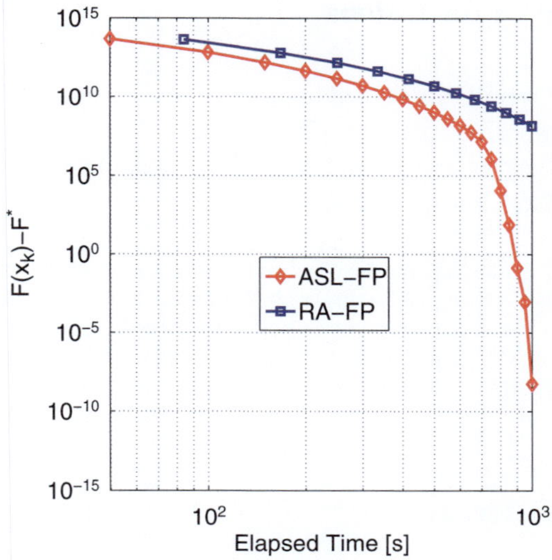

Fig. 6 Evolution of $F(x_k) - F^*$ in time. ASL-FP significantly outperforms RA-FP. The loss F is pushed down by 25 degrees of magnitude in less than 30 min (3TB problem)

Sparse Least Squares (LASSO). Next, we solved an artificial instance of sparse least squares with a matrix of $n = 10^9$ rows and $d = 5 \cdot 10^8$ columns in block-angular form:

$$A = \begin{pmatrix} A_{loc}^{(1)} & 0 & \cdots & 0 \\ 0 & A_{loc}^{(2)} & \cdots & 0 \\ \vdots & \vdots & \ddots & \vdots \\ \hline A_{glob}^{(1)} & A_{glob}^{(2)} & \cdots & A_{glob}^{(C)} \end{pmatrix}. \tag{40}$$

requiring 3 TB to store. Such matrices often arise in stochastic optimization. We used 128 nodes with 4 MPI processes on each node. Each MPI process ran 8 OpenMP threads, giving a total of 4,096 hardware threads. Each node c stored two matrices: $A_{loc}^{(c)} \in \mathbf{R}^{1,952,148 \times 976,562}$ and $A_{glob}^{(c)} \in \mathbf{R}^{500,224 \times 976,562}$. The average number of non-zero elements per row is 175 and 1,000 for $A_{loc}^{(c)}$ and $A_{glob}^{(c)}$, respectively. When communicating $g_k^{(c)}$, only entries corresponding to the global part of $A^{(c)}$ need to be communicated, and hence in RA, a *reduce all* operation is applied to vectors $\delta g_{glob}^{(c)} \in \mathbf{R}^{500,224}$. In ASL, vectors with the same length are sent. The optimal solution x^* has exactly 160,000 nonzero elements. Figure 6 compares the evolution of $F(x_k) - F^*$ for ASL-FP and RA-FP.

9 Conclusions

Overall, distributed algorithms can be both very efficient and easy to implement, when one picks the right approach. The first steps taken by the present authors over the past 2 years seem to have been validated by the considerable interest [7, 11, 27] they have generated.

Acknowledgements The first author was supported by EPSRC grant EP/I017127/1 (Mathematics for Vast Digital Resources) in 2012 and by the EU FP7 INSIGHT project (318225) subsequently. The second author was supported by EPSRC grant EP/I017127/1. The third author was supported by the Centre for Numerical Algorithms and Intelligent Software, funded by EPSRC grant EP/G036136/1 and the Scottish Funding Council.

References

1. Alham, N.K., Li, M., Liu, Y., Hammoud, S.: A MapReduce-based distributed SVM algorithm for automatic image annotation. Comput. Math. Appl. **62**(7), 2801–2811 (2011)
2. Bertsekas, D.P.: Nonlinear Programming, 2nd edn. Athena Scientific, Belmont (1999)
3. Bertsekas, D.P., Tsitsiklis, J.N.: Parallel and Distributed Computation: Numerical Methods. Prentice-Hall Inc., Upper Saddle River (1989)
4. Chang, E.Y., Zhu, K., Wang, H., Bai, H., Li, J., Qiu, Z., Cui, H.: PSVM: parallelizing support vector machines on distributed computers. Adv. Neural Inf. Process. Syst. **20**, 1–18 (2007)
5. Fercoq, O., Richtárik, P.: Accelerated, parallel and proximal coordinate descent. arXiv:1312.5799 (2013)
6. Fercoq, O., Richtárik, P.: Smooth minimization of nonsmooth functions with parallel coordinate descent methods. arXiv:1309.5885 (2013)
7. Fercoq, O., Qu, Z., Richtárik, P., Takáč, M.: Fast distributed coordinate descent for non-strongly convex losses. In: IEEE Workshop on Machine Learning for Signal Processing (2014)
8. Ge, D., Jiang, X., Ye, Y.: A note on the complexity of ℓ_p minimization. Math. Program. **129**(2), 285–299 (2011)
9. Hsieh,C.-J., Chang, K.-W., Lin, C.-J., Sathiya Keerthi, S., Sundararajan, S.: A dual coordinate descent method for large-scale linear SVM. In: Proceedings of the 25th International Conference on Machine Learning, ICML '08, pp. 408–415. ACM, New York (2008)
10. InfiniBand Trade Association: InfiniBand Architecture Specification, vol. 1, Release 1.0 (2005)
11. Jaggi, M., Smith, V., Takáč, M., Terhorst, J., Hofmann, T., Jordan, M.I.: Communication-efficient distributed dual coordinate ascent. In: Advances in Neural Information Processing Systems, vol. 27, 3068–3076 (NIPS 2014) http://papers.nips.cc/book/advances-in-neural-information-processing-systems-27-2014
12. Lee, Y.T., Sidford, A.: Efficient accelerated coordinate descent methods and faster algorithms for solving linear systems. In: 54th Annual Symposium on Foundations of Computer Science. IEEE, New York (2013)
13. Lewis, D.D., Yang, Y., Rose, T.G., Li, F.: Rcv1: A new benchmark collection for text categorization research. J. Mach. Learn. Res. **5**, 361–397 (2004)
14. LIBSVM Data: http://www.csie.ntu.edu.tw/~cjlin/libsvmtools/datasets/binary.html. Accessed 25 Oct 2014
15. Liu, J., Wright, S.J., Ré, C., Bittorf, V.: An asynchronous parallel stochastic coordinate descent algorithm. arXiv:1311.1873 (2013)

16. Lu, Z., Xiao, L.: On the complexity analysis of randomized block-coordinate descent methods. arXiv:1305.4723 (2013)
17. Luo, Z.Q., Tseng, P.: On the convergence of the coordinate descent method for convex differentiable minimization J. Optim. Theory Appl. **72**(1), pp 7–35 (1992) http://link.springer.com/article/10.1007%2FBF00939948?LI=true
18. Natarajan, B.K.: Sparse approximate solutions to linear systems. SIAM J. Comput. **24**(2), 227–234 (1995)
19. Necoara, I., Clipici, D.: Distributed coordinate descent methods for composite minimization. arXiv:1312.5302 (2013)
20. Nesterov, Y.: Introductory Lectures on Convex Optimization. Applied Optimization, vol. 87. Kluwer, Boston (2004)
21. Nesterov, Y.: Efficiency of coordinate descent methods on huge-scale optimization problems. SIAM J. Optim. **22**(2), 341–362 (2012)
22. Niu, F., Recht, B., Ré, C., Wright, S.J.: Hogwild!: a lock-free approach to parallelizing stochastic gradient descent. Adv. Neural Inf. Process. Syst. **24**, 693–701 (2011)
23. OpenMP Architecture Review Board: OpenMP Application Program Interface (2011)
24. Patrascu, A., Necoara, I.: Random coordinate descent methods for ℓ_0 regularized convex optimization. arXiv:1403.6622 (2014)
25. Richtárik, P., Takáč, M.: Efficient serial and parallel coordinate descent methods for huge-scale truss topology design. In: Operations Research Proceedings 2011, pp. 27–32. Springer, New York (2012)
26. Richtárik, P., Takáč, M.: Parallel coordinate descent methods for big data optimization. Mathematical Programming, DOI:10.1007/s10107-015-0901-6 (2012)
27. Richtárik, P., Takáč, M.: Distributed coordinate descent method for learning with big data. arXiv:1310.2059 (2013)
28. Richtárik, P., Takáč, M.: On optimal probabilities in stochastic coordinate descent methods. arXiv:1310.3438 (2013)
29. Richtárik, P., Takáč, M.: Iteration complexity of randomized block-coordinate descent methods for minimizing a composite function. Math. Program. **144**(1–2), 1–38 (2014)
30. Saha, A., Tewari, A.: On the finite time convergence of cyclic coordinate descent methods. SIAM J. Optim. **23**(1), 576–601 (2013)
31. Salleh, N.S.M., Suliman, A., Ahmad, A.R.: Parallel execution of distributed SVM using MPI (CoDLib). In: Information Technology and Multimedia (ICIM), pp. 1–4. IEEE (2011)
32. Scherrer, C., Tewari, A., Halappanavar, M., Haglin, D.: Feature clustering for accelerating parallel coordinate descent. Adv. Neural Inf. Process. Syst. **25**, 28–36 (2012)
33. Shalev-Shwartz, S., Zhang, T.: Stochastic dual coordinate ascent methods for regularized loss. J. Mach. Learn. Res. **14**(1), 567–599 (2013)
34. Shalev-Shwartz, S., Singer, Y., Srebro, N., Cotter, A.: Pegasos: primal estimated sub-gradient solver for SVM. Math. Program. **127**(1), 3–30 (2011)
35. Snir, M., Otto, S., Huss-Lederman, S., Walker, D., Dongarra, J.: MPI-The Complete Reference, Volume 1: The MPI Core, 2nd (revised) edn. MIT Press, Cambridge, MA (1998)
36. Takáč, M., Bijral, A.S., Richtárik, P., Srebro, N.: Mini-batch primal and dual methods for SVMs. J. Mach. Learn. Res. W&CP **28**, 1022–1030 (2013)
37. Tappenden, R., Richtárik, P., Gondzio, J: Inexact coordinate descent: complexity and preconditioning. arXiv:1304.5530 (2013)
38. Tappenden, R., Richtárik, P., Büke, B.: Separable approximations and decomposition methods for the augmented lagrangian. Optim. Methods Softw. arXiv:1308.6774 (2015) Doi:10.1080/10556788.2014.966824
39. Tappenden, R., Richtárik, P., Takáč, M.: On the complexity of parallel coordinate descent. Technical Report ERGO 15-001, The University of Edinburgh (2015) http://arxiv.org/abs/1503.03033
40. Tseng, P.: Convergence of a block coordinate descent method for nondifferentiable minimization. J. Optim. Theory Appl. **109**(3), 475–494 (2001)

41. Tseng, P., Yun, S.: A coordinate gradient descent method for nonsmooth separable minimization. Math. Program. **117**(1), 387–423 (2008)
42. Tseng, P., Yun, S.: Block-coordinate gradient descent method for linearly constrained nonsmooth separable optimization. J. Optim. Theory Appl. **140**, 513–535 (2009)
43. Zhao, P., Zhang, T.: Stochastic optimization with importance sampling. arXiv:1401.2753 (2014)

Notation Glossary

	Optimization problem	
N	Dimension of the optimization variable	(1)
x,h	Vectors in $\mathbf{R}^N$	
F	$F = f + \Omega$ (loss / objective function)	(1)
F^*	Optimal value, we assume $F^* > -\infty$	
f	Smooth convex function ($f : \mathbf{R}^N \to \mathbf{R}$)	(1)
Ω	Convex block separable function ($\Omega : \mathbf{R}^N \to \mathbf{R} \cup \{+\infty\}$)	(1)
	Block structure	
n	Number of blocks	
$[n]$	$[n] = \{1,2,\dots,n\}$ (the set of blocks)	Section 2
N_i	Dimension of block i ($N_1 + \cdots + N_n = N$)	Section 2
U_i	An $N_i \times N$ column submatrix of the $N \times N$ identity matrix	Section 2
$x^{(i)}$	$x^{(i)} = U_i^T x \in \mathbf{R}^{N_i}$ (block i of vector x)	Section 2
$\nabla_i f(x)$	$\nabla_i f(x) = U_i^T \nabla f(x)$ (block gradient of f associated with block i)	Section 2
L_i	Block Lipschitz constant of the gradient of f	(5)
L	$L = (L_1,\dots,L_n)^T \in \mathbf{R}^n$ (vector of block Lipschitz constants)	
w	$w = (w_1,\dots,w_n)^T \in \mathbf{R}^n$ (vector of positive weights)	
$\|x\|_w$	$\|x\|_w = (\sum_{i=1}^n w_i \|x^{(i)}\|_{(i)}^2)^{1/2}$ (weighted norm associated with x)	(3)
Ω_i	i-th component of $\Omega = \Psi_1 + \cdots + \Omega_n$	(7)
$\mu_\Omega(W)$	Strong convexity constant of Ω with respect to the norm $\|\cdot\|_w$	(20)
$\mu_f(W)$	Strong convexity constant of f with respect to the norm $\|\cdot\|_w$	(20)
J	Subset of $\{1,2,\dots,n\}$	
$x_{[Z]}$	Vector in $\mathbf{R}^N$ formed from x by zeroing out blocks $x^{(i)}$ for $i \notin Z$	(2)

	Block samplings	
ω	Degree of partial separability of f	Assumption 1
$\hat{Z}, Z_k$	Distributed block samplings (random subsets of $\{1,2,\ldots,n\}$)	Section 3
C	Number of nodes (partitions)	Section 3
τ	# of blocks updated in 1 iteration within one partition	
$\{P^{(c)}\}_{c=1}^{C}$	Partition of $[n]$ onto C parts	
	Algorithm	
β	Stepsize parameter depending on f and $\hat{Z}$	
$h^{(i)}(x)$	$h^{(i)}(x) = (h(x))^{(i)} = \arg\min_{t\in\mathbf{R}^{N_i}} \langle\nabla_i f(x), t\rangle + \frac{\beta w_i}{2}\|t\|_{(i)}^2 + \Omega_i(x^{(i)}+t)$	(8)

Models for Optimization of Power Systems

Paolo Pisciella, Marida Bertocchi, and Maria Teresa Vespucci

Abstract This chapter provides an overview of possible approaches that can be outlined to model and analyze the decision problems encountered in different stages of power production and delivery. The introduced models can be used for the control of two of the most important activities in power system management: production and transmission. In both cases, we describe how a single producer or an entire system can draw benefits from using optimization techniques for fine-tuning the expansion decisions to be taken. The theoretical basis for the analysis is drawn from different branches of operational research and optimization, ranging from mixed integer linear programming to stochastic programming and bilevel programming.

Keywords Distributed coordinate descent • Empirical risk minimization • Support vector machine • Big data optimization • Partial separability • Huge-scale optimization • Iteration complexity • Expected separable over-approximation • Composite objective • Convex optimization • Communication complexity

1 Management of Electrical Power Systems

Society depends on electricity for almost all its activities, from manufacturing to business to leisure. Electricity is a secondary energy source as it must be generated by converting primary energy sources, such as coal, natural gas, oil, solar, wind, and water streams. Electricity requires an exact match of demand and supply at any time: the limited storage possibilities make it necessary to generate electrical power according to the current load and deliver it by transmission and distribution networks. There is no difference in electricity generated by any two sources, which implies that it is suitable for trading. However, it is characterized by bounds on possibilities of transmission, which implies that there is not a global market.

The power industry has undergone an increasing pressure by governments, large industries, and investors to privatize and deregulate in the quest for efficiency

P. Pisciella • M. Bertocchi • M.T. Vespucci (✉)
Department of Management, Economics and Quantitative Methods, University of Bergamo, Via dei Caniana 2, 24127 Bergamo, Italy
e-mail: paolo.pisciella@unibg.it; marida.bertocchi@unibg.it; maria-teresa.vespucci@unibg.it

M. Al-Baali et al. (eds.), *Numerical Analysis and Optimization*, Springer Proceedings in Mathematics & Statistics 134, DOI 10.1007/978-3-319-17689-5_12

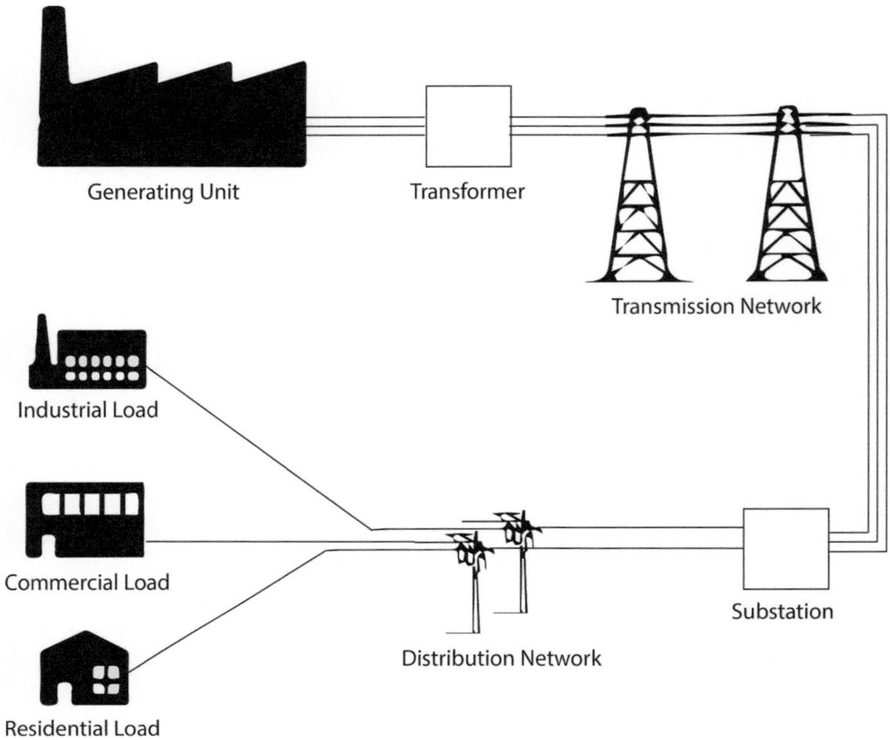

Fig. 1 Structure of an electrical power system

and increase of social welfare. Vertically integrated utilities, which were the past standard structure for electricity generation and delivery, have nowadays assumed a deregulated structure, which is introducing several challenges in planning and operating power systems. The power industry is composed of four main activities: generation, dispatch, transmission, and distribution, see Figure 1. For a detailed explanation of the composition of a Power System and its possible developments, the reader may refer to [49].

Generation is the production of energy from primary sources and represents the first step of the process. Thermal generation plants use either coal or oil; in Combined Cycle plants, two thermal sources in sequence are used, namely gas and steam. Since the 1973 oil crisis there is a growing interest in the possibility of using renewable sources. Because of the technological advances, exploitation of renewable sources has become more and more possible and has been incentivized by the introduction of a regulatory system which subsidizes their usage. A clear example of these rules is the obligation for the Generation Companies (GenCos) to provide a percentage of their total generation using renewable sources. Alternatively, they can purchase an equivalent amount of rights, named Green Certificates, for the part of production not covered by usage of renewable sources. Similarly, GenCos are compelled to the purchase of Emission Permits, in proportion to the amount of CO_2 produced when using thermal sources. In the competition among GenCos the entrance of a new actor is usually limited by the uncertainties on their return on investments; it is also limited by the power that can be injected into the network, therefore, in order to foster competition, major network upgrades may be necessary.

Dispatch is used to balance the electrical system in order to guarantee matching between load and generation at any time: it is based on short-term forecasts of the load and generation and may require the generation units to deviate from their initial production plan, in order to eliminate possible unbalances between demand and supply.

Transmission is the activity of transporting electricity from the generation units to the consumption areas. Activities related to transmission and dispatch are usually considered as a natural monopoly, to be operated on a fair and transparent basis by an independent system operator (ISO). There are two different options for the ownership of the transmission network: in the first model, named ISO-GridCo, the ISO operates the network, whose property is left to many different network companies or the State; in the second model, named TransCo, the ISO is also the owner of the transmission system [7, 37].

Distribution is the last stage of the delivery process to the end user after generation and transmission. It is carried out at lower voltage compared to transmission. This activity can be carried out by different actors, which also collect the usage fees from end users.

In electricity markets exchanges are set by identifying a matching point and the exchange price. Given bids and offers, the market is cleared by the so-called market operator (MO), while guaranteeing the balance between demand and supply. Exchange prices are defined in the day-ahead market matching offers from generators and bids from consumers at each node of the network in order to develop a classic supply-demand equilibrium price. Such match is performed on a hourly basis and calculated separately for areas, according to the extent of congestion and constraints binding the cross areas transmission. The prices calculated at each node of the transmission network are supposed to reflect the marginal cost that the system would bear for a unit load increase in a given location in correspondence of an optimized power flow re-dispatch within the network. For this reason this price is referred to as locational marginal price (LMP). The relatively low possibilities for storage, counterbalanced by fluctuating demand and supply levels imply the need for the TransCo or ISO to coordinate the dispatch of generating units to meet the expected demand across the transmission network. The way electricity flows into the transmission network is determined by physical laws, therefore the amount of losses and congestions in a particular branch of the network will determine the responses that GenCos will provide at each node of the network and, depending on the level of load coverage, the related LMPs.

Before deregulation, most elements of the power industry were heavily regulated. Deregulation meant new challenges and new business structures for the players involved in energy production and delivery. However, despite changes in different structures, market rules, and uncertainties, the underlying requirements for power system operations to be secure, economical, and reliable remain the same [3, 22, 26, 50]. This has increased the need for tools dedicated to offering support to decisions in order to optimize and manage operations both under a single GenCo perspective and under a wide system perspective. The former approach typically pursues the objective of maximizing the profitability stemming from energy productions from various sources. Typical decisions at this level reflect the amount of power to be produced under a given time window, the choice of the fuel sources and possible generation expansion plans encompassing different types of generation units. The latter approach is mainly focused on optimizing the system performance, reducing social costs or maximizing a social welfare function and providing guarantees for reliable system operations, especially for what concerns transmission and dispatch of power flows. This is accomplished by redirecting the electricity flows to nodes where there is more need and withdraw it from nodes where there is more generation capacity. In addition, long-term decisions planning on upgrades of the network are considered at this level in order to avoid congestions and reduce the costs bore by the community.

In this chapter we present three models, each reflecting to a different level of granularity and details the optimization of the various problems related to the electrical power system. We will start introducing a model for supporting the planning of power generation expansion for a single power producer. Then we will shift our focus on the network, introducing two models for grid upgrade, the former based on a centralized perspective, in which the ISO plans the network expansion problem taking as granted the reactions of the power producers injecting power in the network, while the latter takes a decentralized perspective, where decisions on generation expansion is included in the framework as reactions of the GenCos to the decisions taken by the TransCo.

2 A Model for Generation Expansion Planning via Time-Consistent Risk Averse Stochastic Programming

In the literature, models for power generation expansion have been proposed with both a deterministic and a stochastic perspective (see, e.g., [2, 8, 17, 18, 25, 43–47]). In this section we present a decision support model for a power producer who wants to determine the optimal planning for investment in power generation capacity in a long-term horizon (typically, 25 years or more). The power producer operates in a liberalized electricity market, where rules are issued by the Regulatory Authorities with the aim of promoting the development of power production systems with reduced CO_2 emissions. Indeed, CO_2 emission allowances have to be bought by the power producer as a payment for the emitted CO_2. Moreover, the Green Certificate scheme supports power production from renewable energy sources (RES), i.e. by geothermal, wind, biomass, and hydro power plants, and penalizes production from conventional power plants, i.e. CCGT, coal, and nuclear power plants. Every year a prescribed ratio is required between the electricity produced from RES and the total electricity produced. If the ratio attained in a given year is less than the prescribed one, the power producer has to buy Green Certificates in order to satisfy the related constraint; on the contrary, if the realized ratio is greater than the prescribed one, the power producer can sell Green Certificates in the market.

The power producer, assumed to be price-taker, aims at maximizing, the profit over the planning period. Revenues from sale of electricity depend on the electricity market price and on the amount of electricity sold, which is bounded above by the power producer's market share and also depends on the number of operating hours per year of the power plants in the production system. Investment costs depend on the plant rated power and on the investment costs per power unit: typically for thermal power plants, rated powers are higher and unit investment costs are lower than for RES power plants. For conventional power plants, variable generation costs are highly dependent on the fuel prices. Revenues and costs associated with Green Certificate scheme depend on the Green Certificate price, as well as on the actual ratio between production from RES and total annual production realized by the producer every year. Finally, costs for emitting CO_2 depend on the price of the emission allowances, as well as on the amount of CO_2 emitted, that greatly varies among production technologies.

We notice that the evolution of both electricity prices and fuel costs along the planning period are not known at the time when the investment decisions have to be done, therefore a risk is associated with the capacity expansion problem, due to the uncertainty of prices and of demand (see [10]). Uncertainty is included in the model by a multistage scenario tree representing the evolution of the uncertain information on electricity prices and cost of fuels per year. Techniques to measure the quality of using a stochastic multistage model over a deterministic one have attracted some interest in recent research, see, e.g., [23].

The proposed decision support model determines the evolution of the production system along the planning horizon. In order to take into account that power plants greatly differ in terms of construction time and industrial life, the model determines for every technology the number of power plants whose construction has to be started in every year of the planning period. Each new power plant is then available for production when its construction is completed and its industrial life is not ended.

The uncertainty is modeled by a three-stage scenario tree that represents the set Ω of alternative scenarios assumed for the evolution of prices of electricity and costs of fuels along the planning horizon. The scenario tree is defined by a set N of nodes and by the sets B_n of successors (children nodes) of node n, $n \in N$. A leaf node has no successors, i.e. its associated set B_n is empty. A probability value ϕ_n is associated with every node $n \in N$. Each path N_{ω}^{Ω} from the root node to a leaf node corresponds to one scenario $\omega \in \Omega$ and the probability p_{ω} of scenario ω equals the probability of its leaf node. In full generality, the planning horizon is divided in to $|S|$ stages (in our case $|S| = 3$) and every stage $s \in S$ is associated with the set of nodes N_s^S. The set I^n indicates the years of the planning horizon associated with node n.

In order to introduce the model, we define the following sets, parameters and decision variables.

Sets

S	Set of stages in which the planning horizon is divided
Ω	set of scenarios
N	Set of nodes of the scenario tree
N^S_s	Set of nodes in stage $s \in S$
N^{Ω}_{ω}	Set of nodes on the path from the root to a leaf that defines scenario $\omega \in \Omega$
B_n	Set of successors (children nodes) of node $n \in N$
I	Set of years in the planning horizon
I^n	Set of years in node $n \in N$
K^T	Set of thermal power plants owned by the power producer at the beginning of the planning horizon (i.e., in year 0)
K^R	Set of power plants using RES owned in year 0
K	Set of all power plants owned in year 0 (i.e., $K = K^T \cup K^R$)
J^T	Set of candidate technologies for thermal power production
J^R	Set of candidate technologies for power production from RES
J	Set of all candidate technologies (i.e., $J = J^T \cup J^R$)

For instance, in the scenario tree depicted in Figure 2 $N = \{1,2,\ldots,21\}$, $N^S_3 = \{6,\ldots,21\}$ and $N^{\Omega}_4 = \{1,2,9\}$.

Parameters

p_{ω}	(–)	Probability of scenario $\omega \in \Omega$
p^N_n	(–)	Probability of node $n \in N$
ϕ_n	(–)	Conditional probability of reaching node $n \in N$ from its predecessor
$v^J_{j,i,n}$	(k€/GWh)	Variable production cost of a thermal power plant of candidate technology $j \in J^T$ in year $i \in I_t$ in node $n \in N^S_s$, $s \in S$
$v^K_{k,i,n}$	(k€/GWh)	Variable production cost of thermal power plant $k \in K^T$ in year i in node n
$\pi^E_{i,n}$	(k€/GWh)	Market electricity price in year i in node n
$\pi^{GC}_{i,n}$	(k€/GWh)	Green Certificate price in year i in node n
$\pi^{CO_2}_{i,n}$	(k€/t)	CO_2 emission permit price in year i in node n
$\overline{M}_{i,n}$	(GWh)	Demand in year i in node n up to the producer
S_j	(years)	Construction time of a power plant of candidate technology $j \in J$
L^J_j	(years)	Industrial life of a power plant of candidate technology $j \in J$
$\overline{Z}_j$	(–)	Number of sites ready for constructing a power plant of candidate technology $j \in J$
$\overline{P}^J_j$	(MW)	Rated power of a power plant of candidate technology $j \in J$

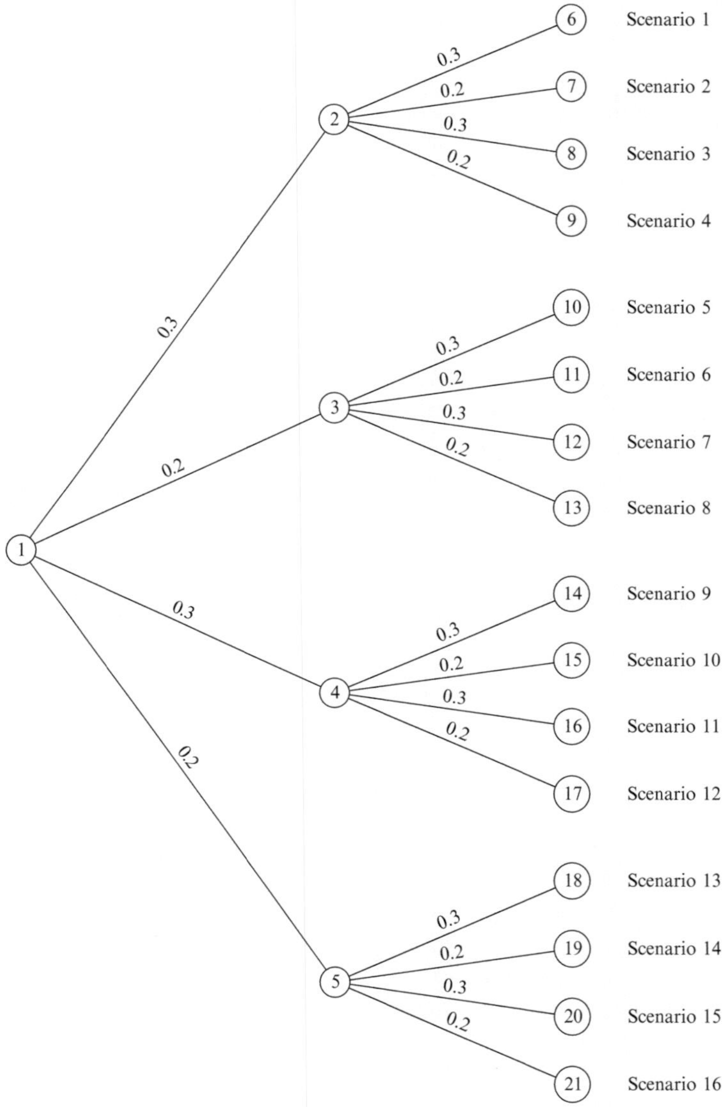

Fig. 2 Example for Scenario tree with assigned probabilities

H_j^J	(h)	Operating hours per year of a power plant of candidate technology $j \in J$
v_j^J	(–)	Percentage of loss of a power plant of technology $j \in J$
$\overline{E}_{j,i}^J$	(GWh)	Maximum energy produced by a power plant of technology $j \in J$ in year $i \in I$
θ_j^J	(t/GWh)	CO_2 emission rate of a thermal power plant of candidate technology $j \in J^T$

C_j	(M€ /MW)	Investment cost of a power plant of candidate technology $j \in J$
R_j	(k€)	Annualized investment cost of a power plant of candidate technology $j \in J$
f_j^J	(k€)	Fixed production cost of a power plant of technology $j \in J$
v_j^J	(k€)	Variable production cost of an RES power plant of candidate technology $j \in J^R$
L_k^K	(years)	Residual life of power plant $k \in K$ owned by the power producer in year 0
$\overline{P}_k^K$	(MW)	Rated power of power plant $k \in K$
H_k^K	(h)	Operating hours per year of power plant $k \in K$
v_k^K	(–)	Percentage of loss of power plant $k \in K$
$\overline{E}_{k,i}^K$	(GWh)	Maximum energy produced by power plant $k \in K$ in year $i \in I$
θ_k^K	(t/GWh)	CO_2 emission rate of thermal power plant $k \in K^T$
f_k^K	(k€)	Fixed production cost of power plant $k \in K$
v_k^K	(k€)	Variable production cost of RES power plant $k \in K^R$
β_i	(–)	Ratio “electricity from RES / total electricity produced” to be attained in year $i \in I$
B	(M€)	Budget available
r	(–)	Interest rate

Decision Variables

$w_{j,i,n}$	(–)	Integer number of power plants of technology $j \in J$ whose construction is to start in year $i \in I^n$ in node $n \in N$
$W_{j,i,n}$	(–)	Number of power plants of technology $j \in J$ available for production in year $i \in I^n$ in node $n \in N$
$E_{j,i,n}^J$	(GWh)	Electricity produced by all power plants of technology $j \in J$ in year i in node n
$E_{k,i,n}^K$	(GWh)	Electricity produced by existing power plant $k \in K$ in year i in node n
$G_{i,n}$	(GWh)	Green Certificates sold ($G_{i,n} \geq 0$) or bought ($G_{i,n} \leq 0$) in year i in node n
$Q_{i,n}$	(t)	CO_2 produced in year i in node n

The decision variables $w_{j,i,n}$, $W_{j,i,n}$, $E_{j,i,n}^J$, $E_{k,i,n}^K$, $G_{i,n}$, and $Q_{i,n}$ are determined so as to

$$\max \sum_{\omega \in \Omega} p_\omega \sum_{n \in N^\Omega_\omega} \sum_{i \in I^n} \frac{1}{(1+r)^i} \left[\pi^E_{i,n} \left(\sum_{j \in J} E^J_{j,i,n} + \sum_{k \in K} E^K_{k,i,n} \right) + \pi^{GC}_{i,n} G_{i,n} - \pi^{CO_2}_{i,n} Q_{i,n} + \right.$$
$$- \sum_{j \in J^T} v^J_{j,i,n} E^J_{j,i,n} - \sum_{j \in J^R} v^J_{j,i,n} E^J_{j,i,n} - \sum_{j \in J} \left(R_j + f^J_j \right) W_{j,i,n} +$$
$$\left. - \sum_{k \in K^T} v^K_{k,i,n} E^K_{k,i,n} - \sum_{k \in K^R} v^K_{k,i,n} E^K_{k,i,n} - \sum_{k \in K} f^K_k \right] \tag{1}$$

subject to

$$\sum_{n \in N^\Omega_\omega} \sum_{i \in I^n} w_{j,i,n} \le \overline{Z}_j\,, \qquad j \in J, \quad \omega \in \Omega \tag{2}$$

$$W_{j,i,n} = \sum_{h \in N^\Omega_\omega} \sum_{l \in I^n \cap [i-S_j-L_j+1, i-S_j]} w_{j,l,h}, \qquad j \in J, \quad i \in I, \quad n \in N^\Omega_\omega, \quad \omega \in \Omega \tag{3}$$

$$\sum_{n \in N^\Omega_\omega} \sum_{i \in I^n} \frac{1}{(1+r)^i} \left(\sum_{j \in J} R_j W_{j,i,n} \right) \le B, \qquad \omega \in \Omega \tag{4}$$

$$0 \le E^J_{j,i,n} \le \overline{E}^J_{j,i} W_{j,i,n}\,, \qquad j \in J\,, \quad \forall i \in I\,, \quad n \in N^\Omega_\omega, \quad \omega \in \Omega\,. \tag{5}$$

$$0 \le E^K_{k,i,n} \le \overline{E}^K_{k,i}\,, \qquad k \in K\,, \quad i \in I \quad n \in N^\Omega_\omega, \quad \omega \in \Omega \tag{6}$$

$$\sum_{j \in J} E^J_{j,i,n} + \sum_{k \in K} E^K_{k,i,n} \le \overline{M}_{i,n}\,, \qquad i \in I\,, \quad n \in N^\Omega_\omega, \quad \omega \in \Omega \tag{7}$$

$$G_{i,n} = \sum_{j \in J^R} E^J_{j,i,n} + \sum_{k \in K^R} E^K_{k,i,n} - \beta_i \left(\sum_{j \in J} E^J_{j,i,n} + \sum_{k \in K} E^K_{k,i,n} \right), \qquad i \in I, \quad n \in N^\Omega_\omega, \quad \omega \in \Omega \tag{8}$$

$$Q_{i,n} = \sum_{j \in J^T} \theta^J_j \cdot E^J_{j,i,n} + \sum_{k \in K^T} \theta^K_k \cdot E^K_{k,i,n}\,, \qquad i \in I, \quad n \in N^\Omega_\omega, \quad \omega \in \Omega \tag{9}$$

$$w_{j,i,n} \in \mathbb{Z}_+\,, \qquad j \in J\,, \quad i \in I \quad n \in N^\Omega_\omega, \quad \omega \in \Omega \tag{10}$$

The cost for investment in new power plants of technology j is annualized and represented by $R_j W_{j,i,n}$ with R_j computed by the usual formula for the periodic payment of an annuity along L^J_j years, i.e.

$$R_j = \frac{C_j P_j^J \cdot r \cdot 1000}{1 - (\frac{1}{1+r})^{L_j^J}}, \qquad j \in J \tag{11}$$

Constraints (2) guarantee that for every candidate technology j the total number of new power plants constructed along the planning horizon is not greater than the number $\overline{Z}_j$ of sites ready for construction of a new power plant, i.e. sites for which all administrative permits have been released. These upper bounds could also be considered as varying from year to year. Constraints (3) determine for every year i and node n the number of new power plants of technology j available for production, i.e. those plants for which both construction is completed and industrial life is not ended. The sum of the actualized annual equivalent investment costs, which depend on the number of new power plants of each technology j available for production in every year i and node n, is required by constraint (4) not to exceed the available budget.

Bounds on electricity production are set by constraints (5) and (6). The annual electricity production obtained by all new power plants of technology j is nonnegative and bounded above by the product of the maximum annual production $\overline{E}_{j,i}^J$ of a power plant of technology j, defined as

$$\overline{E}_{j,i}^J = \frac{1}{1000} \overline{P}_j^J H_j^J (1 - v_j^J), \qquad j \in J\,, \qquad \forall i \in I \tag{12}$$

times the number of new plants of technology j available for production in year i. The annual electricity production of power plant k is nonnegative and bounded above by the maximum annual production $\overline{E}_{k,i}^K$ in year i with

$$\overline{E}_{k,i}^K = \begin{cases} \frac{1}{1000} \overline{P}_k^K H_k^K (1 - v_k^K) & \text{if} \quad i \le L_k^K \\ 0 & \text{if} \quad i > L_k^K. \end{cases} \tag{13}$$

Parameters H_j^J and H_k^K take into account possible plant breakdown and maintenance. Notice that for some technologies, if selected, a lower bound to the annual electricity production could be imposed in order to take into account technical limitations.

The electricity generated in year i cannot exceed the power producer's demand at node n in year i, as expressed in constraint (7). The amount of electricity $G_{i,n}$, for which in year i under node n the corresponding Green Certificates are bought, if $G_{i,n} \le 0$, or sold, if $G_{i,n} \ge 0$, is defined by constraints (8), where β_i is the ratio, required in year i, between the electricity produced from RES and the total electricity produced.

The amount $Q_{i,n}$ of CO_2 emissions that the power producer must pay for in year i under node n is defined by constraint (9), where θ_k^K and θ_j^J are the CO_2 emission rates of thermal power plant $k \in K^T$ and thermal power plant of candidate technology $j \in J^T$, respectively.

A risk averse extension of the model (1)–(10) can be obtained by using a risk measure that takes into account the potential losses of each decision. When maximizing profits a widely accepted risk measure is the conditional value at risk (CVaR) of profits, i.e. the barycentral value of profits in the worst α percentile of cases, see [33] and [34]. We define the multi-stage CVaR in a nested fashion, following the approach featured on Philpott et al. [27] and Rudloff et al. [39] which iteratively solve a convex combination of performance and risk in the last stage, using it as the performance measure for the previous stage. This approach has been borrowed from dynamic programming and it has been used by several authors in the stochastic programming framework (see, e.g., [27, 38, 39]). At every stage s two auxiliary variables, d_n^s and η_s, are defined, where η_s plays the same role as the VaR at the optimal value of a two-stage CVaR constrained problem. We also denote F_n^N as the discounted value of profits in node n. Finally we denote parameters α_s and ρ_s as the stage-wise confidence parameter and weighting factor of performance and risk, respectively. We use a weighting factor ranging between zero and one, where zero denotes risk neutrality, which corresponds to neglecting risk, whilst one defines complete aversion towards risk. Usage of a convex combination of the present value of expected profits and CVaR is a standard procedure when striking a balance between risk and performance (see, e.g., [10]) and it turns out to be a particularly suitable approach when performance and risk are measured in the same units (see, for example, [6, 10, 45–48]).

We have that our definition at stage three is given by

$$d_m^3 \geq 0 \qquad d_m^3 \geq \eta_{3n} - F_m^N \qquad n \in N_2^S, \quad m \in B_n \tag{14}$$

then for each node of stage two we define

$$d_n^2 \geq 0 \qquad d_n^2 \geq \eta_2 - K_n^N \qquad n \in N_2^S \tag{15}$$

where, for each node of stage two we have

$$K_n^N = F_n^N + (1-\rho_3)\left(\sum_{m \in B_n} \phi_m F_m^N\right) + \rho_3\left(\eta_{3n} - \frac{1}{\alpha_3}\sum_{m \in B_n} \phi_m d_m^3\right) \qquad n \in N_2^S \tag{16}$$

which defines the objective function at the third stage conditioned to reaching node $n \in N_2^S$ in stage 2. Objective function in stage one is given by

$$F_1^S + (1-\rho_2)\left(\sum_{n \in N_2^S} \phi_n K_n^N\right) + \rho_2\left(\eta_2 - \frac{1}{\alpha_2}\sum_{n \in N_2^S} \phi_n d_n^2\right) \tag{17}$$

Time inconsistency is one of the main drawbacks of using CVaR in a multi-stage framework. Let the optimal solution of the risk averse multi-stage stochastic problem be computed, let the optimal decisions be assigned to the corresponding variables on the path from the root to node n in an intermediate stage and let the problem be solved for the subtree emanating from node n. If the optimal values of the subtree problem coincide with those computed on the complete problem, the solution is time consistent, otherwise it is time inconsistent. In other words, an optimal policy is time consistent if and only if the future planned decisions are actually going to be implemented. This property does not generally hold for risk averse optimality problems. However, the CVaR modeling approach used in this framework satisfies time consistency as it is characterized by properties stated in [38], Theorem 1. The model can be used to analyze how the production mix changes with different amounts of budget available to the GenCo. Let us assume that the GenCo can install new plants pertaining to four different technologies: Combined Cycle (CCGT), Coal, Wind, and Nuclear. Years from 1 to 5 define the first stage of the model, years from 6 to 10 define the second stage, and years from 11 to 25 define the third stage. The decision to be taken is related to the expansion plan of years 1–5. The power producer will consider how the decision taken for these 5 years will impact on future possible decisions depending on how the uncertainty on energy prices and fuel costs will unveil. This recursive decisions are taken in years 6–10, and for each recursive decision related to a given realization of the random events another set of possible reactions are considered depending on how prices will evolve in years 11–25. Scenarios are generated to account for 1,296 possible final joint realizations of the random variables. For further details on the implementation, refer to [30].

We observe the effects of an increasing budget availability on the installation of new power plants. For low amounts of budget, Combined Cycle is preferred over Coal because of its higher and more stable productivity. When budget availability increases, investments move towards wind technology, which has lower operational cost but must be installed more extensively, and finally to coal. The effects of budget increase on the Net Present Value of profits displays a monotonically increasing behavior.

For what concerns the effects of shifts on risk aversion no production uncertainty for renewable sources is considered. Therefore the model considers wind as a production source yielding a higher level of profit stability, even though it comes at the expense of a lower average production capacity. In the considered model, coal is the most profitable technology but, at the same time, it bears a very uncertain pattern for operational costs, while CCGT entails a more stable costs profile. As risk aversion becomes higher, investments in coal are postponed to future years, in order to exploit the knowledge on the unveiling scenarios, while investments in CCGT increase. Net Present Value of profits decreases as the relative importance of risk increases.

Possible other approaches to tackle the considered problem could encompass the usage of a two-stage stochastic structure for the optimization problem. Such approach would allow for consideration of various risk measures in order to define different viewpoints to model the risk preferences of the power producer.

3 A Centralized Framework for Power Generation Capacity Expansion

Planning the expansion of power generation capacity is not only important under a single investor perspective. Under a system perspective requirements on the introduction of renewable sources, pollution control, and network stability have an impact on how the additional capacity will be composed and its optimal geographical distribution. In this section we describe a deterministic multiperiod centralized model for Power Generation Capacity Expansion aiming at determining the optimal generation mix of the entire power system. With perfect competition among producers power demand is satisfied at the lowest aggregated operational and investment costs, given the legal requirements on production from renewable energy sources, as well as CO_2 emission allowance costs. The system perspective, which takes into account cumulated costs and how the production of a given generation unit influences the production level of other generation units, requires the network structure to be taken into account explicitly. In other words, it is not enough to take into account the demand for each generation unit regardless of which part of the network such demand comes from, but interconnections between nodes need to be modeled explicitly, in order to understand how the global demand is reallocated amongst the different generation units as the structure of the production capacity changes. The model has many similarities and some major difference to the one introduced in the previous section. The main differences are: the objective function defined as a cost minimization, power demand constraints in peak hours, and the modeling of power flows. As for the previous model, this approach considers investment costs depending on the plant rated power and on the investment costs per power unit. The regulator also wants to minimize possible costs linked to Green Certificates and Emissions Permits for CO_2 externalities. The following notation is introduced.

Sets

I	Set of years in the planning horizon
Z	Set of nodes in the network
L	Set of connecting lines
J^T	Set of candidate technologies for thermal power production
J^R	Set of candidate technologies for power production from RES
J	Set of all candidate technologies (i.e., $J = J^T \cup J^R$)
J_z	Set of all candidate technologies in node z
K^T	Set of thermal power plants owned by the power producer at the beginning of the planning horizon (i.e., in year 0)
K^R	Set of power plants using RES owned in year 0
K	Set of all power plants owned in year 0 (i.e., $K = K^T \cup K^R$)
K_z	Set of all power plants owned in year 0 in node z

Parameters

S_j	(years)	Construction time of a power plant of candidate technology $j \in J$
L_j^J	(years)	Industrial life of a power plant of candidate technology $j \in J$
$\overline{N}_{j,z}$	(–)	Upper bound on power plants of candidate technology $j \in J$ that can be installed in area $z \in Z$
$\overline{P}_j^J$	(GW)	Rated power of a power plant of candidate technology $j \in J$
H_j^J	(h)	Operating hours per year of a power plant of candidate technology $j \in J$
v_j^J	(–)	Percentage of loss of a power plant of technology $j \in J$
$\overline{E}_{j,i}^J$	(GWh)	Maximum energy produced by a power plant of technology $j \in J$ in year $i \in I$
θ_j^J	(t/GWh)	CO_2 emission rate of a thermal power plant of candidate technology $j \in J^T$
C_j	(M€/MW)	Investment cost of a power plant of candidate technology $j \in J$
R_j	(k€)	Annualized investment cost of a power plant of candidate technology $j \in J$
f_j^J	(k€)	Fixed production cost of a power plant of technology $j \in J$
B	(M€)	Budget available
L_k^K	(years)	Residual life of power plant $k \in K$ owned by the power producer in year 0
$\overline{P}_k^K$	(GW)	Rated power of power plant $k \in K$
H_k^K	(h)	Operating hours per year of power plant $k \in K$
v_k^K	(–)	Percentage of loss of power plant $k \in K$
$\overline{E}_{k,i}^K$	(GWh)	Maximum energy produced by power plant $k \in K$ in year $i \in I$
θ_k^K	(t/GWh)	CO_2 emission rate of thermal power plant $k \in K^T$
f_k^K	(k€)	Fixed production cost of power plant $k \in K$
β_i	(–)	Ratio "electricity from RES / total electricity produced" to be attained in year $i \in I$
$\sigma_{z,l}$		Power transfer distribution factor related to link $l \in L$ and to node $z \in Z$
r	(–)	Interest rate
$D_{z,i}^P$	(GW)	Power load in the peak hour of period i in zone z
D_i^E	(GWh)	Energy load of period i
$\overline{TR}_l$	(MW)	Maximum capacity of transmission line l
$\underline{TR}_l$	(MW)	Minimum capacity of transmission line l

Decision Variables

$w_{j,i}$	(–)	Integer number of power plants of technology $j \in J$ whose construction is to start in year $i \in I^s$
$W_{j,i}$	(–)	Number of power plants of technology $j \in J$ available for production in year $i \in I$
$E^J_{j,i}$	(GWh)	Electricity produced by all power plants of technology $j \in J$ in year i
$P^J_{j,i,z}$	(GW)	Power produced by a power plant of candidate technology $j \in J$ at time $i \in I$ from node $z \in Z$
$E^K_{k,i}$	(GWh)	Electricity produced by existing power plant $k \in K$ in year i
$P^K_{k,i,z}$	(GW)	power produced by an existing power plant of technology $k \in K$ at time $i \in I$ from node $z \in Z$
G_i	(GWh)	Green Certificates sold ($G_{i,n} \geq 0$) or bought ($G_i \leq 0$) in year i
Q_i	(t)	CO_2 produced in year i
$TR_{l,i}$	(MW)	Power flow on transmission line l in period i

The model is as follows:

$$\min \sum_{i \in I} \frac{1}{(1+r)^i} \left[-\pi_i^{GC} G_i + \pi_i^{CO_2} Q_i + \sum_{k \in K^T} v^K_{k,i} E^K_{k,i} + \sum_{k \in K^R} v^K_{k,i} E^K_{k,i} + \sum_{k \in K} f^K_k + \right.$$

$$\left. + \sum_{z \in Z} \sum_{j \in J_z} \left(R_j + f^J_j\right) W_{j,i,z} + \sum_{j \in J^T} v^J_{j,i} E^J_{j,i} + \sum_{j \in J^R} v^J_{j,i} E^J_{j,i} \right] \tag{18}$$

subject to

$$\sum_{i \in I} w_{j,i,z} \leq \overline{N}_{j,z} \,, \qquad j \in J, \quad z \in Z \tag{19}$$

$$W_{j,i,z} = \sum_{l \in I \cap \left[i - S_j - L_j + 1, i - S_j\right]} w_{j,l,z} \qquad z \in Z, \quad j \in J_z, \quad i \in I \tag{20}$$

$$\sum_{z \in Z} \sum_{i \in I} \frac{1}{(1+r)^i} \left(\sum_{j \in J} R_j W_{j,i,n} \right) \leq B \tag{21}$$

$$0 \leq E^J_{j,i} \leq \sum_{z \in Z_j} \overline{E}^J_{j,i} W_{j,i,z} \qquad j \in J, \quad i \in I \,. \tag{22}$$

$$0 \leq E^K_{k,i} \leq \overline{E}^K_{k,i} \,, \qquad k \in K, \quad i \in I \tag{23}$$

$$0 \leq P^J_{j,i,z} \leq \overline{P}^J_{j,i} W_{j,i,z} \,, \qquad z \in Z, \quad j \in J_z, \quad i \in I \tag{24}$$

$$0 \leq P^K_{k,i,z} \leq \overline{P}^K_{k,i}\,, \qquad z \in Z, \quad k \in K_z, \quad i \in I \tag{25}$$

$$TR_{l,i} = \sum_{z \in Z} \sum_{k \in K_z} \sigma_{z,l} P^K_{k,i,z} + \sum_{z \in Z} \sum_{j \in J_z} \sigma_{z,l} P^J_{j,i,z} - \sum_{z \in Z} \sigma_{z,l} D^P_{z,i} \quad l \in L, \quad i \in I \tag{26}$$

$$\underline{TR}_l \leq TR_{l,i} \leq \overline{TR}_l \qquad l \in L, \quad i \in I \tag{27}$$

$$\sum_{j \in J} E^J_{j,i} + \sum_{k \in K} E^K_{k,i} = D^E_i \qquad i \in I \tag{28}$$

$$\sum_{z \in Z} \sum_{j \in J_z} P^J_{j,i,z} + \sum_{z \in Z} \sum_{k \in K_z} P^K_{k,i,z} = \sum_{z \in Z} D^P_{z,i} \qquad i \in I \tag{29}$$

$$G_i = \sum_{j \in J^R} E^J_{j,i} + \sum_{k \in K^R} E^K_{k,i} - \beta_i \left(\sum_{j \in J} E^J_{j,i} + \sum_{k \in K} E^K_{k,i} \right) \qquad i \in I \tag{30}$$

$$Q_i = \sum_{j \in J^T} \theta^J_j E^J_{j,i} + \sum_{k \in K^T} \theta^K_k E^K_{k,i} \qquad i \in I \tag{31}$$

$$w_{j,i,z} \in \mathbb{Z}_+ \qquad z \in Z, \quad j \in J_z, \quad i \in I. \tag{32}$$

Parameter R_j is defined by (11) and $\overline{E}^J_{j,i}$ and $\overline{E}^K_{k,i}$ are defined by (12) and (13), respectively. The objective function (18) accounts for providing electricity to consumers at the minimum system total cost. Similarly to the previous model, constraints (20) determine the number of new power plants of technology j available for production in every year i, i.e. those plants for which both construction is completed and industrial life is not ended. Bounds (22)–(25) impose restrictions on total energy from each technology source and on power from plants at each node. Constraint (26) defines the power flows taking into account the power transfer distribution factors, that define how much of the power generated in all the nodes connected to a given link flows in the link. Bounds on power flows in every line l are defined by constraints (27). Constraints (28) and (29) define the balance for the total energy in the system and for the total power delivered in peak hours of year i. The model can be used to define the generation capacity expansion necessary to satisfy the load of the entire system at minimum cost and to determine the impact of different geographical distributions of generation plants on possible congestion of transmission lines. Results for a test case based on the Italian power system show how the optimal solution depends on the available budget. For lower levels of budget a mix of coal and wind is preferred, as it does not need high installation investments. Even though coal has a low operating cost, it also provides a low efficiency. For an increasing amount of budget, wind and coal technologies are steadily replaced by Combined Cycle and nuclear. Combined Cycle features higher investment and operational costs, but has a higher efficiency compared to coal, while nuclear has the highest investment costs but relatively low operational costs. The test case assumed that the total yearly load is shared with different proportions between the four main market areas (north 40 %, center 25 %, south 20 %, Sardinia 5 %, and Sicily 10 %). In correspondence of a very high budget a nuclear plant is installed in every area, which allows to cover the entire national load at minimum cost. In Sicily and Sardinia the nuclear plants will produce much more energy than the local demand and power needs to be transmitted to other areas, therefore inducing congestions in the lines connecting Sicily and Sardinia to the other market areas. Summarizing, this model can help in determining system expansion plans at minimum cost and whether the adoption of these plans can be prevented by possible congestions of the existing transmission lines.

4 A Leader-Followers Model of Power Transmission Capacity Expansion in a Market Driven Environment

In this section we discuss a deterministic multiperiod model for analysis of transmission grid expansion planning with competitive generation capacity planning in electricity markets. The model considers a TransCo paradigm for the ownership and operations of the network. The purpose of the model is to provide a tool for the definition of the optimal grid upgrading program in a market driven environment. Interplay between TransCo and several independent GenCos is treated as a leader-followers Stackelberg game. Such game is expressed as the following sequential decision pattern: the TransCo decides on the best possible upgrades of transmission lines at time zero and the GenCos modify their production plans over time and potential capacity expansion at time zero accordingly, reaching an equilibrium together with the MO, which clears the market providing new LMP. GenCos' reactions lead to a power production equilibrium that is properly taken into account within the TransCo decision problem, which receives back new LMPs and planned production used to determine the aggregated social costs, defined as the weighted sum of costs paid by consumers and negative aggregated GenCo profits. Problem of coordination of transmission expansion planning with competitive generation capacity planning in electricity markets has been addressed by a quite large strand of literature (see, e.g., [10, 16, 19, 21, 24, 41]), and usually solved by means of different techniques, from plain usage of Linear Programs [12, 15], to hierarchic approaches [1, 13, 14, 20, 31, 32, 40, 42] to Benders' decomposition methods aiming at computing shadow prices for LMPs and introduce cuts to prevent GenCos and transmission companies expanding capacity in an uncoordinated manner [35, 36]. The model introduced in this section tackles the coordination problem using bilevel programming techniques [4, 11]. Our aim is to find a method to obtain a global optimal solution for the TransCo, given the responses provided by GenCos. LMPs are obtained by including first order conditions of the market clearing problem into the TransCo problem.

The model defines a sequential game between three players: the TransCo, a group of GenCos and the MO. The model is structured in two different, interrelated decision levels that represent the sequential nature of the decisions up to each player. Namely, the TransCo will take its decision on the transmission structure as first mover, then each GenCo will decide on power production levels and potential investments according to the choice made by the TransCo. Bids provided by each GenCo are collected and sorted by a MO, which clears the market and sets the LMPs according to load and bids submitted by GenCos.

Power is delivered to consumers spread over different nodes having different load and power production capacity, which in turn depend on the amount of existing and candidate generation units. The problem refers to a medium/long time horizon, which we discretize in years. The modeling framework is displayed in Figure 3. The two-level Stackelberg game involves only the network planner as the upper level player and a group of power generating companies, whose bid-ask mechanism with the consumers is mediated by a MO, as the lower level. The TransCo aims at minimizing a social cost function, defined as the weighted sum of the total costs up to the consumers and the negative sum of the GenCos aggregated profit. The TransCo takes a decision on installation of new power transmission lines l which, together with the existing transmission lines, will define bounds for new capacities for the transmission corridors at time t. These bounds are used by the MO to define how much power can be transferred between nodes in order to cover the peak hour load of period t. Depending on the relative importance of consumers and GenCos, congestions might or might not occur, leading to several different LMPs or just one, respectively.

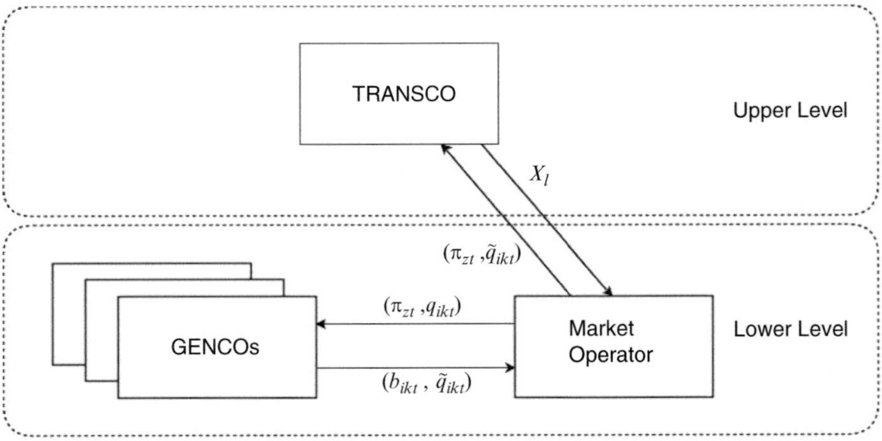

Fig. 3 Interdependencies between transmission company, generator companies and MO

Bids are sent to the MO by the GenCos in form of a pair $(b_{ikt}, \tilde{q}_{ikt})$ defining the price bid at time t from generator k belonging to GenCo i and the related quantity, respectively. The MO, given the power flow capacities defined by the TransCo, will define the LMPs π_{zt} for node z at time t and the accepted quantity q_{ikt} of power generator i. GenCos aim at maximizing their profit by deciding how much power to supply and whether to open new generation units. We assume that GenCos do not influence LMPs through strategic bidding, so that their bids simply reflect a mark-up on their marginal costs. The problem solved by the TransCo is described in the following subsection

4.1 The TransCo Problem

TransCo aims at minimizing the weighted sum of aggregated costs up to consumers and negative GenCo profits given the expansion investment budget. If the consumers have a high importance in the aggregated social cost function, the TransCo will add new lines to reduce congestion and let GenCos with low costs place a bid in different areas, lowering the value of the accepted bids and, consequently, the electricity prices. The problem is formalized as follows.

Sets

T	Set of periods t (in every period we consider the peak hour load)
I	Set of producers i
Z	Set of nodes z
L^E	Set of existing transmission lines l
L^C	Set of candidate transmission lines l
$K^E_{i,z}$	Set of existing technologies k belonging to producer i in node z
$K^C_{i,z}$	Set of candidate technologies k of producer i in node z

Parameters

$C_{z,t}$	(MW)	Load in node z in period t
f^L_l	(€)	Investment cost for opening line l
H	(€)	Total budget for lines expansion
$c_{i,k}$	(€/MWh)	Generation marginal cost of technology k for producer i
α		Weighting factor, ranging between 0 and 1, measuring the relative importance of consumers and producers in the social cost function minimized by the TransCo.

Decision Variables

$\pi_{z,t}$	(€/MWh)	LMP in node z at time t;
$\tilde{q}_{i,k,t}$	(MW)	Power produced by GenCo i from generation unit k at time t;
X_l		Binary variable taking value 1 if line l is built.

The TransCo model is as follows:

$$\min_{\pi_{z,t},\tilde{q}_{i,k,t},X_l} \alpha \sum_{t\in T}\sum_{z\in Z} C_{z,t}\pi_{z,t} + \\ -(1-\alpha)\sum_{t\in T}\delta^{-t}\sum_{z\in Z}\left(\pi_{z,t}\sum_{k\in K^E_{i,z}\cup K^C_{i,z}}\tilde{q}_{i,k,t} - \sum_{k\in K^E_{i,z}\cup K^C_{i,z}} c_{i,k}\tilde{q}_{i,k,t}\right) \tag{33}$$

subject to

$$\sum_{l\in L^C} f^L_l X_l \le H \tag{34}$$

$$(\pi_{z,t},\tilde{q}_{i,k,t}) \in \Omega(X) \qquad z\in Z,\quad k\in K^E_{i,z}\cup K^C_{i,z},\quad t\in T,\quad i\in I \tag{35}$$

$$X_l \in \{0,1\} \qquad l\in L^C \tag{36}$$

with X defining the vector whose components are X_l, $l\in L^C$.

The objective function is the convex combination of total cost paid by the consumers $\sum_{t\in T}\sum_{z\in Z} C_{z,t}\pi_{z,t}$ and the negative sum of the discounted operations profits obtained by the GenCos $\sum_{t\in T}\delta^{-t}\sum_{z\in Z}\left(\pi_{z,t}\sum_{k\in K^E_{i,z}\cup K^C_{i,z}}\tilde{q}_{i,k,t} - \sum_{k\in K^E_{i,z}\cup K^C_{i,z}} c_{i,k}\tilde{q}_{i,k,t}\right)$. Inequality (34) is the budget constraint to investment cost for lines expansion. $\Omega(X)$ represents the space of joint solutions of problems solved by the MO and the GenCos, which are introduced in the following two subsections, parametrized by vector X. Such set contains the possible equilibria for the involved GenCos and for the MO. We have considered that the main objective of the TransCo is securing a viable power transmission by removing congestion between market areas and, more generally, reducing social costs. As such, investment costs have only been considered as a constraint, without modeling the effects in the objective function. The TransCo problem is solved by considering the reactions of two actors: the MO and a group of GenCos. In what follows we define the problems of these two groups of actors.

4.2 The Market Operator Problem

The main task of the MO is matching energy demand and supply at each time point and determine Locational Marginal Prices. Let us introduce the following notation

Parameters

$b_{i,k,t}$	(€ /MWh)	Price of sell bid of generation unit k belonging to GenCo i in period t
$A_{z,l}$		Element of the incidence matrix of the transmission network assuming value 1 if power in the transmission corridor l flows towards node z and -1 if the flow is directed towards the opposite direction
$C_{z,t}$	(MW)	Load in node z in period t
B_l		Susceptance of line l
X_l		Boolean parameter assuming value 1 if a line l is built
$\tilde{q}_{i,k,t}$	(MW)	Power produced by generator k belonging to GenCo i at time t
$\overline{TR}_{l,t}$	(MW)	Maximum capacity of transmission line l in period t
$\underline{TR}_{l,t}$	(MW)	Minimum capacity of transmission line l in period t
$\overline{\theta}_z$		Maximum voltage angle value for node z;
$\underline{\theta}_z$		Minimum voltage angle value for node z;

Decision Variables

$q_{i,k,t}$	(MW)	Accepted bid for technology k of producer i in period t
$TR_{l,t}$	(MW)	Power flow on transmission line l in period t
$\theta_{z,t}$		Voltage angle for terminal node in node z in period t

Market clearing conditions for the perfect competitive system considered for a group of similar producers is given by the solution of the problem

$$\min_{q_{i,k,t},TR_{l,t}} \sum_{i\in I} \sum_{t\in T} \sum_{z\in Z} \sum_{k\in K^E_{i,z}\cup K^C_{i,z}} b_{i,k,t} q_{i,k,t} \tag{37}$$

subject to

$$\sum_{i\in I} \sum_{k\in K^E_{i,z}\cup K^C_{i,z}} q_{i,k,t} + \sum_{l\in L^E\cup L^C} A_{z,l} TR_{l,t} = C_{z,t} \qquad z\in Z, \quad t\in T \tag{38}$$

$$0 \le q_{i,k,t} \le \tilde{q}_{i,k,t} \qquad z\in Z, \quad k\in K^E_{i,z}\cup K^C_{i,z}, \quad t\in T, \quad i\in I \tag{39}$$

$$TR_{l,t} = B_l \left(\sum_{z\in Z} A_{z,l}\theta_{z,t} \right) X_l \qquad l\in L^C, \quad t\in T \tag{40}$$

$$TR_{l,t} = B_l \left(\sum_{z\in Z} A_{z,l}\theta_{z,t} \right) \qquad l\in L^E, \quad t\in T \tag{41}$$

$$\underline{TR}_{l,t} \leq TR_{l,t} \leq \overline{TR}_{l,t} \qquad l \in L^E \cup L^C, \quad t \in T \tag{42}$$

$$\underline{\theta}_z \leq \theta_{z,t} \leq \overline{\theta}_z \qquad z \in Z, \quad t \in T \tag{43}$$

$$TR_{l,t} \in \Re \qquad l \in L^E \cup L^C, \quad t \in T \tag{44}$$

$$\theta_{z,t} \in \Re \qquad z \in Z, \quad t \in T \tag{45}$$

Solution of the introduced problem determines the accepted quantities minimizing the sum of the quantities times their bid prices. Constraint (38) ensures market balance between demand and supply in each zone and its dual variable represents the zonal LMP. Constraint (39) defines the upper bound for the accepted production from each generator: the MO cannot accept more than what has been produced. Each transmission line of a power network transmits power from its sending node to its receiving node. The amount of transmitted power is proportional between the difference of the voltage angles of these nodes. The principle underlying this relation is similar to the one that stands between the flow of water through a pipeline connecting two water tanks, with the flow level proportional to the difference of height between the two tanks. This constant of proportionality is called *susceptance* and denoted by B_l (see [9]). This relation is modeled by constraints (40) and (41), which represent the power flow through candidate and existing lines. Since a line l connects only two nodes, the sums in (40) and (41) will only consider the voltage angles related to such two nodes. In other words, the coefficient $A_{z,l}$ will be zero if link l is not connected to node z, otherwise it will take value 1 or -1 if the link models an inflow or an outflow from node, z respectively. Finally $\underline{\theta}_z$ and $\overline{\theta}_z$ of the slack node z are set to zero. Notice that the power flow equation (40) for the candidate line is multiplied by a boolean parameter, which sets the transmission to zero when no line is built in the considered transmission corridor.

4.3 *The GenCo Problem*

At the same level as the MO is the set of GenCos. These actors aim at maximizing their own profit by submitting bids $(b_{i,k,t}, \tilde{q}_{i,k,t})$ to the MO and defining their optimal expansion plan according to the grid structure. Notice how the structure of the GenCo problem is simplified compared to the model shown in the previous section. This simplification is made to include the GenCo problem as a part of an equilibrium model whose goal is determining the upgrade of the transmission network. This requires describing the GenCo problem with a lower level of details in order to maintain the model tractable. The problem of the i-th GenCo involves the following notation:

Parameters

$\pi_{z,t}$	(€/MWh)	Locational Marginal Price in node z at time t
δ		Discounting factor
$c_{i,k}$	(€/MWh)	Generation cost of technology k for producer i
$f^G_{i,k}$	(€)	Investment cost of technology k for producer i
$\Gamma^C_{i,k}$	(MW)	Capacity of candidate technology k of producer i
$\Gamma^E_{i,k}$	(MW)	Capacity of existing technology k of producer i
$q_{i,k,t}$	(MW)	Accepted bid for technology k of producer i in period t

Decision Variables

$Y_{i,k}$		Binary variable set to 1 if producer i activates candidate generation unit k
$\tilde{q}_{i,k,t}$	(MW)	Power produced by generator k belonging to GenCo i at time t

Every GenCo has a capacity limit for both existing and candidate generators. Such limits are expressed by the constraints

$$\tilde{q}_{i,k,t} \leq \Gamma^E_{i,k} \qquad z \in Z, \quad k \in K^E_{i,z}, \quad t \in T$$

and

$$\tilde{q}_{i,k,t} \leq \Gamma^C_{i,k} Y_{i,k} \qquad z \in Z, \quad k \in K^C_{i,z}, \quad t \in T$$

respectively.

Finally, the quantity sold by each GenCo cannot be larger than the quantity accepted by the MO, i.e.

$$\tilde{q}_{i,k,t} \leq q_{i,k,t} \qquad z \in Z, \quad k \in K^E_{i,z} \cup K^C_{i,z}, \quad t \in T.$$

The decision problem of GenCo i is therefore the following:

$$\max_{\tilde{q}_{i,k,t}, Y_{i,k}} \sum_{t \in T} \delta^{-t} \sum_{z \in Z} \left(\pi_{z,t} \sum_{k \in K^E_{i,z} \cup K^C_{i,z}} \tilde{q}_{i,k,t} - \sum_{k \in K^E_{i,z} \cup K^C_{i,z}} c_{i,k} \tilde{q}_{i,k,t} \right) - \sum_{z \in Z} \sum_{k \in K^C_{i,z}} f^G_{i,k} Y_{i,k} \tag{46}$$

subject to

$$\tilde{q}_{i,k,t} \leq \Gamma^C_{i,k} Y_{i,k} \qquad z \in Z, \quad k \in K^C_{i,z}, \quad t \in T \tag{47}$$

$$\tilde{q}_{i,k,t} \leq \Gamma_{i,k}^{E} \qquad z \in Z, \quad k \in K_{i,z}^{E}, \quad t \in T \tag{48}$$

$$0 \leq \tilde{q}_{i,k,t} \leq q_{i,k,t} \qquad z \in Z, \quad k \in K_{i,z}^{E} \cup K_{i,z}^{C}, \quad t \in T \tag{49}$$

$$Y_{i,k} \in \{0,1\} \qquad k \in K_{i,z}^{C} \tag{50}$$

In this problem, $\pi_{z,t}$ represents the LMP in node z in period t and is defined as the shadow price associated with constraint (38).

The solution of each GenCo problem changes according to the bid level accepted by the MO in each node and the resulting Locational Marginal Price. The bound $q_{i,k,t}$ on maximum accepted bid is defined in constraint (49) and can be understood as a demand constraint at time t for power output up to generator k belonging to GenCo i. This latter in turn depends on the possibility of transferring electricity between different nodes. This mechanism establishes a hierarchical relation between the decision up to each GenCo in terms of MW to provide to the market and potential new installments and the decision taken by the TransCo in terms of opened transmission lines. According to this relation we will refer to the GenCo problem as $GP_i(\pi_{z,t}, q_{i,k,t})$, where $\pi_{z,t}$ and $q_{i,k,t}$ are defined by the MO as a sequential response to the decisions taken by the TransCo and as a concurrent response to the bids offered by the GenCos.

The TransCo model can be solved using an approach based on the k-*th* best algorithm [5] properly modified to allow treatment of binary variables in the lower level (see [28]). The model can be used to study how budget availability or different importance of consumers and producers can change the upgrade policy up to the TransCo.

We tested our model on the classical 6-bus example from Garver (see [29] for the details of the implementation).

First, we focus on how different budget availability up to the TransCo can influence the components of the social cost function, namely profitability of the GenCos and social costs up to consumers. Then we move our attention on how the profitability of GenCos and social costs change when we shift the weighting factor α, which measures the relative importance of consumers and producers in the aggregate social cost function.

Results for different budget availability are displayed in Figures 4 and 6. The weighting factor parameter α is set to 0.5, Figure 6 (left) shows us two distinct effects of a budget increase. For smaller increases the effect is a convergence of average (over periods) Locational Marginal Prices from different nodes, whilst for larger increases the converged prices follow a common path downwards. This has an impact on GenCos' profits as explained later on.

Producers with larger availability of power plants with low marginal production costs increase the production, eroding market shares from their competitors. Some GenCos will experience a decrease of power sold as consumers can buy from cheaper producers, while other GenCos will have an increase in sales and some of them (A2 and A4 in Figure 4 left) will install new generation units with lower

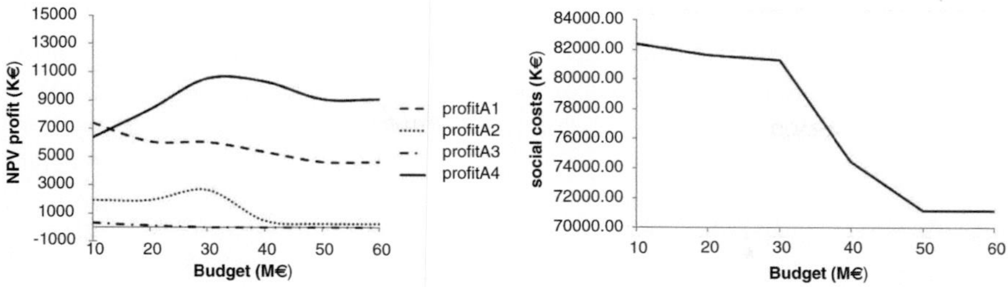

Fig. 4 Net Present Value of profits and total social costs for different levels of Budget (weighting factor set to 0.5)

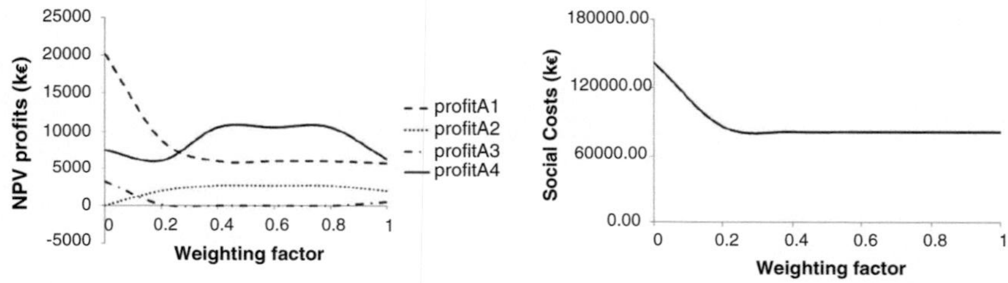

Fig. 5 Net Present Value of profits and total social costs for different choices of the weighting factor (budget set to 30 M€)

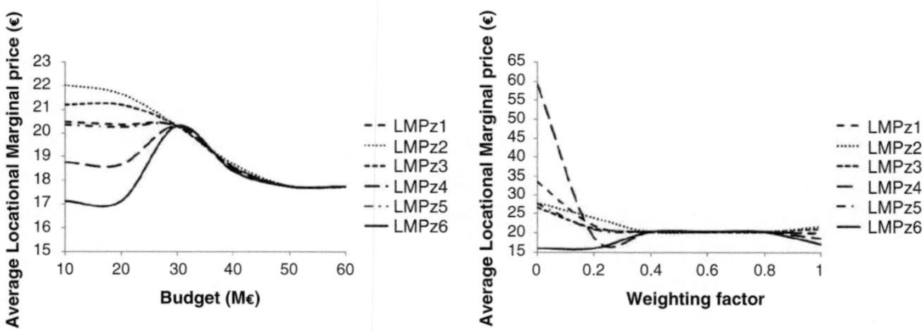

Fig. 6 Locational Marginal Prices for different choices of budget (weighting factor set to 0.5) and weighting factor (budget set to 30 M€)

marginal costs compared to the competition. As a consequence of congestion removal, prices will converge to a common point as it is shown in Figure 6 (left). Also the consumers will benefit of such upgrade as they can buy energy for lower prices. With increasing importance of consumers, the TransCo will eliminate the large part of congestions allowing prices to decrease in all nodes of the grid. This will decrease the overall GenCos' profits, as it can be seen on the second portion of the curves in Figure 4 (left), but it will lower to a greater extent social costs, as it can be seen in Figure 4 (right). To analyze details on how power exchanges between nodes are modified by different grid configurations see, [29].

The effects of varying the weighting factor are displayed in Figures 5 and 6 (right). We have performed the analyses setting a value of 30 Million Euros for the TransCo budget and shifted the values of α. The effect on the grid is an increasing level of congestion removal. This implies an initial convergence of prices over different nodes, as shown in Figure 6 (right) which, in turn allows more efficient GenCos to increase their market share by opening new generation units with lower marginal production costs. When the weighting factor approaches the unit value (i.e., only consumers are important in the social cost function) the TransCo will select a grid expansion plan which further penalizes GenCos as some zones will have lower LMPs. This entails a profit decrease for some GenCos. Possible variations to the introduced model could consist in the introduction of uncertainty related to some parameters such as load or possible failures of lines, in order to describe the level of security of the network. Network security should also be addressed through the management of the $N-1$ security issue, in order to evaluate how the stability of the network persists when one or more elements fail to operate.

5 Conclusions

In this chapter we have analyzed the benefits of optimization techniques in planning power system capacity expansions, both under a single producer perspective and under a system-wide perspective. Different approaches have been considered, depending on the nature and the goals of the problem to be analyzed and solved. At a high level of granularity single producer perspective is considered. When there is a single decision maker with high control on all the model's required input data it is possible to consider a decentralized approach focusing on the decisions to be taken by the single producer. This approach has the undeniable benefit of allowing a very detailed input structure, considering the main sources of uncertainty and accounting for risk. Nevertheless, power system expansion planning is operated by a multitude of actors. Therefore in the long run the reality of decision making is highly distributed and the optimization problem solved by the single producer will not be able to consider other agents' decisions. In such cases it can be useful to extend the viewpoint and also consider a system-wide approach to capacity expansion decisions planning.

When considering a multi-player environment one needs to define whether to consider a centralized approach, where a single regulator takes decisions on the best expansion plan on a system perspective or consider an equilibrium approach where each player takes her best decision reacting to other players' decisions. The centralized approach has been used to introduce the second model, which considers the problem of power generation capacity expansion in a centralized system-wide perspective, whilst the equilibrium-based approach has been used to set the focus on the problem of transmission expansion, which is formulated as a Stackelberg game in order to take into account the reactions of the producers to different grid configurations. When the focus is moved from the single producer to the system-wide interplay between several actors, the modeling and computational difficulties tend to grow. This implies that each part of the equilibrium model is described with a lower level of detail compared to the single producer case.

References

1. Aguado, J.A., de la Torre, S., Contreras, J., Conejo A.J., Martínez A.: Market-driven dynamic transmission expansion planning. Electr. Power Syst. Res. **82**(1), 88–94 (2012)
2. Albornoz, V.M., Benario, P., Rojas, M.E.: A two-stage stochastic integer programming model for a thermal power system expansion, Int. Trans. Oper. Res. **11**, 243–257 (2004)
3. Bacon, R.W.: Privatization and reform in the global electricity supply industry. Annu. Rev. Energy Environ. **20**, 119–143 (1995)
4. Bard, J.F.: Practical Bilevel Optimization. Nonconvex Optimization and its Applications, vol. 30. Dordrecht, Kluwer Academic (1998)
5. Bialas, W.F., Karwan, M.H., Sourie, J.-C.: On Two-Level Optimization. IEEE Trans. Autom. Control **1**, 211–214 (1982)
6. Bjorkvoll, T., Fleten, S.E., Nowak, M.P., Tomasgard, A., Wallace, S.W.: Power generation planning and risk management in a liberalized market. IEEE Porto Power Tech. Proc. **1**, 426–431 (2001)
7. Bompard, E., Invernizzi, A., Napoli, R.: Transmission expansion in the competitive environment: the Italian case. In: Proceedings 2005 IEEE Power Tech Conference (2005)
8. Booth, R.R.: Optimal generation planning considering uncertainty. IEEE Trans. Power Syst. **PAS-91**(1), 70–77 (1972)
9. Castillo, E., Conejo, A.J., Pedregal, P., García, R., Alguacil, N.: Building and Solving Mathematical Programming Models in Engineering and Science. Wiley, New York (2011)
10. Conejo, A.J., Carrion, M., Morales, J.M.: Decision Making Under Uncertainty in Electricity Market. International Series in Operations Research and Management Science. Springer Science+Business Media, New York (2010)
11. Fortuny-Amat, J., McCarl, B.: A representation and economic interpretation of a two-level programming problem. J. Oper. Res. Soc. **32**, 783–792 (1981)
12. Galloway, C.D., Garver, L.L., Kirchmayer, L.K., Wood, A.J.: Generation-transmission expansion planning. In: Proceedings of Power Systems Computation Conference, pt. 5, Stockholm (1966)
13. Garcés, L.P., Conejo, A.J., Garcia-Bertrand, R., Romero, R.: A bilevel approach to transmission expansion planning within a market environment. IEEE Trans. Power Syst. **24**(3), 1513–1522 (2009)
14. Garcés, L.P., Romero, R., Lopez-Lezama, J.M.: Market-driven security-constrained transmission network expansion planning transmission and distribution conference and exposition: Latin America (T&D-LA). In: 2010 IEEE/PES, pp. 427–433 (2010)
15. Garver, L.L.: Transmission Network Estimation using Linear Programming. IEEE Trans. Power Apparat. Syst. **89**, 1688–1697 (1970)
16. Genesi, C., Marannino, P., Siviero, I., Zanellini, F., Carlini, E.M., Pericolo, P.P.: Coordinated transmission and generation planning to increase the electricity market efficiency. In: XVI Power Systems Computation Conference (PSCC 2008), Glasgow (2008)
17. Genesi, C., Marannino, P., Montagna, M., Rossi, S., Siviero, I., Desiata, L., Gentile, G.: Risk management in long term generation planning. In: 6th International Conference on the European Energy Market, pp. 1–6 (2009)
18. Han, S., Lee, J., Kim, T., Park, S., Kim, B.H.: The development of the generation expansion planning system using multi-criteria decision making rule. In: The International Conference on Electrical Engineering 2009, (2009)

19. Hashimoto, H.: A spatial nash equilibrium model. In: Harker, P. (ed.) Spatial Price Equilibrium: Advances in Theory, Computation and Application, pp. 20–40. Springer, Berlin (1985)
20. Hesamzadeh, M.R., Hosseinzadeh, N., Wolfs, P.J.: A leader-followers model of transmission augmentation for considering strategic behaviours of generating companies in energy markets. Int. J. Electr. Power Energy Syst. **32**(5), 358–367 (2010)
21. Hobbs, B.E.: Linear complementarity models of Nash-Cournot competition in bilateral and POOLCO power markets. IEEE Trans. Power Syst. **16**(2), 194–202 (2001)
22. International Energy Agency: Electricity Market Reform: An IEA Handbook. OECD/IEA, Paris (1999)
23. Maggioni, F., Wallace, S.W.: Analyzing the quality of the expected value solution in stochastic programming. Ann. Oper. Res. **200**(1), 37–54 (2012)
24. Ng, S.K.K, Zhong, J., Lee, C.W.: A game-theoretic study of the strategic interaction between generation and transmission expansion planning. In: Power Systems Conference and Exposition, p. 10. IEEE, Seattle (2009)
25. Noonan, F., Giglio, R.J.: Planning electric power generation: a nonlinear mixed integer model employing benders decomposition. Manag. Sci. **23**(9), 946–956 (1977)
26. Patterson, W.: Transforming Electricity. Earthscan.203, London (1999)
27. Philpott, A.B., de Matos, V.L.: Dynamic sampling algorithms for multi-stage stochastic programs with risk aversion. Eur. J. Oper. Res. **218**(2), 470–483 (2012)
28. Pisciella, P., Bertocchi, M., Vespucci, M.T.: A bilevel programing approach to modeling of power transmission capacity planning. Statistica e Applicazioni, 79–94 (2013). ISSN:1824-6672
29. Pisciella, P., Bertocchi, M., Vespucci, M.T.: A leader-followers model of power transmission capacity expansion in a market driven environment. Comput. Manag. Sci. (Special Issue on Optimisation Methods and Applications in the Energy Sector) (2014). doi: 10.1007/s10287-014-0223-9
30. Pisciella, P., Vespucci, M.T., Bertocchi, M., Zigrino, S.: A time consistent risk averse three-stage stochastic mixed integer optimization model for power generation capacity expansion. Energy Econ. (2014). doi: 10.1016/j.eneco.2014.07.016
31. Pozo, D., Sauma, E.E., Contreras, J.: A three-level static MILP model for generation and transmission expansion planning. IEEE Trans. Power Syst. **28**(1), 202–210 (2013)
32. Pozo, D., Contreras, J., Sauma, E.E.: If you build it, he will come: anticipative power transmission planning. Energy Econ. **36**, 135–146 (2013)
33. Rockafellar, R.T., Uryasev, S.: Optimization of conditional value-at-risk. J. Risk **2**, 21–41 (2000)
34. Rockafellar, R.T., Uryasev, S.: Conditional value-at-risk for general loss distributions. J. Bank. Finance **26**, 1443–1471 (2002)
35. Roh, J.H., Shahidehpour, M., Fu, Y.: Market-based coordination of transmission and generation capacity planning. IEEE Trans. Power Syst. **22**(4), 1406–1419 (2007)
36. Roh, J.H., Shahidehpour, M., Wu, L.: Market-based generation and transmission planning with uncertainties. IEEE Trans. Power Syst. **24**(3), 1587–1598 (2009)
37. Rosellon, J.: Different approaches towards electricity transmission expansion. Rev. Netw. Econ. **2**(3), 238–269 (2003)
38. Ruczcynski, A.: Risk averse dynamic programming for Markov decision processes. Math. Program. **125**, 235–261 (2010)
39. Rudloff, B., Street, A., Valladao, D.: Time consistency and risk averse dynamic decision models: interpretation and practical consequences. Internal Research Reports 17 (2011)
40. Sauma, E.E., Oren, S.S.: Proactive planning and valuation of transmission investments in restructured electricity markets. J. Regul. Econ. **30**, 261–290 (2006)
41. Sauma, E.E., Oren, S.S.: Economic criteria for planning transmission investment in restructured electricity markets. IEEE Trans. Power Syst. **22**(4), 1394–1405 (2007)
42. Shan, J., Ryan, S.M.: Capacity expansion in the integrated supply network for an electricity market. IEEE Trans. Power Syst. **26**(4), 2275–2284 (2011)

43. Shiina, T., Birge, J.R.: Multi-stage stochastic programming model for electric power capacity expansion problem. Jpn. J. Ind. Appl. Math. **20**, 379–397 (2003)
44. Stoughton, N.M., Chen, R.C., Lee, S.T.: Direct construction of optimal generation mix. IEEE Trans. Power Syst. **99**(2), 753–759 (1980)
45. Vespucci, M.T., Bertocchi, M., Innorta, I., Zigrino, S.: A stochastic model for the single producer capacity expansion problem in the Italian electricity market. Technical Report DIIMM, 6, University of Bergamo (2011)
46. Vespucci, M.T., Bertocchi, M., Innorta, I., Zigrino, S.: Models for the generation expansion problem in the Italian electricity market. Technical Report DMSIA, 2, University of Bergamo (2011)
47. Vespucci, M.T., Bertocchi, M., Zigrino, S., Escudero, L.F.: Stochastic optimization model for power generation expansion planning with risk management. European Energy Market (EEM13) 10th International Conference Proceedings, Stockholm, 27–31 May 2013, pp. 1–8 (2013). doi: 10.1109-EEM.2013.6607352
48. Vespucci, M.T., Bertocchi, M., Innorta, I., Zigrino, S.: A stochastic model for investments in different technologies for electricity production in the long period. CEJOR J. **22**(2), 407–426 (2014). doi: 10.1007-s10100-013-0317-4
49. Vv.Aa.: The Future of The Electric Grid. An Interdisciplinary MIT Study. Massachusetts Institute of Technology, Cambridge (2011)
50. World Energy Council.: The Benefits and Deficiencies of Energy Sector Liberalisation, vol. 1. World Energy Council, London (1998)

On Chubanov's Method for Solving a Homogeneous Inequality System

Kees Roos

Abstract We deal with a recently proposed method of Chubanov for solving linear homogeneous systems with positive variables. Our first aim is to show that the performance of this method can be improved by a slight modification of Chubanov's so-called Basic Procedure. In theory this results in at least the same decrease of the merit function used by Chubanov, but both in theory and in practice the decrease may be much faster. Theoretical evidence for the speed-up follows from a lemma, whereas some numerical experiments provide convincing computational evidence. We also present a complete, somewhat simplified analysis of Chubanov's Main Algorithm, thereby including also some numerical experiments.

Keywords Linear homogeneous systems • Algorithm • Polynomial-time

1 Introduction

Let A be an $m \times n$ matrix with $m < n$, and $\operatorname{rank}(A) = m$. Recently Chubanov [2–4] presented a new algorithm that finds in polynomial time a solution of the system

$$Ax = 0, x > 0, \tag{1}$$

or establishes that no such solution exists. In the algorithm the author uses a nonzero vector $y \geq 0$ that is updated in each iteration and eventually serves to decide which of the two cases occurs. If a positive solution exists, then such a solution can be obtained from y.

A crucial tool in Chubanov's approach is a result showing that as long as no solution of (1) has been found, a "better" y can be constructed. Eventually this leads to a "small" vector y, which induces a "cut" of the form $x_k \leq \frac{1}{2}$ for some index k. In the proof of this result he uses Farkas's lemma. Now it is well-known that this lemma is equivalent to the duality theorem for linear optimization; both results are far from trivial but each of them can be easily derived from the other [1]. The

K. Roos (✉)
Delft University of Technology, Delft, Netherlands
e-mail: c.roos@tudelft.nl

M. Al-Baali et al. (eds.), *Numerical Analysis and Optimization*, Springer Proceedings in Mathematics & Statistics 134, DOI 10.1007/978-3-319-17689-5_13

first result in our paper is an elementary proof of the mentioned result that does not depend on the Farkas lemma, which makes the analysis independent of the existing theory of linear optimization (LO).

In hindsight, duality theory helps to understand the role of the vector y in Chubanov's approach. For this we recall a variant of Farkas's lemma that is due to Stiemke [14]. It states that (1) has no solution if and only if the system

$$A^T u \geq 0, \quad A^T u \neq 0 \tag{2}$$

has a solution. Now one has $y = A^T u$ for some u if and only if $P_A y = 0$, where P_A denotes the orthogonal projection onto the null space of A. It follows that system (2) has a solution if and only if the system

$$P_A y = 0, \quad y \geq 0, \quad y \neq 0 \tag{3}$$

has a solution. Chubanov's algorithm can be viewed as a systematic search method for a vector y satisfying (3). It will be convenient to call any such vector a *dual feasible vector.*

Since (3) is homogeneous in y and $y \neq 0$, we may restrict the search to vectors y such that $e^T y = 1$, where e denotes the all-one vector. If during this search it happens that $P_A y > 0$, then $z = P_A y$ is a positive solution of (1). This follows because $AP_A = 0$, whence $Az = 0$. If this happens, we call the vector y *primal feasible.*

On the other hand, if y is not primal feasible then there must exist an index k such that $z_k \leq 0$. In that case it becomes natural to look for a new y' such that $\|P_A y'\| < \|P_A y\|$. This is exactly what the so-called Basic Procedure (BP) of Chubanov does, and [2, Lemma 2.1] shows how such an y' can be found.

Of course, if (1) has a positive solution, then there is no y satisfying (3). A clever finding of Chubanov is to stop the BP when a y has been found such that

$$2\sqrt{n}\|P_A y\| \leq \max(y), \quad 0 \neq y \geq 0, \tag{4}$$

where $\max(y) := \max_i(y_i)$. In that case the vector y is said to be *small.* As Chubanov showed, this happens after at most $4n^3$ iterations, which makes his BP strongly polynomial. Any small vector y gives rise to a cut for problem (1) of the form $x_k \leq \frac{1}{2}$, where k is such that $y_k = \max(y)$. Hence it can be used to reduce problem (1) to a problem similar to (1), with A replaced by AD. Here D denotes the identity matrix I with I_{kk} replaced by $\frac{1}{2}$. Thus Chubanov's Main Algorithm (MA) in [2, 4] replaces A by AD and then calls the BP again. If this yields a positive solution x for the new system, then Dx is a positive solution of (1); otherwise, the BP will generate a new vector y satisfying (4), and so on.

If A has integer (or rational) entries, then the number of calls of the BP is polynomially bounded by the size of the matrix A, as follows by using a classical result of Khachiyan [9] that gives a positive lower bound on the positive entries of a solution of a linear system of equations. As a result the algorithm solves problem (1) in polynomial time [2, Theorem 2.1].

The aim of this short paper is twofold. We want to present the main ideas behind Chubanov's algorithm, thereby including a relatively simple analysis of his complexity result. The outline is as follows. We start in the next section by presenting an improved version of Chubanov's BP. It does not improve the complexity result in [2], but it speeds up implementations of Chubanov's original version drastically. This has been acknowledged in [4], which is an updated version of Chubanov [2], and is confirmed by numerical results that we present in Section 5. In Section 2.1 we recall Chubanov's original method of generating cuts for problem (1) from a small vector y generated by the BP. In [4] a more general method for generating cuts has been discussed, which we present in Section 2.2. This method is used in some of the numerical experiments in Section 5. It turns out that it reduces the iteration number, but not necessarily the time required by the algorithm. Section 2.3 contains the modified BP, and its analysis. Section 3 shows how the cuts generated by the BP can be used to obtain a polynomial-time method for solving (1). This method is more formally described in the MA, which is presented in Section 4, where we include some lemmas that reduce the iteration bound by a factor n.

In Section 5 we report some numerical experiments, and we conclude with some comments in Section 6.

2 Improved Basic Procedure

Let N_A denote the null space of A, i.e., $N_A := \{x \,:\, Ax = 0\}$, and

$$P_A := I - A^T \left(AA^T\right)^{-1} A.$$

Note that our assumption $\operatorname{rank}(A) = m$ implies that the inverse of AA^T exists. It is obvious that $AP_A = 0$ and since P_A is symmetric, also $P_A A^T = 0$. If $x \in N_A$, then $P_A x = x$. On the other hand, if $x \in N_A^{\perp}$, then $x = A^T y$ for some y, whence $P_A x = P_A A^T y = 0$. These properties resemble the well-known fact that P_A is the orthogonal projection of $\mathbf{R}^n$ onto N_A. It follows that x satisfies $Ax = 0$ if and only if $P_A x = x$, and x is feasible for (1) if and only if $P_A x = x > 0$.

2.1 Three Simple Lemmas

A simple result—crucial for the approach of Chubanov's paper [2, 4]—is the following lemma.

Lemma 1. *If $P_A y > 0$, for an arbitrary vector y, then $z = P_A y$ solves (1).*

Proof. This follows from $Az = AP_Ay = 0$ and $z > 0$. □

The next lemma consists of the easy part of the aforementioned lemma of Stiemke [14].

Lemma 2. *If $P_Ay = 0$ holds for some nonzero $y \geq 0$, then (1) is infeasible.*

Proof. Let $P_Ay = 0$ and $0 \neq y \geq 0$. Suppose that x is feasible for (1). Then $P_Ax = x > 0$. Since $x > 0$ and $0 \neq y \geq 0$, we have $y^Tx > 0$. We now may write

$$0 < y^Tx = y^TP_Ax = x^TP_Ay = 0.$$

This contradiction proves the lemma. □

Before stating the third lemma we note that if x is feasible for (1), then also $x' = x/\max(x)$ is feasible for (1), and this solution belongs to the unit cube, i.e., $x' \in [0,1]^n$. Hence, (1) is feasible if and only if the following system has a solution:

$$Ax = 0, \;\; x \in (0,1]^n. \tag{5}$$

The next lemma shows that if y satisfies $0 \neq y \geq 0$ and (4), then it gives rise to a cut. More precisely, it induces an inequality that cuts off half of the unit cube in (5).

Lemma 3. *Let y satisfy $0 \neq y \geq 0$ and (4), and let j be such that $y_j = \max(y)$. Then every feasible solution of (5) satisfies $x_j \leq \frac{1}{2}$.*

Proof. Let x be feasible for (5). This means that $P_Ax = x \in [0,1]^n$, which implies that $\|x\| \leq \sqrt{n}$. Hence, using (4) and the Cauchy–Schwarz inequality, we may write

$$y_jx_j \leq y^Tx = y^TP_Ax = x^TP_Ay \leq \|x\| \, \|P_Ay\| \leq \|x\| \, \frac{\max(y)}{2\sqrt{n}} \leq \tfrac{1}{2}\max(y) = \tfrac{1}{2}y_j.$$

Since $y_j > 0$, it follows that $x_j \leq \frac{1}{2}$. □

It will be convenient to call y *small* if (4) holds, and *large* otherwise. Of course, this terminology is relative to the (current) matrix A. Note that $\|P_Ay\| > 0$ if y is large. Moreover, if y is a dual feasible vector as defined by (3), then $P_Ay = 0$ and hence y is small. Recall that y is primal feasible if $P_Ay > 0$ and in that case $z = P_Ay$ is a solution of (1), by Lemma 1. So we have shown that (1) has a solution if y is primal feasible and it has no solution if y is dual feasible.

For future use we also state the following result.

Lemma 4. *Let $0 \neq y \geq 0$ and $2n\sqrt{n}\,\|P_Ay\| \leq e^Ty$. Then y is small.*

Proof. By using $e^Ty \leq n\max(y)$, we may write

$$2\sqrt{n}\,\|P_Ay\| \leq \frac{e^Ty}{n} \leq \max(y), \tag{6}$$

which implies (4). □

2.2 More Cut-Generating Vectors

In the previous section it turned out that any small vector y induces a cut $x_j \leq \frac{1}{2}$, for some j, for problem (5). Recently it was established by Chubanov [4] that also large vectors y can serve this purpose. Fixing k he considered the LO-problem

$$\max\{x_k \,:\, Ax = 0, x \in [0,1]^n\}.$$

The dual problem is

$$\min\left\{e^T w \,:\, A^T v + w \geq e_k, w \geq 0\right\} = \min\left\{e^T [e_k - u]^+ \,:\, P_A u = 0\right\}.$$

The above equality uses again that $u = A^T v$ if and only if $P_A u = 0$. Since P_A is a projection matrix we have $P_A^2 = P_A = P_A^T$. Given y and $z = P_A y$, we therefore have $P_A(y - z) = P_A(y - P_A y) = 0$. Hence, if $y_k > 0$, we may take $u = \frac{y-z}{y_k}$. It then follows from the Duality Theorem for Linear Optimization that

$$x_k \leq e^T \left[e_k - \frac{y-z}{y_k}\right]^+. \tag{7}$$

In particular we have

$$e^T \left[e_k - \frac{y-z}{y_k}\right]^+ \leq \frac{1}{2} \quad \Rightarrow \quad x_k \leq \frac{1}{2}. \tag{8}$$

Criterion (8) for the cut $x_k \leq \frac{1}{2}$ is weaker than (4) in the sense that if (4) gives rise to a cut, then so does (8). This follows from $e_k - \frac{y}{y_k} \leq 0$. Hence if $y_k = \max(y)$ then

$$e^T \left[e_k - \frac{y-z}{y_k}\right]^+ \leq \frac{e^T [z]^+}{y_k} \leq \frac{\sqrt{n}\|[z]^+\|}{y_k} \leq \frac{\sqrt{n}\,\|z\|}{y_k} = \frac{\sqrt{n}\,\|P_A y\|}{\max(y)}.$$

Even tighter cuts can obtained in a simpler way, without using the Duality Theorem for Linear Optimization. Let u be such that $P_A u = 0$. If x is feasible, then $Ax = 0$, which implies $P_A x = x$. Therefore, $u^T x = u^T P_A x = (P_A u)^T x = 0$. Also using $0 \leq x \leq e$ we may write

$$x_k = e_k^T x = x^T (e_k - u) \leq e^T [e_k - u]^+. \tag{9}$$

Defining $v = y - z$ we have $P_A v = 0$, as we saw above. By substituting $u = \frac{v}{y_k}$ in (9) we obtain (7). More generally we may substitute $u = \alpha v$, yielding $x_k \leq q(\alpha)$ for every $\alpha \in \mathbf{R}$, where the function $q(\alpha)$ is defined by

$$q(\alpha) := e^T [e_k - \alpha v]^+ = [1 - \alpha v_k]^+ + \sum_{i \neq k} [-\alpha v_i]^+, \quad \alpha \in \mathbf{R}.$$

One may easily verify that $q(\alpha)$ is a nonnegative piecewise linear convex function with a breakpoint at $\alpha = 0$ and if $v_k \neq 0$ another breakpoint at $\alpha = \frac{1}{v_k}$. The breakpoint at $\alpha = 0$ yields the void inequality $x_k \leq q(0) = 1$. So only the breakpoint at $\alpha = \frac{1}{v_k}$ is of interest, which yields the inequality

$$x_k \leq \sum_{i \neq k} \left[\frac{-v_i}{v_k} \right]^+. \tag{10}$$

Of course, this new cut is nonvoid only if the right-hand side is less than 1, but then it is always at least as tight as (7). The theoretical analysis below is based on the weakest cut we have found, namely in Lemma 3. In the computational part of the paper we also use the new cuts and demonstrate numerically that they are superior to the cuts in Lemma 3.

To conclude this section we point out that the vector v has a nice geometric interpretation: whereas z is the orthogonal projection of y onto the null spaxe of A, v is the orthogonal projection of y onto the row space of A.

2.3 Search for a Small Vector

Note that inequality (4) is homogeneous in y. Since y is nonzero, without loss of generality we may assume that $e^T y = 1$. In the sequel we always assume that y is nonnegative and normalized in this way.

Assuming that y is large relative to a given matrix A, we present in this section a simple algorithm that generates a new vector y that is either primal feasible or dual feasible or small, in at most $4n^3$ iterations. The algorithm closely resembles the BP in [2, 4] but deviates at one particular point.

It is also convenient to use the vector z defined by $z = P_A y$. Note that if $z = 0$ then y is dual feasible, which by Lemma 2 implies that (1) is infeasible. On the other hand, by Lemma 1, if $z > 0$, then z satisfies (1), and we are also done. So, if z is such that the status of (1) is not yet decided, then $z \neq 0$ and at least one component z is not positive. In that case we may find a nonempty set K of indices such that

$$\sum_{k \in K} z_k \leq 0.$$

Denoting the k-th column of P_A as p^k, we have $p^k = P_A e_k$, where e_k denotes the k-th unit vector. We define

$$e_K := \frac{1}{|K|} \sum_{k \in K} e_k, \quad p_K := P_A e_K = \frac{1}{|K|} \sum_{k \in K} p^k. \tag{11}$$

Note that $0 \neq e_K \geq 0$, and $e^T e_K = 1$. If $p_K = 0$ ($p_K > 0$), then e_K is dual (primal) feasible and we are done. Hence, we may assume that $p_K \neq 0$. Using again that P_A is a projection matrix we obtain $P_A z = P_A^2 y = P_A y = z$. This implies $z^T p^k = z^T P_A e_k = z^T e_k = z_k$ for each k. Thus we obtain

$$z^T p_K = \frac{1}{|K|} \sum_{k \in K} z^T p^k = \frac{1}{|K|} \sum_{k \in K} z_k \leq 0.$$

As a consequence, in the equation

$$\|z - p_K\|^2 = \left(\|z\|^2 - z^T p_K\right) + \left(\|p_K\|^2 - z^T p_K\right) \tag{12}$$

the two bracketed terms are both positive, because z and p_K are nonzero and $z^T p_K \leq 0$. Therefore, we may define a new y-vector, denoted by $\tilde{y}$, according to

$$\tilde{y} = \alpha y + (1 - \alpha) e_K, \quad \alpha = \frac{\|p_K\|^2 - z^T p_K}{\|z - p_K\|^2} = \frac{p_K^T (p_K - z)}{\|z - p_K\|^2}. \tag{13}$$

Because of (12), α is well-defined and $\alpha \in (0,1)$. Since $y \geq 0$ and $e_K \geq 0$, we may conclude that $\tilde{y} \geq 0$ and, since $e^T y = e^T e_K = 1$, also $e^T \tilde{y} = 1$.

The transformation (13) from y to $\tilde{y}$ is the key element in Algorithm 1. It iterates (13) until y is small or primal feasible or dual feasible.

The only difference from the BP of Chubanov is that he always takes a singleton for K. Usually larger sizes of K are possible; as we will see below, this may speed up the BP. Theoretically this statement is justified by the following lemma, which generalizes [2, Lemma 2.1] and [4, Lemma 2.1].

Lemma 5. *Let $z \neq 0$ and $p_K \neq 0$. With $\tilde{z} := P_A \tilde{y}$, one has*

Algorithm 1: $[y, z, \text{case}]$ = BASIC PROCEDURE(P_A, y)

INITIALIZE: $z = P_A y$; case = 0
while $2\sqrt{n}\|z\| > \max(y)$ **and** case = 0 **do**
  if $z > 0$ **then**
    case = 1 (y is primal feasible); **return**
  else
    if $z = 0$ **then**
      case = 2 (y is dual feasible); **return**
    else
      find $K \neq \emptyset$ such that $\sum_{k \in K} z_k \leq 0$
      $\alpha = p_K^T (p_K - z) / \|z - p_K\|^2$
      $y = \alpha y + (1 - \alpha) e_K$
      $z = \alpha z + (1 - \alpha) p_K$ $(= P_A y)$

$$\frac{1}{\|\tilde{z}\|^2} \geq \frac{1}{\|z\|^2} + |K|. \tag{14}$$

Proof. We have

$$\tilde{z} = \alpha P_A y + (1-\alpha) P_A e_K = \alpha z + (1-\alpha) p_K = p_K + \alpha(z - p_K).$$

Hence,

$$\|\tilde{z}\|^2 = \alpha^2 \|z - p_K\|^2 + 2\alpha p_K^T (z - p_K) + \|p_K\|^2.$$

The value of α that minimizes this expression is given in (13). It follows that

$$\|\tilde{z}\|^2 = \|p_K\|^2 - \frac{\left[p_K^T(z-p_K)\right]^2}{\|z-p_K\|^2} = \frac{\|p_K\|^2\|z\|^2 - (z^T p_K)^2}{\|p_K\|^2 + \|z\|^2 - 2z^T p_K} \leq \frac{\|p_K\|^2\|z\|^2}{\|z\|^2 + \|p_K\|^2},$$

where we used $z^T p_K \leq 0$. Since P_A is a projection matrix, $\|P_A e_K\| \leq \|e_K\|$. So we may write

$$\|p_K\|^2 = \|P_A e_K\|^2 \leq \|e_K\|^2 = \left\| \frac{1}{|K|} \sum_{k \in K} e_k \right\|^2 = \frac{1}{|K|^2} \left\| \sum_{k \in K} e_k \right\|^2 = \frac{|K|}{|K|^2} = \frac{1}{|K|}.$$

It follows that

$$\frac{1}{\|\tilde{z}\|^2} \geq \frac{1}{\|z\|^2} + \frac{1}{\|p_K\|^2} \geq \frac{1}{\|z\|^2} + |K|, \tag{15}$$

as desired. □

Theorem 1. *After at most $4n^3$ iterations the BP yields a vector y that is either small (case = 0) or primal feasible (case = 1) or dual feasible (case = 2).*

Proof. As before, we assume that $e^T y = 1$, $y \geq 0$, and $z = P_A y$. If y is small, then Algorithm 1 requires only 1 iteration. Otherwise y is large, which by Lemma 4 implies that

$$\frac{1}{\|z\|^2} < 4n^3.$$

If during the execution of Algorithm 1 it happens that $z > 0$ or $z = 0$, then the BP immediately stops. Otherwise, since $|K| \geq 1$, each iteration of the while loop increases $1/\|z\|^2$ by at least 1. Hence, after at most $4n^3$ executions of the while loop the algorithm yields a vector y that is primal feasible (case $= 1$) or dual feasible (case $= 2$) or such that $1/\|z\|^2 \geq 4n^3$. In the last case y is small (case $= 0$). □

Since each execution of the while loop requires at most $O(n)$ time, the BP will require at most $O(n^4)$ time. It must be mentioned that for solving (1) one needs to call the BP several times. Surprisingly, Chubanov was able to show that on average the BP then will require only $O(n^3)$ time. We deal with this in the next section. For some computational results, we refer to Section 5.1.

3 Exploration of Cuts from Small Vectors

Let y be a small vector relative to A, and let D denote the diagonal matrix that arises from the identity matrix I when replacing I_{kk} by $\frac{1}{2}$, and where k is such that $y_k = \max(y) := \max_i(y_i)$. We conclude from Lemma 3 that (5) has a solution if and only if the system

$$Ax = 0, \; D^{-1}x \in (0,1]^n \tag{16}$$

has a solution. Obviously this is the case if and only if the system

$$ADx' = 0, \; x' \in (0,1]^n, \tag{17}$$

has a solution. Indeed, if x satisfies (16), then $x' = D^{-1}x$ satisfies (17) and conversely. Hence we have reduced the problem of finding a solution of (5) to finding a solution of (17), and this is in essence the same problem, except that the matrix A is replaced by AD.

Suppose that we can find a vector y that is small relative to AD. Then we obtain a second diagonal matrix, D_1 say, with one of its diagonal entries 1/2, and the remaining diagonal entries equal to 1, such that (17) has a positive solution if and only if the system

$$ADD_1x'' = 0, \; x'' \in (0,1]^n.$$

has a solution. Note that DD_1 is a diagonal matrix with either two entries on the diagonal equal to 1/2 and all other entries 1, or one entry 1/4 and all other entries 1.

Proceeding in this way, after T steps we may conclude that finding a solution of (5) is equivalent to finding a solution of a system of the form (16), where D is a diagonal matrix whose diagonal entries are nonpositive powers of 2, say $D_{ii} = 2^{-t_i}$, with

$$T = \sum_{i=1}^{n} t_i.$$

Yet we observe that if a (positive) solution exists then T cannot become too large. This can be understood by noting that the elements on the diagonal of D are upper bounds for the respective entries x_i in feasible solutions of (16). Hence we may proceed as follows. We may restate (16) in the following way:

$$Ax = 0, \quad 0 < x_i \leq 2^{-t_i}, \quad 1 \leq i \leq n. \tag{18}$$

Any solution of the system (5) will also satisfy (18), and vice versa. Obviously the closure of the feasible region of (18) is convex and bounded. Hence every feasible x is a convex combination of basic feasible solutions. According to Khachiyan's result [9] there exists a positive number τ such that $1/\tau = O(2^{\text{size}(A)})$ with the property that the positive coordinates of the basic feasible solutions of this system are bounded from below by τ. Here $\text{size}(A)$ denotes the binary size of matrix A [9, 12, 13]. Now let x be a (positive) solution of (18) and i an arbitrary index. There must exist a vertex solution x' whose i-th coordinate is positive and hence at least equal to τ. Then $\alpha x + (1-\alpha)x'$ is positive for $\alpha \in (0,1)$, whereas its i-th coordinate becomes larger than or equal to τ if α approaches zero. We conclude that we must have $\tau \leq 2^{-t_i}$ for each i. It follows that

$$\tau^n \leq \prod_{i=1}^{n} 2^{-t_i} = 2^{-\sum_{i=1}^{n} t_i} = 2^{-T},$$

which leads to the following upper bound for T:

$$T \leq n \log_2 \frac{1}{\tau} = O(n \cdot \text{size}(A)). \tag{19}$$

We can now describe the idea underlying Chubanov's algorithm. His Main Algorithm (MA) repeatedly calls the BP. If the vector y generated by the BP is primal or dual feasible, then the MA stops. Otherwise y generates a cut; in that case, the BP is called again with the current A replaced by AD, where D incorporates all cuts generated so far. The above reasoning makes clear that the number of calls of the BP will be not larger than $O(n \cdot \text{size}(A))$. Each iteration of the BP requires $O(n)$ time. Since the number of iterations of the BP is $O(n^3)$, we conclude that solving (1) requires at most $O(n^5 \cdot \text{size}(A))$ time. In the next section we describe the MA in more detail, as well as the version of the BP that was used by Chubanov. We also demonstrate how a more careful analysis, which in essence is also due to Chubanov, reduces the above bound on the time-complexity by a factor n.

4 Chubanov's Main Algorithm

In essence Chubanov's algorithm generates a sequence of pairs $(d^{(i)}, y^{(i)})$, where $y^{(i)}$ is large relative to $AD^{(i)}$, with $D^{(i)} = \text{diag}\,(d^{(i)})$, and $e^T y^{(i)} = 1$. The vectors $d^{(i)}$ are constructed such that $x \leq d^{(i)}$ for each i. The sequence starts with $d^{(0)} = e$ and $y^{(0)}$

Algorithm 2: $[\bar{y}, y, z, J, \text{case}]$ =CHUBANOV'S BASIC PROCEDURE(P_A, y)

1: INITIALIZE: $\bar{y} = 0$; $z = P_A y$; $J = \emptyset$; case = 0
2: **while** $2\sqrt{n}\|z\| > \max(y)$ **and** case = 0 **do**
3:     **if** $z > 0$ **then**
4:         case = 1 (y is primal feasible); **return**
5:     **else**
6:         **if** $z = 0$ **then**
7:             case = 2 (y is dual feasible); **return**
8:         **else**
9:             $\bar{y} = y$
10:             find K such that $\sum_{k \in K} z_k \leq 0$
11:             $\alpha = p_K^T(p_K - z)/\|z - p_K\|^2$
12:             $y = \alpha y + (1-\alpha)e_K$
13:             $z = \alpha z + (1-\alpha)p_K \ (= P_A y)$
14: **if** case = 0 **then**
15:     find a nonempty set J such that $J \subseteq \{j : y_j = \max(y)\}$

such that $y^{(0)} > 0$ and $e^T y^{(0)} = 1$. If $y^{(i)}$ is primal feasible, i.e., $z^{(i)} := P_{AD^{(i)}} y^{(i)} > 0$ for some $i \geq 0$, then $x = D^{(i)} z^{(i)}$ is a positive solution for (1). On the other hand, if $z^{(i)} = 0$, then $y^{(i)}$ is dual feasible, and hence a certificate for infeasibility of (1). Otherwise we construct $y^{(i+1)}$ by using (a slightly extended version of) the BP, as in Algorithm 2, with the pair $(P_{AD^{(i)}}, y^{(i)})$ as input.

The extension concerns the new variables $\bar{y}$ and J, and the fact that the initial vector y is now input. The reason for the latter is that we want to explore knowledge gathered in the vector y during a previous call of the BP in the next call. More precisely, if the BP starts with a large vector y that is not primal or dual feasible, then it generates a small vector. We start the next call of the BP with the last large vector that was constructed during the last call. This is the vector $\bar{y}$ defined in line 9. As we will see later, this makes sense. It reduces the average number of iterations of the BP by a factor n. If the BP generates a small vector y, then J is a set of indices, each of which gives rise to a cut that halves the feasible region. Whereas Chubanov takes for J always a singleton, we assume that $|J| = O(1)$.

We are now ready to analyze the Main Algorithm (MA), Algorithm 3. In this algorithm, A_J denotes the submatrix of A consisting of the columns indexed by the elements in the index set J. The notations d_J and y_J are defined in a similar way. The output of the algorithm is a 4-tuple $[x, y, d, \text{case}]$, where the vector d is such that $D = \text{diag}(d)$ and case $\in \{1, 2, 3\}$. The meaning of these three cases is as follows:

case = 1: x is a solution of (1);

case = 2: y is a certificate for infeasibility of (1);

case = 3: d is a certificate for infeasibility of (1), due to Khachiyan's result.

When at the start of the BP the vector y is primal or dual feasible, then the BP needs only one iteration with output case = 1 or case = 2. Then the MA will stop.

Algorithm 3: $[x, y, d, \text{case}] = \text{MAIN ALGORITHM}(A, \tau)$

INITIALIZE: $d = e$; $y = e/n$; $x = 0$; case $= 0$;
while *case* $= 0$ **do**
  $P_A = I - A^T(AA^T)^{-1}A$
  $[\bar{y}, y, z, J, \text{case}] = \text{Basic Procedure}(P_A, y)$
  if case $= 0$ **then**
    $d_J = d_J/2$
    if $\min(d_J) < \tau$ **then**
      case $= 3$
    else
      $A_J = A_J/2$
      if $\bar{y} \neq 0$ **then**
        $y = \bar{y}$
      $y_J = y_J/2$
      $y = y/e^T y$
if case $= 1$ **then**
  $D = \text{diag}(d)$
  $x = Dz$

So let us assume that at the start of the BP y is not primal or dual feasible. Then there are still two situations: y is either small or large. If y is small, the BP also requires only one iteration. In that case y and z are not changed by the BP. Moreover, at termination of the BP we have $\bar{y} = 0$ and the set J will be not empty. Hence the MA will modify the current d and y by dividing their coordinates d_j, y_j, with $j \in J$, by 2. As a consequence, the next call of the BP has input $(P_{AD}, Dy/e^T Dy)$, with $D = \text{diag}(d)$, with d and y updated as described. Following [2], when the BP requires only one iteration, we call the corresponding iteration of the MA a *fast* iteration. Note that a fast iteration occurs if and only if the BP yields $\bar{y} = 0$.

Next we focus on the case where y is large at the start of the BP, but not primal or dual feasible. This will give rise to a so-called *slow* iteration of the MA. The BP then outputs a small vector y, and the large vector $\bar{y}$ that was generated just before y. It also gives $z = P_A y$ and the index set J consisting of all j such that $y_j = \max(y)$. Note that y gives rise to new cuts $x_j \leq 1/2, j \in J$. The MA updates d and A accordingly. But now the MA replaces y by $\bar{y}$ before dividing its coordinates in J by 2, and hence the next call of the BP uses as input $(P_{AD}, D\bar{y}/e^T D\bar{y})$, with D as before.

An important question is whether $y' = D\bar{y}/e^T D\bar{y}$ is small or large relative to AD. This question is hard to answer, but the next lemma provides some useful information. At the end of the proof of this lemma it becomes clear why we use the large vector $\bar{y}$ instead of the small vector y. In this lemma D is a diagonal matrix whose diagonal entries are positive and at most 1. The proof is essentially the same as the proof of Lemma 3.2 in [2].

Lemma 6. *Let $\bar{y}$ be large relative to A, and $y' = D\bar{y}/e^T D\bar{y}$, with D as just defined. If $\bar{z} = P_A \bar{y}$ and $z' = P_{AD} y'$, then*

$$\frac{1}{\|\bar{z}\|^2} - \frac{1}{\|z'\|^2} < 8\,|Q|\,n^2,$$

where $Q = \{q\,:\, D_{qq} < 1\}$.

Proof. We start by proving $\|P_{AD}D\bar{y}\| \le \|P_A\bar{y}\|$. The projection matrix P_{AD} is

$$P_{AD} = I - DA^T(AD^2A^T)^{-1}AD.$$

Since $P_{AD}DA^T = 0$ it follows that

$$P_{AD}D\bar{y} = P_{AD}\left(D\bar{y} - DA^T v\right), \quad \forall v.$$

Since P_{AD} is a projection matrix, it does not increase the length of a vector. Therefore,

$$\|P_{AD}D\bar{y}\| \le \|D\bar{y} - DA^T v\| = \|D\left(\bar{y} - A^T v\right)\| \le \|\bar{y} - A^T v\|,$$

where the last inequality follows because the entries on the diagonal of D are positive and less than or equal to 1. Now taking $v = (AA^T)^{-1}A\bar{y}$ we get $\bar{y} - A^T v = P_A\bar{y}$, whence we obtain $\|P_{AD}D\bar{y}\| \le \|P_A\bar{y}\|$.

From the definition of y' it follows that

$$\|P_{AD}y'\| \le \frac{\|P_A\bar{y}\|}{e^T D\bar{y}}.$$

Since $e^T\bar{y} = 1$, and $D_{qq} = 1$ if $q \notin Q$, we may write

$$e^T D\bar{y} = \sum_{q=1}^{n} D_{qq}\bar{y}_q \ge \sum_{q\notin Q} D_{qq}\bar{y}_q = \sum_{q\notin Q} \bar{y}_q = 1 - \sum_{q\in Q} \bar{y}_q.$$

It follows that

$$\frac{1}{\|P_{AD}y'\|^2} \ge \frac{(e^T D\bar{y})^2}{\|P_A\bar{y}\|^2} \ge \frac{\left(1 - \sum_{q\in Q}\bar{y}_q\right)^2}{\|P_A\bar{y}\|^2} \ge \frac{1 - 2\sum_{q\in Q}\bar{y}_q}{\|P_A\bar{y}\|^2}.$$

By rearranging terms, we get

$$\frac{1}{\|P_A\bar{y}\|^2} - \frac{1}{\|P_{AD}y'\|^2} \le \frac{2\sum_{i\in Q}\bar{y}_i}{\|P_A\bar{y}\|^2} \le \frac{2\,|Q|\max(\bar{y})}{\|P_A\bar{y}\|^2}.$$

Finally, since $\bar{y}$ is large we have $2\sqrt{n}\,\|P_A\bar{y}\| > \max(\bar{y}) \ge 1/n$. As a consequence we may write

$$\frac{\max(\bar{y})}{\|P_A\bar{y}\|^2} = \frac{\max(\bar{y})}{\|P_A\bar{y}\|}\frac{1}{\|P_A\bar{y}\|} < 2\sqrt{n}\cdot 2n\sqrt{n} = 4n^2$$

and the lemma follows. □

We are ready to derive an improved upper bound for the total number of iterations, as stated in the following theorem. The proof is as in [2], but adapted to the case where $|J| = O(1)$ (instead of $|J| = 1$).

Theorem 2. *Assuming* $O(|J|) = 1$, *the total number of BP-iterations is* $O\left(n^3 \log_2 \tau^{-1}\right)$.

Proof. As before, let T denote the number of iterations performed by the MA. Moreover, let the i-th of these MA-iterations be followed by n_i iterations of the BP. Then the total number of BP-iterations is

$$N = \sum_{i=1}^{T} n_i. \tag{20}$$

We denote the values of $z = P_A y$ at the start and end of the j-th BP-iteration during the i-th MA-iteration as z_{ij} and $\hat{z}_{ij}$, respectively.

For a fast MA-iteration we have $n_i = 1$. Next we consider a slow MA-iteration. Then we may write

$$n_i = \sum_{j=1}^{n_i} 1 = 1 + \sum_{j=1}^{n_i-1} 1 \le 1 + \sum_{j=1}^{n_i-1}\left(\frac{1}{\|\hat{z}_{ij}\|^2} - \frac{1}{\|z_{ij}\|^2}\right),$$

where the inequality is due to Lemma 5, since $|K| \ge 1$. Since $\hat{z}_{ij} = z_{i,j+1}$ for $1 \le j < n_i$, we get

$$n_i \le 1 + \sum_{j=1}^{n_i-1}\left(\frac{1}{\|z_{i,j+1}\|^2} - \frac{1}{\|z_{ij}\|^2}\right) = 1 + \frac{1}{\|z_{i,n_i}\|^2} - \frac{1}{\|z_{i1}\|^2}.$$

Let S denote the number of slow iterations, and let the corresponding indices be i_s, with $1 \le s \le S$, such that

$$1 \le i_1 < i_2 < \ldots < i_S \le T.$$

Then we obtain

$$N \le \sum_{i=1}^{T} 1 + \sum_{s=1}^{S}\left(\frac{1}{\|z_{i_s,n_{i_s}}\|^2} - \frac{1}{\|z_{i_s,1}\|^2}\right) = T + \sum_{s=1}^{S}\left(\frac{1}{\|z_{i_s,n_{i_s}}\|^2} - \frac{1}{\|z_{i_s,1}\|^2}\right).$$

By rearranging the terms in the last sum we obtain

$$N \leq T + \frac{1}{\|z_{i_S,n_{i_S}}\|^2} - \frac{1}{\|z_{i_1,1}\|^2} + \sum_{s=1}^{S-1}\left(\frac{1}{\|z_{i_s,n_{i_s}}\|^2} - \frac{1}{\|z_{i_{s+1},1}\|^2}\right).$$

Omitting the third term in the last expression, and also using that the last y-vector generated by the BP is small, we obtain

$$N \leq T + 4n^3 + \sum_{s=1}^{S-1}\left(\frac{1}{\|z_{i_s,n_{i_s}}\|^2} - \frac{1}{\|z_{i_{s+1},1}\|^2}\right).$$

Now consider the bracketed term in the last sum. This term contains the last z-vector in MA-iteration i_s and the first z-vector in MA-iteration i_{s+1}. If other iterations occur between these two slow iterations, these are fast iterations, and their indices i satisfy $i_s < i < i_{s+1}$. For the moment we may safely assume that $D = I$ at the end of MA-iteration i_s. Since the BP does not change the vector y during fast iterations, the y vector at the start of MA-iteration i_{s+1} has the form $D\bar{y}/e^T D\bar{y}$, where $\bar{y}$ is the large vector generated by the last BP-iteration during MA-iteration i_s, and where D is the product of matrices $D_i = \text{diag}\,(d_i)$, where each d_i arises from the all-one vector by replacing the J_i-entries by $\frac{1}{2}$, where $i_s \leq i < i_{s+1}$. But this means that Lemma 6 applies. Hence we obtain

$$\frac{1}{\|z_{i_s,n_{i_s}}\|^2} - \frac{1}{\|z_{i_{s+1},1}\|^2} \leq 8n^2\,|Q|,$$

where Q is the set of entries in d smaller than 1.

If each J_i is a singleton, for $i_s \leq i < i_{s+1}$, then $|Q|$ equals at most $i_{s+1} - i_s$. But we allow J_i to be larger, though not larger than $O(1)$. In that case we obtain $|Q| = O(i_{s+1} - i_s)$. Substituting this we obtain

$$N \leq T + 4n^3 + 8n^2 O\left(\sum_{s=1}^{S-1}(i_{s+1} - i_s)\right) = T + 4n^3 + 8n^2 O\,(i_S - i_1).$$

Since $i_S - i_1 < T$ we get

$$N \leq T + 4n^3 + 8n^2 O\,(T).$$

Finally, using (19), we get the following upper bound for the total number of iterations:

$$N \leq n\log_2 \tau^{-1} + 4n^3 + 8n^2 O\left(n\log_2 \tau^{-1}\right) = O\left(n^3 \log_2 \tau^{-1}\right). \qquad \square$$

Since each BP-iteration requires $O(n)$ time, the time-complexity for the BP becomes $O\left(n^4 \log_2 \tau^{-1}\right)$. We also need to take into account the time needed for the computation of P_A during each iteration of the MA. This can be done in $O(n^3)$

time [6, 12], and even faster with the Sherman-Morrisen-Woodbury formula [8]. Hence the MA needs about the same time as the BP or less, which means that the total time-complexity of the algorithm is $O\left(n^4 \log_2 \tau^{-1}\right) = O\left(n^4 \text{size}(A)\right)$.

5 Computational Results

5.1 Computational Results for the BP

In order to illustrate the effect of allowing the set K to be larger than a singleton we refer to Figures 1 and 2. These figures were obtained by applying Algorithm 1 to a randomly generated matrix A of size 25×50. In that case the algorithm needs 256 iterations if we take for K the singleton k for which z_k is minimal. On the other hand, if we take for K the set consisting of all indices k for which $z_k \leq 0$, the number of iterations is only 25.

The black graphs in Figures 1 and 2 show, respectively, the behavior of $1/\|z\|^2$ and $\sqrt{n}\,\|z\|/\max(y)$ during the course of the algorithm if K is a singleton and the grey graphs do the same when we take for K the larger set. Figure 1 makes clear that if K is a singleton the increase in $1/\|z\|^2$ is on average 80, which is much larger than 1, as guaranteed by Lemma 5. But if K consists of all k for which $z_k \leq 0$, the increase per iteration is on average 2200, which is also much larger than guaranteed by Lemma 5.

To compare Algorithm 1 and the original BP [2, 4] further we also present Table 2. Each line gives the average number of iterations, the average time (in seconds) and the average size of the set K, for a class of 100 randomly generated problems with matrices A of size $m \times n$ as given in the first two columns. The elements of A were randomly chosen in the intervals $[-100, 100]$, and uniformly distributed. In all cases we used $y = e/n$ as initial vector, where e denotes the all-one vector.

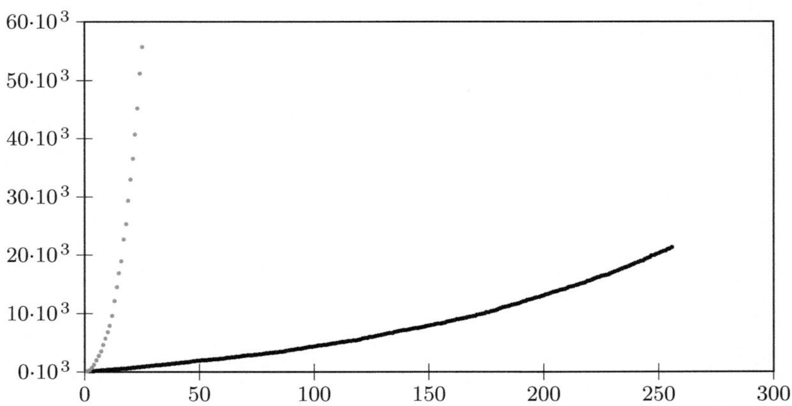

Fig. 1 Typical behavior of $\frac{1}{\|z\|^2}$ for $K = \{k\}$ (*black*), where $z_k = \min(z)$, and for $K = \{k : z_k \leq 0\}$ (*grey*)

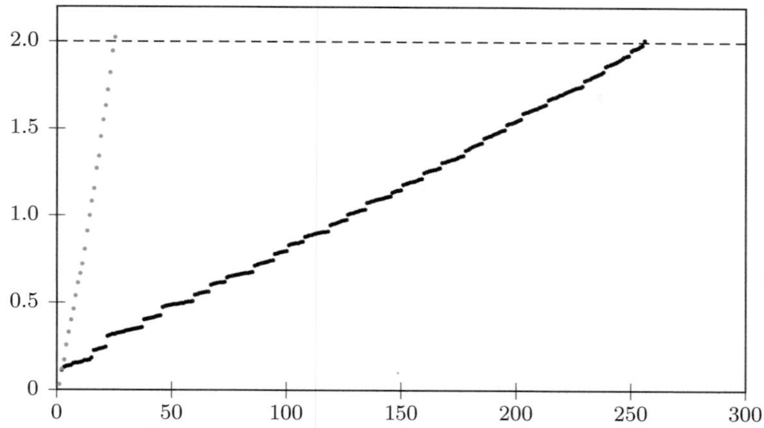

Fig. 2 Typical behavior of $\frac{\max(y)}{\sqrt{n}\|z\|}$ for $K = \{k\}$, where $z_k = \min(z)$ (*black*), and for $K = \{k : z_k \leq 0\}$ (*grey*)

Table 1 Comparison of the original and the new basic procedure

		$K = \{k\}, z_k = \min(z)$			$K = \{k : z_k \leq 0\}$			Improvement factor	
m	n	Iterations	Seconds	$\|K\|$	Iterations	Seconds	$\|K\|$	Iterations	Seconds
5	10	14.8	0.0004	0.8	7.1	0.0003	2.0	2.1	1.7
25	50	650.4	0.0213	1.0	88.2	0.0038	8.2	7.4	5.6
125	250	47,965.1	3.0959	1.0	1,607.3	0.1991	32.5	29.8	15.6
625	1,250	997,796.8	1,546.3305	1.0	4,320.1	36.1645	228.3	231.0	42.8

We conclude from Table 1 that in the new approach the average size of the set K is substantially larger than 1. Moreover, as expected, the improvement factor for the average number of iterations is about the same as the average size of K. Finally, this factor, as well as the improvement factor for the computational time, increases with the size of A. For further computational evidence in favor of the new approach, we refer to [4].

It should be noted that it is not at all excluded that for the new approach worst-case examples exist where $|K| = 1$ in every iteration. But it can easily be understood that the occurrence of this worst-case behavior will be a rare event. In fact, during our experiments we encountered this event only for (very) small sizes of A, with $n \leq 10$. On the other hand, it might be worth investigating if it is possible to derive an estimate for the average size of K in the new approach.

We repeated the same experiment with exactly the same set of randomly generated problems as before, but now using the cuts (10) in Section 2.2. Table 2 shows the results. Comparing this table with Table 1 we see that in all cases substantially smaller iteration numbers and computational times arise. The last two columns in both tables demonstrate a similar effect of using larger sets K.

Note that the number of iterations decreases drastically when using the larger sets K. This does not yield a similar decrease in computational time, however. This

Table 2 Comparison of the original and the new basic procedure, with cuts from (10)

		$K = \{k\}, z_k = \min(z)$				$K = \{k : z_k \leq 0\}$				Improvement factor	
m	n	Iterations	Seconds	$\lvert K\rvert$	$\lvert J\rvert$	Iterations	Seconds	$\lvert K\rvert$	$\lvert J\rvert$	Iterations	Seconds
5	10	1.2	0.0001	0.8	5.3	1.2	0.0001	1.7	5.5	1.0	1.1
25	50	25.1	0.0015	1.0	5.9	9.7	0.0007	8.6	9.5	2.6	2.4
125	250	2,446.6	0.1316	1.0	11.7	207.1	0.0193	37.3	20.6	11.8	6.8
625	1,250	64,392.6	6.1867	1.0	39.1	1,295.8	2.0168	206.6	110.8	49.7	3.1

is due to the fact that the computation of p_K requires $O(|K|n)$ time, which is a factor $|K|$ larger than when K is a singleton. We verified that if $m = 625$ then this computation is responsible for about 80 % of the time needed per iteration. This explains the smaller reduction in computational time.

In Table 2 we also show the average number of cuts generated by the BP, i.e., the average size of the set J. Since we checked for every index k if (10) yields a nonvoid cut, these computations require $O(n^2)$ time. This also results in a negative effect on the time required by the algorithm. By limiting the size of J, as in Theorem 2, we might prevent this negative effect.

The above results, as well as the results in the next section, were obtained using Matlab (version R2014a) on a Windows 7 desktop (Intel(R) Core(TM) i3 CPU, 3.2 GHz), with 8 GB RAM. For the computation of the projection matrix P_A we used the Matlab commands

```
[m,n] = size(A);
[Y,R] = qr(A',0);
P = eye(n) - Y*Y'.
```

5.2 *Computational Results for the MA*

Using the same set of random problems as in the previous section we also ran the MA. The results are presented in Table 3. Besides the usual quantities we include a column that gives the two-norm of the error $\|Ax\|$ in the solution, after rescaling x such that $e^T x = 1$. To compare computing times with a well-known solver we also solved all problems with Sedumi [11, 15]; the average solution times for Sedumi are given in the last column. To obtain these results we used the variant of the BP with $K = \{k : z_k \leq 0\}$ and with the tightest possible cuts, as in (10). The table indicates that on average the Projection Method of Chubanov—equipped with the improvements developed in this paper—is the winner.

It should be mentioned that the Sedumi times are rather stable, whereas the behavior of Chubanov's Projection Method is rather unpredictable. This is made clear in Table 4, which shows the range of the solution times for Chubanov's method and Sedumi, respectively.

Table 3 Computational behavior of the MA

Size(A)		Iterations		Accuracy	Sizes K and J		Time (s)	
m	n	MA	BP	$\|\|Ax\|\|$	$\|K\|$	$\|J\|$	Chubanov	Sedumi
5	10	1.8	3.3	1.3e−14	1.9	5.3	0.0008	0.0167
25	50	3.4	39.1	1.1e−13	10.1	18.2	0.0036	0.0333
125	250	4.4	928.8	8.5e−13	39.5	72.3	0.1091	0.5375
625	1,250	6.2	4,590.0	1.2e−12	231.4	536.0	9.0458	43.0655

Table 4 Range of solution times for Chubanov's method and Sedumi

Size(A)		Range of solution times	
m	n	Chubanov	Sedumi
5	10	[0.0001, 0.0119]	[0.0099, 0.0255]
25	50	[0.0006, 0.0267]	[0.0191, 0.0534]
125	250	[0.0142, 1.0401]	[0.2606, 0.7549]
625	1,250	[1.0911, 239.1962]	[27.5789, 89.5803]

6 Conclusion

The BP of Chubanov arises from Algorithm 1 by using for the set K in line 9 a single index with $z_k \leq 0$; a natural choice is to take k such that z_k is minimal. In this paper we allow K to be larger, e.g., the set of all k such that $z_k \leq 0$. We have shown, both theoretically and computationally, that the new approach outperforms the BP of Chubanov. Apparently this is because the average size of the set K in the new approach is substantially larger than 1, at least in our test problems.

From our experiments we conclude that Chubanov's MA, when equipped with the new BP and the new cut criterion, is competitive with Sedumi.

It remains as a topic for further research to find out if more can be said on the behavior of the size of the set K. If we could show that on average the size of K is a certain fixed fraction of the dimension n, this might open the way to improving the iteration bound of the MA.

Finally, as Chubanov mentions in [4], his BP resembles a procedure proposed by Von Neumann, which has been described by Dantzig in [5]. This Von Neumann algorithm has been elaborated further in [7] and more recently in [10]. It may be a subject for further research to investigate if the idea developed in the current paper can also be used to speed up Von Neumann's procedure.

Acknowledgements We thankfully acknowledge valuable comments of an anonymous referee that not only made the paper more readable but also helped to reduce the time for computing the matrix P_A. Thanks are also due to Tomonari Kitahara (Tokyo Inst. of Technology) for the correction of some typos in an earlier version.

References

1. Broyden, C.G.: A simple algebraic proof of Farkas's lemma and related theorems. Optim. Methods Softw. **8**(3/4), 185–199 (1998)
2. Chubanov, S.: A polynomial relaxation-type algorithm for linear programming. http://www.optimization-online.org/DB_FILE/2011/02/2915.pdf (2012)
3. Chubanov, S.: A strongly polynomial algorithm for linear systems having a binary solution. Math. Program. Ser. A **134**(2), 533–570 (2012)
4. Chubanov, S.: A polynomial projection algorithm for linear programming. http://www.optimization-online.org/DB_FILE/2013/07/3948.pdf (2013)
5. Dantzig, G.B.: An ε-precise feasible solution to a linear program with a convexity constraint in $1/\varepsilon^2$ iterations, independent of problem size. Technical Report SOL 92-5, Systems Optimization Laboratory, Department of Operations Research, Stanford University, Stanford, Oct 1992
6. Edmonds, J.: Systems of distinct representatives and linear algebra. J. Res. Nat. Bur. Stand. Sect. B **71B**, 241–245 (1967)
7. Epelman, M., Freund, R.M.: Condition number complexity of an elementary algorithm for computing a reliable solution of a conic linear system. Math. Program. Ser. A **88**(3), 451–485 (2000)
8. Golub, G.H., Van Loan, C.F.: Matrix Computations, 2nd edn. Johns Hopkins University Press, Baltimore (1989)
9. Khachiyan, L.G.: A polynomial algorithm in linear programming. Dokl. Akad. Nauk SSSR **244**, 1093–1096 (1979). Translated into English in Sov. Math. Dokl. **20**, 191–194
10. Li, D., Terlaky, T.: The duality between the perceptron algorithm and the Von Neumann algorithm. In: Zukuaga, L., Terlaky, T. (eds.) Modelling and Optimization: Theory and Applications, pp. 113–136. Springer Science+Business, New York (2013)
11. Pólik, I.: Addendum to the Sedumi User Guide. Version 1.1. http://www.sedumi.ie.lehigh.edu/wp-content/sedumi-downloads/SeDuMi_Guide_11.pdf (2005)
12. Renegar, J.: A polynomial-time algorithm, based on Newton's method, for linear programming. Math. Program. **40**, 59–93 (1988)
13. Schrijver, A.: Theory of Linear and Integer Programming. Wiley, New York (1986)
14. Stiemke, E.: Über positive Lösungen homogener linearer Gleichungen. Math. Ann. **76**, 340–342 (1915)
15. Sturm, J.F.: Using Sedumi 1.02, a Matlab toolbox for optimization over symmetric cones. http://www.plato.asu.edu/ftp/usrguide.pdf (2001)

List of NAOIII-2014 Conference Participants

#	Name	Affiliation
Organizing Committee		
1	Mehiddin Al-Baali	DOMAS, SQU
2	Asma Al-Ghassani	DOMAS, SQU
3	Magda Al-Hinai	DOMAS, SQU
4	Mohamed Al-Lawati	DOMAS, SQU
5	Nasser Al-Salti	DOMAS, SQU
6	Easwaran Balakrishnan	DOMAS, SQU
7	Lucio Grandinetti	Calabria University, Italy
8	Bernhard Heim	German Univeristy of Technology, Oman
9	Humaid Khalfan	Euclid Center, UAE
10	Issam A.R. Moghrabi	Gulf University for Science and Technology, Kuwait
11	Anton Purnama	DOMAS, SQU
12	Syed S. Ridwan	Caledonian College of Engineering, Oman
13	Muhammad I. Syam	UAE University, UAE
14	Chefi Triki	Department of Mechanical and Industrial Engineering, SQU
15	Tsukasa Yashiro	DOMAS, SQU
16	Saiful I. Zaman	DOMAS, SQU
Invited Speakers		
17	Paul Armand	CNRS et Universite de Limoges, France
18	Oleg Burdakov	Linkoping University, Sweden
19	John C. Butcher	University of Auckland, New Zealand
20	Andrew C. Conn	IBM Research Center, USA
21	Yu-Hong Dai	Chinese Academy of Science, China
22	Jacques Desrosiers	HEC Montreal and GERAD, Canada
23	Iain Duff	Rutherford Appleton Laboratory, UK
24	David M. Gay	AMPL Optimization Inc, USA
25	Michael Hintermuller	Humboldt-Universitaet zu Berlin, Germany

(continued)

M. Al-Baali et al. (eds.), *Numerical Analysis and Optimization*, Springer Proceedings in Mathematics & Statistics 134, DOI 10.1007/978-3-319-17689-5

#	Name	Affiliation
26	Nezam Mahdavi-Amiri	Sharif University of Technology, Iran
27	Dominique Orban	Ecole Polytechnique de Montreal, Canada
28	Amiya Kumar Pani	India Institute of Technology Bombay, India
29	Martin Reed	University of Bath, UK
30	Peter Richtarik	University of Edinburgh, UK
31	Cornelis Roos	Delft University of Technology, The Netherlands
32	Ekkehard W. Sachs	University of Trier, Germany
33	Michael Saunders	Stanford University, USA
34	Tamas Terlaky	Lehigh University, USA
35	Maria Teresa Vespucci	Bergamo University, Italy
36	Hong Wang	University of South Carolina, USA
37	Andrew Wathen	Oxford University, UK
Other Participants		
38	Mahmoud Abd El-Gelil	Department of Civil and Architectural Engineering, SQU
39	Mohmmad Ayaz Ahmad	University of Tabuk, Saudi Arabia
40	Afaq Ahmad	Department of Electrical and Computer Engineering, SQU
41	Muhammad Idrees Ahmad	DOMAS, SQU
42	Waleed K. Ahmed	UAE University, UAE
43	Mohammed Saleh Ahmed	DOMAS, SQU
44	Fahir Talay Akyildiz	Gaziantep University, Turkey
45	Nasser A. Al Azri	Department of Mechanical and Industrial Engineering, SQU
46	Masood Alam	Foundation Program Unit, SQU
47	Said Al-Arimi	Master Student, DOMAS, SQU
48	Munira Al-Balushi	Student, DOMAS, SQU
49	Moza Al-Balushi	Student, DOMAS, SQU
50	Khadija Al-Balushi	Student, DOMAS, SQU
51	Latifa Alblushi	Master Student, DOMAS, SQU
52	Rahma Al-Busaidi	Student, Department of Physics, SQU
53	Nida Al-Chalabi	Department of Computer Science, SQU
54	Mahmoud Al-Hashami	Student, DOMAS, SQU
55	Amaal Al-Hashimy	Department of Computer Science, SQU
56	Fatema Al-Hatimi	Student, DOMAS, SQU
57	Shamsaa Al-Hatimi	Student, DOMAS, SQU
58	Mohammed Al-Hatmi	Master Student, DOMAS, SQU
59	Majid Ali	DOMAS, SQU
60	Ahmed A. Al-Kasbi	Master Student, DOMAS, SQU
61	Kamel Al-Khaled	DOMAS, SQU
62	Amal Al-Kharusi	Student, DOMAS, SQU
63	Mahmood Al-Kindi	Department of Mechanical and Industrial Engineering, SQU
64	Fatma Al-Kindi	Master Student, DOMAS, SQU
65	Rashid Al-Kiyumi	Student, College of Engineering, SQU
66	Mohammed Al-Lawatia	Student, College of Engineering, SQU
67	Faisal Al-Malki	Taif University, Saudi Arabia
68	Khalid S.M. Al-Mashrafi	PhD Student, DOMAS, SQU
69	Mariam Al-Maskari	Student, DOMAS, SQU
70	Waad Al-Mazroui	Student, DOMAS, SQU
71	Qasem Al-Mdallal	UAE University, UAE
72	Huda Almemari	Master Student, DOMAS, SQU

(continued)

#	Name	Affiliation
73	Abdullrahman A. Al-Muqbali	Master Student, DOMAS, SQU
74	Fatma Al-Musalhi	Foundation Program Unit, SQU
75	Tafool Alojaili	Student, DOMAS, SQU
76	Sami Al-Riyami	Diwan Royal Court
77	Amal Al-Saidi	Master Student, DOMAS, SQU
78	Abdul-Sattar J. Al-Saif	Department of Mathematics, Basrah University, Iraq
79	Tahani Al-Sariri	DOMAS, SQU
80	Amina Al-Sawaai	DOMAS, SQU
81	Mohammed Alshahrani	King Fahd University of Petroleum and Minerals, Saudi Arabia
82	Aysha Al-Shamsi	UAE University, UAE
83	Hamed Al-Shamsi	DOMAS, SQU
84	Ziyad Al-Sharawi	DOMAS, SQU
85	AbdulAdheem Al-Soodinay	University of Nizwa, Oman
86	Hamed Al-Sunaidi	Master Student, DOMAS, SQU
87	Khalid Alzebdeh	Department of Mechanical and Industrial Engineering, SQU
88	Nadjadji Anwar	Institute of Technology Sepuluh November (ITS), Indonesia
89	Dilnawaz Anwar	Foundation Program Unit, SQU
90	Muhammad Ashfaq	Foundation Program Unit, SQU
91	Muhammad Ashfaq	Foundation Program Unit, SQU
92	Medhat H.A. Awadalla	Department of Electrical and Computer Engineering, SQU
93	Isa Abdullah Baba	Near East University, Norther Cyprus
94	Elena E. Berdysheva	German Univeristy of Technology, Oman
95	Abdelkader Boudi	University of Bechar, Algeria
96	Messaoud Boulbrachene	DOMAS, SQU
97	Pallath Chandran	DOMAS, SQU
98	Nabil Channouf	College of Economics and Political Science, SQU
99	Boumediene Chentouf	DOMAS, SQU
100	Zouaoui Chikr Elmezouar	University of Bechar Algeria
101	Yassir Dinar	DOMAS, SQU
102	Atsu Dorvlo	DOMAS, SQU
103	Ibrahim Dweib	Department of Computer Science, SQU
104	Tayfour El-Bashir	DOMAS, SQU
105	Ibrahim A. Eltayeb	DOMAS, SQU
106	Godfrey Engwau	Foundation Program Unit, SQU
107	Afifa Essefi	Foundation Program Unit, SQU
108	Rudolf Fleischer	German Univeristy of Technology, Oman
109	Jurgen Garloff	University of Applied Sciences, HTWG Konstanz, Germany
110	Ahmed F. Ghaleb	Department of Mathematics, Cairo University at Giza, Egypt
111	Hachemi Glauoi	University of Bechar, Algeria
112	Rim Gouai-Zarrad	American University of Sharjah, UAE
113	Sanjiv Gupta	DOMAS, SQU
114	Hyung-Tae Ha	Gachon University, South Korea
115	Iftikhar Haider	Foundation Program Unit, SQU

(continued)

#	Name	Affiliation
116	Iftikhar Haider	Foundation Program Unit, SQU
117	Said Hakima	University of Ouargla, Algeria
118	Munawar Hameed	Oman College of Management and Technology, Oman
119	Joachim Heinze	Springer, Germany
120	Afzal Husain	Department of Mechanical and Industrial Engineering, SQU
121	Robert Ibatullin	Moskow Academy of Water Transport, Russia
122	Sofiya Ibatullina	Bashkir Academy of Public Administration and Management, Russia
123	Jai Prakash Jaiswal	Maulana Azad National Institute of Technology, Bhopal, India
124	Tariq Jamil	Department of Electrical and Computer Engineering, SQU
125	Riyadh M.K. Jasim	Dr. B.A. Marthwada University, Aurangabad, India
126	Damian Kajunguri	Foundation Program Unit, SQU
127	Aref Kamal	DOMAS, SQU
128	Samir Karaa	DOMAS, SQU
129	Gabor Kassay	Babes-Bolyai University, Romania
130	Sebti Kerbal	DOMAS, SQU
131	Sareh Keshavarzi	Shiraz University of Medical Sciences, Iran
132	Mohammed S. Khan	DOMAS, SQU
133	Qamar Khan	DOMAS, SQU
134	Lazhar Khriji	Department of Electrical and Computer Engineering, SQU
135	Judith Kreuzer	German University of Technology, Oman
136	Edamana Krishnan	DOMAS, SQU
137	Alfiya Kurmangaleeva	Bashkir Academy of Public Administration and Management, Russia
138	Kenneth K. Kwikiriza	Foundation Program Unit, SQU
139	Abdolmajid Lababpour	National Institute of Genetic Engineering and Biotechnology, Iran
140	Ivan S. Latif	University of Salahaddin, Erbil, Iraq
141	Shamil Makhmutov	DOMAS, SQU
142	Marina Makhmutova	DOMAS, SQU
143	Jasbir S. Manhas	DOMAS, SQU
144	Mihaela Miholca	Babes-Bolyai University, Romania
145	Ahmed M. Mohammed	Foundation Program Unit, SQU
146	Kamel Nafa	DOMAS, SQU
147	Hanifa M. Nasir	University of Peradeniya, Sri Lanka
148	Amar Oukil	Department of Operations Management and Business Statistics, SQU
149	M. Reza Peyghami	K.N. Toosi University of Technology, Iran
150	Sujan Piya	Department of Mechanical and Industrial Engineering, SQU
151	Maitree Podisuk	Kasem Bundit University, Thailand
152	Mansur M. Rahman	DOMAS, SQU

(continued)

#	Name	Affiliation
153	Medhat Rakha	DOMAS, SQU
154	Florian Rupp	German University of Technology, Oman
155	Nirmal C. Sacheti	DOMAS, SQU
156	Said Mohammed Said	University of Ouargla, Algeria
157	Ahmad Sana	Department of Civil and Architectural Engineering, SQU
158	Khadija Semhi	Department of Earth Science, SQU
159	Vinay Singh	School of Engineering and Technology, Nagaland University, India
160	Irina Skhomenko	Foundation Program Unit, SQU
161	Ilya Spitkovsky	New York University Abu Dhabi, UAE
162	Nasser H. Sweilam	Cairo University, Faculty of Science, Egypt
163	Anwar Tabouk	Student, DOMAS, SQU
164	Delfim F.M. Torres	University of Aveiro, Portugal
165	Stefan Veldsman	DOMAS, SQU
166	Muhammad Waheed	Foundation Program Unit, SQU
167	Gerald Wanjala	DOMAS, SQU
168	Mohamed Yasin	Department of Computer Science, SQU
169	Muhammad Ziad	DOMAS, SQU

SQU ج س ق
NAO III
Muscat 2014 مسقط
كلية العلوم
The Third Internatinal
conference on
Numerical Analysis
and Optimization
Theory, Algorithms, Applications, and Technology
جامعة السلطان قابوس
مسقط - سلطنة عمان
Sultan Qaboos University
Muscat - Sultanate of Oman
January كانون الثاني
6 - 9, 2014
المؤتمر الدولي الثاني حول
التحليل العددي
والحلول المثلى
طرق ونظريات وتطبيقات وتكنولوجيا

MIX
Papier aus verantwortungsvollen Quellen
Paper from responsible sources
FSC® C105338
FSC
www.fsc.org

If you have any concerns about our products,
you can contact us on
ProductSafety@springernature.com

In case Publisher is established outside the EU,
the EU authorized representative is:
Springer Nature Customer Service Center GmbH
Europaplatz 3, 69115 Heidelberg, Germany

Printed by Libri Plureos GmbH
in Hamburg, Germany